COMPUTER-AIDED DESIGN,
ENGINEERING, AND MANUFACTURING
SYSTEMS TECHNIQUES AND APPLICATIONS

VOLUME
V

THE DESIGN OF MANUFACTURING SYSTEMS

COMPUTER-AIDED DESIGN,
ENGINEERING, AND MANUFACTURING
SYSTEMS TECHNIQUES AND APPLICATIONS

VOLUME
V

THE DESIGN OF MANUFACTURING SYSTEMS

EDITOR
CORNELIUS LEONDES

CRC Press
Boca Raton   London   New York   Washington, D.C.

Library of Congress Cataloging-in-Publication Data

Catalog record is available from the Library of Congress.

No claim to original U.S. Government works
International Standard Book Number 0-8493-0997-2
Printed in the United States of America 1 2 3 4 5 6 7 8 9 0
Printed on acid-free paper

Preface

A strong trend today is toward the fullest feasible integration of all elements of manufacturing, including maintenance, reliability, supportability, the competitive environment, and other areas. This trend toward total integration is called concurrent engineering. Because of the central role information processing technology plays in this, the computer has also been identified and treated as a central and most essential issue. These are the issues that are at the core of the contents of this volume.

This set of volumes consists of seven distinctly titled and well-integrated volumes on the broadly significant subject of computer-aided design, engineering, and manufacturing: systems techniques and applications. It is appropriate to mention that each of the seven volumes can be utilized individually. In any event, the great breadth of the field certainly suggests the requirement for seven distinctly titled and well-integrated volumes for an adequately comprehensive treatment. The seven volume titles are:

1. Systems Techniques and Computational Methods
2. Computer-Integrated Manufacturing
3. Operational Methods in Computer-Aided Design
4. Optimization Methods for Manufacturing
5. The Design of Manufacturing Systems
6. Manufacturing Systems Processes
7. Artificial Intelligence and Robotics in Manufacturing

The contributors to this volume clearly reveal the effectiveness and great significance of the techniques available and, with further development, the essential role that they will play in the future. I hope that practitioners, research workers, students, computer scientists, and others on the international scene will find this set of volumes to be a unique and significant reference source for years to come.

Cornelius T. Leondes
Editor

Editor

Cornelius T. Leondes, B.S., M.S., Ph.D., is an Emeritus Professor at the School of Engineering and Applied Science, University of California, Los Angeles. Dr. Leondes has served as a member or consultant on numerous national technical and scientific advisory boards. He has served as a consultant for numerous Fortune 500 companies and international corporations, published over 200 technical journal articles, and edited and/or co-authored over 120 books. Dr. Leondes is a Guggenheim Fellow, Fulbright Research Scholar, and Fellow of IEEE. He is a recipient of the IEEE Baker Prize, as well as its Barry Carlton Award.

Contributors

Shabbir Ahmed
Georgia Institute of Technology
Atlanta, Georgia

Venkat Allada
University of Missouri-Rolla
Rolla, Missouri

Saifallah Benjaafar
University of Minnesota
Minneapolis, Minnesota

Dietrich Brandt
University of Technology (RWTH)
Aachen, Germany

Jože Duhovnik
University of Ljubljana
Ljubljana, Slovenia

Placid M. Ferreira
University of Illinois at Urbana-
   Champaign
Urbana, Illinois

Necdet Geren
University of Çukurova
Adana, Turkey

Klaus Henning
University of Technology (RWTH)
Aachen, Germany

Bao Sheng Hu
Xi'an Jiaotong University
Xi'an, China

Mark A. Lawley
Purdue University
West Lafayette, Indiana

T. Warren Liao
Louisiana State University
Baton Rouge, Louisiana

Spyros A. Reveliotis
Georgia Institute of Technology
Atlanta, Georgia

Nikolaos V. Sahinidis
University of Illinois at Urbana-
   Champaign
Urbana, Illinois

Inga Tschiersch
University of Technology (RWTH)
Aachen, Germany

Ke Yi Xing
Xidian University
Xi'an, China

Roman Žavbi
University of Ljubljana
Ljubljana, Slovenia

Contents

1

Long-Range Planning of Chemical Manufacturing Systems

Shabbir Ahmed
Georgia Institute of Technology

Nikolaos V. Sahinidis[1]
*University of Illinois
at Urbana-Champaign*

1.1   Introduction

Recent years have witnessed increasingly growing awareness for long-range planning in all sectors. Companies are concerned more than ever about long-term stability and profitability. The chemical process industries is no exception. New environmental regulations, rising competition, new technology, uncertainty of demand, and fluctuation of prices have all led to an increasing need for decision policies that will be "best" in a dynamic sense over a wide time horizon. Quantitative techniques have long established their importance in such decision-making problems. It is, therefore, no surprise that there is a considerable number of papers in the literature devoted to the problem of long-range planning in the processing industries. It is the purpose of this chapter to present a summary of recent advances in this area and to suggest new avenues for future research.

The chapter is organized in the following manner. Section 1.2 presents the long-range planning problem. Section 1.3 discusses deterministic models and solution strategies. Models dealing with uncertainty are discussed in Section 1.4. Finally, some recommendations for future research and concluding remarks are presented in Section 1.5.

---

[1]Address all correspondence to this author (e-mail: nikos@uiuc.edu).

1.2   The Long-Range Planning Problem

Let us consider a plant comprising several processes to produce a set of chemicals for sale. Each process intakes a number of raw materials and produces a main product along with some by-products. Any of these main or by-products could be the raw materials for another process. We, thus, have a list of chemicals consisting of the main products or by-products that we wish to sell as well as ingredients necessary for the production of each chemical. We might then contemplate the in-house production of some of the required ingredients, forcing us to consider another tier of ingredients and by-products. The listing continues until we have considered all processes which may relate to the ultimate production of the products initially proposed for sale. At this point, the final list of chemicals will contain all raw materials we consider purchasing from the market, all products we consider offering for sale on the market, and all possible intermediates. The plant can then be represented as a network comprised of nodes representing processes and the chemicals in the list, interconnected by arcs representing the different alternatives that are possible for processing, and purchases to and sales from different markets.

The process planning problem then consists of choosing among the various alternatives in such way as to maximize profit. Once we know the prices of chemicals in the various markets and the operating costs of processes, the problem is then to decide the operating level of each process and amount of each chemical in the list to be purchased and sold to the various markets. The problem in itself grows combinatorially with the number of chemicals and processes and is further complicated once we start planning over multiple time periods.

Let us now consider the operation of the plant over a number of time periods. It is reasonable to expect that prices and demands of chemicals in various markets would fluctuate over the planning horizon. These fluctuations along with other factors, such as new environmental regulations or technology obsolescence, might necessitate the decrease or complete elimination of the production of some chemicals while requiring an increase or introduction of others. Thus, we have some additional new decisions variables: capacity expansion of existing processes, installation of new processes, and shut down of existing processes. Moreover, owing to the broadening of the planning horizon, the effect of discount factors and interest rates will become prominent in the cost and price functions. Thus, the planning objective should be to maximize the net present value instead of short-term profit or revenue. This is the problem to which we shall devote our attention. The problem can be stated as follows: assuming a given network of processes and chemicals, and characterization of future demands and prices of the chemicals and operating and installation costs of the existing as well as potential new processes, we want to find an operational and capacity planning policy that would maximize the net present value. We shall now present a general formulation of this problem for a planning horizon consisting of a finite number of time periods.

General Formulation

The following notation will be used throughout.

Indices

i The set of NP processes that constitutes the network ($i = 1, NP$).
j The set of NC chemicals that interconnect the processes ($j = 1, NC$).
l The set of NM markets that are involved ($l = 1, NM$).
t The set of NT time periods of the planning horizon ($t = 1, NT$).

Variables

E_{it} Units of expansion of process i at the beginning of period t.
P_{jlt} Units of chemical j purchased from market l at the beginning of period t.
Q_{it} Total capacity of process i in period t. The capacity of a process is expressed in terms of its main product.

S_{jlt} Units of chemical j sold to market l at the end of period t.

W_{it} Operating level of process i in period t expressed in terms of output of its main product.

Functions

$INVT_{it}(E_{it})$ The investment model for process i in period t as a function of the capacity installed or expanded.

$OPER_{it}(W_{it})$ The cost model for the operation of process i over period t as a function of the operating level.

$SALE_{jlt}(S_{jlt})$ The sales price model for chemical j in market l in period t as a function of the sales quantity.

$PURC_{jlt}(P_{jlt})$ The purchase price model for chemical j in market l in period t as a function of the purchase quantity.

$\psi_{ij}^{O}(W_{it})$ The mass balance model for the output chemical j from process i as a function of the operating level.

$\psi_{ij}^{I}(W_{it})$ The mass balance model for the input chemical j for process i as a function of the operating level.

Parameters

a_{jlt}^{L}, a_{jlt}^{U} Lower and upper bounds for the availability (purchase amount) of chemical j from market l in period t.

d_{jlt}^{L}, d_{jlt}^{U} Lower and upper bounds for the demand (sale amount) of chemical j in market l in period t.

With this notation, a general model for long-range process planning can be formulated as follows.

Model GP

$$\max NPV = \sum_{t=1}^{NT} \left\{ \sum_{i=1}^{NP} [- INVT_{it}(E_{it}) - OPER_{it}(W_{it})] \right.$$

$$\left. + \sum_{j=1}^{NC} \sum_{l=1}^{NM} [SALE_{jlt}(S_{jlt}) - PURC_{jlt}(P_{jlt})] \right\} \tag{1.1}$$

subject to

$$Q_{it} = Q_{it-1} + E_{it} \qquad i = 1NP \quad t = 1, NT \tag{1.2}$$

$$W_{it} \leq Q_{it} \qquad i = 1NP \quad t = 1, NT \tag{1.3}$$

$$\sum_{l=1}^{NM} P_{jlt} + \sum_{i=1}^{NP} \psi_{ij}^{O}(W_{it}) = \sum_{l=1}^{NM} S_{jlt} + \sum_{i=1}^{NP} \psi_{ij}^{I}(W_{it}) \qquad j = 1NC \quad t = 1, NT \tag{1.4}$$

$$a_{jlt}^{L} \leq P_{jlt} \leq a_{jlt}^{U} \qquad j = 1, NC; \quad l = 1NM \quad t = 1, NT \tag{1.5}$$

$$a_{jlt}^{L} \leq S_{jlt} \leq a_{jlt}^{U} \qquad j = 1, NC; \quad l = 1NM \quad t = 1, NT \tag{1.6}$$

$$E_{it}, Q_{it}, W_{it} \geq 0 \qquad i = 1NP \quad t = 1, NT \tag{1.7}$$

The objective function is to maximize the difference between the sales revenues of the final products and the investment, operating, and raw material costs. Constraint (1.2) in the preceding formulation defines the total capacity available at period t as a sum of capacity available in period $t - 1$ and the capacity expansion at the beginning of period t. The parameter Q_{i0} represents the initial capacity, that is, at $t = 0$. The condition that the operating level of any process cannot exceed the installed capacity is modeled by constraint (1.3). Eq. (1.4) expresses mass balances for chemicals across processes and markets. For each chemical, in each time period, the total amount purchased from all markets plus the total amount produced from all processes must equal the total amount sold to all markets and the total amount consumed by all processes. This balance must be satisfied independently for each time period because it is assumed that no inventories are carried from one period to another. The inclusion of inventory quantities is straightforward by introducing new variables I_{jt} denoting inventory of chemical j at the end of period t. The mass balance constraint (1.4) would then become

$$\sum_{l=1}^{NM} P_{jlt} + \sum_{i=1}^{NP} \psi_{ij}^{O}(W_{it}) + I_{jt-1} = \sum_{l=1}^{NM} S_{jlt} + \sum_{i=1}^{NP} \psi_{ij}^{I}(W_{it}) + I_{jt} \qquad (1.8)$$

The cost of carrying inventory then can be included in the objective, and constraints on inventory capacities can also be included. Eqs. (1.5) and (1.6) express the upper and lower bounds of the amount of each chemical that can be purchased from or sold to each market in each time period. Eq. (1.7) is the nonnegativity constraint.

In subsequent sections, we shall see a number of variants of this general model.

1.3   Deterministic Models

Uncertainty is an integral part of long-range planning. Future prices, demands, and operating and investment costs cannot be known with complete certainty beforehand. However, because of the inherent complexity of the problem even without uncertain factors, deterministic models serve to provide valuable insight. These models assume that uncertainty has been dealt with in the forecast of future demands, prices, and costs. Once these forecasts are available, the problem at hand can be formulated as a mathematical model and solved using various optimization techniques. This section describes deterministic models and solution strategies. Models dealing more explicitly with uncertainty are treated in Section 1.4.

An MILP Model

Under the assumption of linear mass balances in the processes and fixed charge cost models, the general model **GP** reduces to a mixed integer linear program (MILP), with capacity decision variables taking 0–1 integer values. Sahinidis and Grossman [14] present the general MILP model for long-range planning in a chemical process industry which is described next.

Assumptions

The MILP model under consideration has been developed under the following assumptions

1. A network of existing as well as potential new processes and chemicals is given.
2. Forecasts for prices and demands of chemicals, as well as investment and operating costs over the planning horizon, are given.
3. The planning horizon consists of a finite number of time periods.
4. Linear models are assumed for mass balances across processes. The functions $\psi_{ij}^{O}(W_{it})$ and $\psi_{ij}^{I}(W_{it})$ in Eq. (1.4) are replaced by linear proportionality constants.
5. Fixed charge cost models are used to model investment costs. These models assume a fixed charge associated with installation of new capacity and a variable charge proportional to the capacity

installed. Linear models are used for operational costs, and sale and purchase prices. Hence, the functions $INVT_{it}(E_{it})$, $OPER_{it}(W_{it})$, $SALE_{jlt}(S_{jlt})$, and $PURC_{jlt}(P_{jlt})$ in the objective (1.1) are replaced by appropriate linear functions.

6. No inventories are carried over time periods because the length of each period is assumed to be rather long.

Supplementary Notation

The notation used here is the same as in the section on general formulation with the inclusion of the following parameters and variables.

Parameters

Γ_{jlt}, γ_{jlt}	Forecasted buying and selling prices of chemical j in market l in period t.
μ_{ij}^{I}, μ_{ij}^{O}	Input and output proportionality constants for chemical j in process i.
α_{it}	Per unit expansion cost for process i at the beginning of period t.
β_{it}	Fixed cost of establishing or expanding process i at the beginning of period t.
δ_{it}	Unit production cost to operate process i during period t.
E_{it}^{L}, E_{it}^{U}	Lower and upper bounds for the capacity expansion of process i in period t.

All cost and price parameters are discounted at the specific interest rate and include the effect of taxes in the net present value.

Variables

y_{it} A 0–1 integer variable. If process i is expanded during period t, then $y_{it} = 1$, else $y_{it} = 0$.

The Model

Model P

$$\text{max } NPV = \sum_{t=1}^{NT}\left\{-\sum_{i=1}^{NP}(\alpha_{it}E_{it} + \beta_{it}y_{it} + \delta_{it}W_{it})\right.$$

$$\left. + \sum_{j=1}^{NC}\sum_{l=1}^{NM}(\gamma_{jlt}S_{jlt} - \Gamma_{jlt}P_{jlt})\right\} \tag{1.9}$$

subject to

$$\sum_{l=1}^{NM}P_{jlt} + \sum_{t=1}^{NP}\mu_{ij}^{O}W_{it} = \sum_{l=1}^{NM}S_{jlt} + \sum_{t=1}^{NP}\mu_{ij}^{I}W_{it} \qquad j = 1, NC \quad t = 1, NT \tag{1.10}$$

$$y_{it}E_{it}^{L} \leq E_{it} \leq y_{it}E_{it}^{U} \qquad i = 1, NP \quad t = 1, NT \tag{1.11}$$

$$y_{it} \in \{0, 1\} \qquad i = 1, NP \quad t = 1, NT \tag{1.12}$$

and constraints (1.2), (1.3), and (1.5) to (1.7).

Note that the investment model in the objective (1.9) has been replaced by a fixed charge model, and the operating costs and sale and purchase prices are linear. Also, the mass balance equations (1.10) are linear. Eq. (1.11) is the bounding constraint for capacity expansions. A zero value of y_{it} forces the capacity expansion of process i at period t to zero. If the binary variable equals 1, then the capacity expansion is performed within prescribed bounds. Eq. (1.12) expresses the binary nature of variable y_{it}.

The preceding model can be further extended by including the following additional constraints.

1. Limit on the number of expansions of some processes:

$$\sum_{t=1}^{NT} y_{it} \le NEXP(i) \qquad i \in I' \subset \{1, 2, ..., NP\} \tag{1.13}$$

where $NEXP(i)$ is the maximum allowable number of expansions of process i and I' is the set of indices for the process whose expansions are to be limited.

2. Limit on the capital available for investment during some time periods:

$$\sum_{i=1}^{NP} (\bar{\alpha}_{it} E_{it} + \bar{\beta}_{it} y_{it}) \le CI(t) \qquad t \in T' \subset \{1, 2, ..., NT\} \tag{1.14}$$

where $\bar{\alpha}_{it}$ and $\bar{\beta}_{it}$ are the nondiscounted cost coefficients corresponding to period t, $CI(t)$ is the capital available in period t and T' is the set of indices for the periods in which the investment is to be limited.

The case of shut down of an existing plant results when the variable W_{it} takes a value of zero after a given time period t. The economics of shut down can be modeled by the inclusion of the following variables and constraints.

Variables

ϵ_{it} A 0–1 integer variable. If process i is to be shut down at the beginning of time period t, then $\epsilon_{it} = 1$, else $\epsilon_{it} = 0$.

CP_{it} Available plant capacity of process i at time of shut down t. If process i is not shut down at time t, then $CP_{it} = 0$.

Constraints

$$CP_{it} \le Q_{it-1} \qquad\qquad i = 1, NP \quad t = 1, NT + 1 \tag{1.15}$$

$$CP_{it} \le \epsilon_{it} U_{it} \qquad\qquad i = 1, NP \quad t = 1, NT + 1 \tag{1.16}$$

$$Q_{it} \le CP_{it} + (1 - \epsilon_{it}) U_{it-1} \qquad i = 1, NP \quad t = 1, NT + 1 \tag{1.17}$$

$$y_{ik} \le 1 - \epsilon_{it} \qquad\qquad i = 1, NP \quad t = 1, NT + 1 \quad k \ge t \tag{1.18}$$

$$W_{it} \le (1 - \epsilon_{it})(U_{ik} + E_{ik}^{U}) \qquad i = 1, NP \quad t = 1, NT + 1 \quad k \ge t \tag{1.19}$$

$$\sum_{t=1}^{NT+1} \epsilon_{it} = 1 \qquad i = 1, NP \tag{1.20}$$

where $U_{it} = \sum_{k=1}^{t-1} E_U^{it}$. When process i has been decided to be shut down at the beginning of period t, that is, $\epsilon_{it} = 1$, then constraints (1.15) and (1.17) will enforce $CP_{it} = Q_{it}$. Constraint (1.16) will force $CP_{it} = 0$ for as long as $\epsilon_{it} = 0$. According to (1.18) and (1.19) no expansion or production is allowed after shutdown. Finally, constraint (1.20) is due to the fact that a process can be shut down only once. A penalty (or scrap value) can be included in the objective as the shut down costs.

Solution Strategies

Model **P** presented in the previous section is a mixed integer linear program and typically can be solved by a branch and bound algorithm using linear programming (LP) relaxations. This is the basic algorithm employed by most commercial codes (e.g., LINDO, ZOOM, OSL, CPLEX, SCICONIC) for these types of problems. The branch and bound algorithm for mixed integer linear programs with integer variables restricted to binary values is described next.

The Branch and Bound Algorithm

The main idea for the branch and bound algorithm for solving model **P** is that of *divide and conquer*. The original problem is *branched* into two subproblems by selecting one of the integer variables and restricting it to 0 and 1 in the two subproblems, respectively. The resulting subproblems can be further subdivided by selecting another branching variable. In this way, the original problem is partitioned in the form of a binary tree, where the root node denotes the original problem and each subsequent node represents an easier subproblem. Because the binary variables can assume only 0–1 values and the nodes of the tree represent all possible combinations of these values, the optimal solution of **P**, if it exists, must be in one of the nodes. Note, however, that this tree can have have $2^{NP \times NT}$ nodes, where $NP \times NT$ is the number of binary variables (y_{it}). To avoid complete enumeration of the tree, we have to determine whether a given node is required to be partitioned further. We can then prune the nodes that need not be further examined, thus reducing computational effort. The process is called *fathoming* and is achieved by solving an LP relaxation of the MILP subproblem in this node. The LP relaxation is obtained by replacing each integrality constraint $y_{it} \in \{0, 1\}$ in the MILP by the continuous constraints $0 \leq y_{it} \leq 1 \; \forall i, t$. If the relaxed subproblem is infeasible, this indicates that the MILP in this node is also infeasible. If the relaxed solution is no better than the best previously obtained integer solution, we infer that the current node will produce children nodes with inferior solutions than those we have at present. If the relaxed solution is integral, than it is optimal for the current MILP subproblem, and we need not examine this node further. Moreover, if this integral solution is better than the best previous integral solution, we have a new incumbent solution at hand for the original problem **P**. Thus, if any of the preceding conditions is satisfied for a node, we need not partition it further and the node is deemed *fathomed*. The algorithmic statement of the method is now presented.

0 Let $S_0 = \{P_0 = \mathbf{P}\}$, $\widehat{NPV} = -\infty$ and $k = 0$.

1 If $S_k = \emptyset$ STOP, $\widehat{NPV}$ is the optimal *NPV*.

2 Select a subproblem (node) $P_k \in S_k$.

3 Relax P_k by replacing the integrality constraints $y_{it} \in \{0, 1\}$ by $0 \leq y_{it} \leq 1$ and solve it as a LP. Let (NPV^k, y_{it}^k) be the corresponding solution. If P_k is infeasible, by convention $NPV^k = -\infty$. Proceed as follows.

   a. If $NPV^k \leq \widehat{NPV}$, then set $S_{k+1} = S_k \backslash P_k$.

   b. If $NPV^k > \widehat{NPV}$ and $y_{it}^k \in \{0, 1\} \forall i, t$, then let $S_{k+1} = S_k \backslash P_k$ and $\widehat{NPV} = NPV^k$.

   c. If $NPV^k > \widehat{NPV}$ and $y_{it}^k \notin \{0, 1\} \forall i, t$, then select a branching variable $y_{i_k t_k}$ where $(i_k, t_k) \in \{i, t: y_{it}^k \notin \{0, 1\}\}$ and construct the following subproblems from problem P_k:

$$P_k^0 = P_k \cap \{y_{i_k t_k} = 0\}$$

$$P_k^1 = P_k \cap \{y_{i_k t_k} = 1\}$$

Set $S_{k+1} = S_k \backslash P_k \cup \{P_k^0, P_k^1\}$.

Set $k \leftarrow k + 1$ and GOTO 1.

The preceding method might be computationally expensive for the planning model because, for realistic problems, the number of variables is large. For example, a network with 40 processes, 50 chemicals, 2 markets, and 5 time periods would involve 200 binary variables and approximately 1000 continuous variables and

1200 constraints. Moreover, because most of the alternative combinations of the binary variables in the model are feasible, a large number of nodes (subproblems) might have to be examined before the optimal can be found. Therefore, there is a clear incentive to develop efficient computational strategies to reduce the solution effort. One possible means is to add valid constraints to the model in order to reduce the gap between the MILP and LP solutions. Alternatively, other algorithms may be applied to take advantage of the special structure of the model. Some of these techniques are described later on.

Bounding and Integer Constraints

Assume that there are no limitations on the capital investment, that is, constraint (1.14) is not included. Then, a simple bound that can be included in model **P** is given by

$$NPV \geq \max\{LB_1, LB_2\} \tag{1.21}$$

where LB_1 and LB_2 are lower bounds on NPV obtained as follows.

LB_1 relaxed LP solution of **P** with nonzero binaries set to one.
LB_2 relaxed LP solution of **P** with the binary variables of the first period set to one and with the corresponding capacities set to the minimum required in order to serve demand during all subsequent time periods.

Additional computational gains can be obtained by including constraint (1.13), where $NEXP(i)$ can be obtained by solving the following MILP for each i

$$NEXP(i) = \max \sum_{t=1}^{NT} y_{it} \tag{1.22}$$

subject to

$$\sum_{t=1}^{NT} (\alpha_{it} E_{it} + \beta_{it} y_{it}) \leq \alpha_{imax} Q_{imax} + \beta_{imax}$$

$$\sum_{t=1}^{NT} E_{it} = Q_{imax}$$

$$0 \leq E_{it} \leq U y_{it} \qquad t = 1, NT$$

$$y_{it} \in \{0, 1\} \qquad t = 1, NT$$

where U is a large positive quantity.

The preceding problem calculates the maximum number of expansions whose cost is less than or equal to the maximum cost (in the worst-case sense) of any given expansion. The first constraint implies that the cost of the expansions cannot exceed the investment cost of process i at maximum capacity Q_{imax} with the "worst" coefficients, α_{imax} and β_{imax}. Owing to the discount factors, these coefficients usually correspond to period 1. Q_{imax} is the minimum capacity required to serve maximum possible demand. This demand can be found by maximizing the operating level subject to the material balance constraints. From the solution of the preceding small-scale MILPs, constraint (1.13) can be added to the general model.

Both constraints (1.13) and (1.21) help in reducing the gap between the relaxed LP and MILP solutions so as to decrease the computational effort of the branch and bound method. However, for large-scale problems, these provisions may not be sufficient. Furthermore, when Q_{imax} exceeds E_{i1}^{U}, the problem in (1.22) will often underestimate the maximum number of expansions.

Strong Cutting Planes

The main idea behind the method of strong cutting planes for the general MILP model with the capital investment constraint (1.14) is as follows. First, a solution to the LP relaxation of the model is found. Then, by exploiting the network substructure of the model, a *separation problem* is solved to generate additional valid inequalities which attempt to chop off the solution point from the space of the LP relaxation polyhedron. The process is repeated, thereby reducing the gap between the MILP and its LP relaxation.

A network substructure in the model **P** can be identified by making the following substitution.

$$
\left.
\begin{aligned}
x_{it} &= \bar{\alpha}_{it} E_{it} + \bar{\beta}_{it} y_{it} \\
l_{it} &= \bar{\alpha}_{it} E_{it}^{L} + \bar{\beta}_{it} \\
u_{it} &= \bar{\alpha}_{it} E_{it}^{U} + \bar{\beta}_{it}
\end{aligned}
\right\}
\qquad i = 1, NP \qquad\qquad t = 1, NT
$$

With these, we obtain the following substructure for each process in each time period

$$
S_{it} = \left\{ (x, y): \sum_{i=1}^{NP} x_{it} \leq CI(t),\ l_{it} y_{it} \leq x_{it} \leq \bar{u}_{it} y_{it},\ y_{it} \in \{0, 1\} \right\}
$$

For this substructure, two families of valid inequalities have been derived.

1. The simple generalized flow cover inequality [14]:

$$
\sum_{i \in C_t} [x_{it} + (u_{it} - \lambda_t)^{+}(1 - y_{it})] \leq CI(t) \tag{1.23}
$$

2. The extended generalized flow cover inequality [14]:

$$
\sum_{i \in C_t} [x_{it} + (u_{it} - \lambda_y)^{+}(1 - y_{it})] + \sum_{i \in R_t} [x_{it} + (\bar{u}_{it} - \lambda_{it})^{+}(1 - y_{it})] \leq CI(t) \tag{1.24}
$$

where, the notation Φ^{+} stands for $\max(0, \Phi)$ and $C_t \subset \{1,2,\ldots,NP\}$ is a generalized flow cover, that is,

$$
\lambda_t = \sum_{i \in C_t} u_{it} - CI(t) > 0
$$

and

$$
R_t \subset \{1, 2, \ldots, NP\} \qquad \bar{u}_{it} = \max(u_t, u_{it});
$$

$$
\bar{u}_t = \max_{i \in C_t} \{u_{it}\} \quad \text{and} \quad \bar{u}_t > \lambda_t > 0
$$

Then, the *separation problem* to determine the cover C_t for a given relaxed LP solution (x^{*}, y^{*}) for each time period t is

$$
\max \zeta_t = \sum_{i=1}^{NP} \{-(1 - y_{it}^{*}) z_i\} \tag{1.25}
$$

subject to

$$\sum_{i=1}^{NP} u_{it} z_i > CI(t) \qquad z_i \in \{0, 1\} \quad i = 1, NP$$

where $z_i = 1$ if $i \in C_t$; $z_i = 0$, otherwise. The violated inequalities (1.23) and (1.24) are derived whenever $\zeta_t > -1$. The indices i that are included in the set R_t must satisfy the condition: $x_{it}^* - (\bar{u}_{it} - \lambda_t)^+ y_{it}^* \geq 0$. The cutting plane algorithm is then as follows.

0 Solve the LP relaxation of **P**. Set $NPV' = NPV$ (optimum from the relaxed LP).
1 For each time period t, solve the separation problem (1.25), to determine the cover C_t and add the violated inequalities (1.23) and (1.24) to the current MILP formulation.
2 Solve the new LP relaxation. If $(NPV' - NPV)/NPV'$ is greater than a prescribed tolerance, then set $NPV' = NPV$ and repeat steps 1 and 2. Otherwise, start the branch and bound procedure or any other algorithm to find the optimum to the current formulation.

The advantage of this type of algorithm is that no attempt is made to generate all facets of the 0–1 polyhedron. Instead, cuts are added at each iteration to reduce the relaxation gap. However, the cuts are generated from an isolated part of the model and might not always be deep enough to completely eliminate the relaxation gap.

Benders Decomposition

Benders decomposition is a standard decomposition technique in which the MILP problem is solved through a sequence of LP subproblems in the "easy" variables and MILP master problem in the "complicating" variables [3]. The subproblems provide lower bounds to the net present value while the master problem provides upper bounds. The definition of the LP subproblems and the MILP master problem depends on the partitioning of the variables. The natural choice for this partitioning is

1. Complicating variables for the master problem: $u = [y_{it}]$.
2. Remaining variables for the LP subproblems: $v = [E_{it}, P_{jlt}, Q_{it}, S_{jlt}, W_{it}]$.

However, with this variable partition, the master problem is often too relaxed. In order to strengthen the bounds predicted by the master problem, the variable partitioning can be redefined as follows:

1. Complicating variables for the master problem: $u = [y_{it}, Q_{it}, E_{it}]$.
2. Remaining variables for the LP subproblems: $v = [P_{jlt}, S_{jlt}, W_{it}]$.

For this choice of complicating variables, the Benders decomposition algorithm can then be stated as follows:

0 Select u^1; set $NPV^U = +\infty$, $NPV^L = -\infty$, $k = 1$.
1 a. Fix the complicating variables at u^k and solve the LP relaxation of **P** without constraints (1.2), (1.7), and (1.11), to determine the solutions NPV^k and v^k.
   b. Update the lower bounds by setting $NPV^L = \max\{NPV^L, NPV^k\}$.
2 Solve the following MILP master problem to get u^{k+1}.

$$NPV^U = \max_{u,\mu} \mu$$

subject to

$$\mu \leq NPV(u, v^r) + \sum_{i=1}^{NP} \sum_{t=1}^{NT} \rho_{it}^r (W_{it} - Q_{it}) \qquad r = 1, k$$

and constraints (1.2), (1.7), (1.11), and (1.12), and $\mu \in R^1$. Here, $NPV(u, v^r)$ is the NPV function in Eq. (1.9) with variables v^r fixed, and ρ_{it}^r are the Lagrange multipliers for constraints (1.3).

3 If $NPV^L = NPV^U$, stop. Otherwise, set $k \leftarrow k + 1$ and go to step 1.

Numerical experiments carried out by Sahinidis and Grossmann [14] suggest that for large problems, Benders decomposition seems to hold little promise, since the bounds determined are not very tight and a large number of subproblems need to be solved. A better strategy for large-scale models of the type **P** appeared to be a combination of integer cuts, strong cutting plane generation, and branch and bound.

Reformulation to Exploit Lot Sizing Substructures

Aiming at improving the LP bounds, Sahinidis and Grossmann [13] present a reformulation of the general model **P** by exploiting the lot sizing substructures. To see this substructure, let us fix all the chemical flows (W_{it}, P_{jlt}, S_{jlt}) in the network in such a way that the mass balance constraint (1.10) is satisfied for all time periods. Then, every process can be isolated from the rest of the network and the planning problem for each process i becomes: "Find the cheapest capacity expansion sequence for process i (E_{it}, $t = 1, NT$) which will allow the production of the specified operating level and flow of chemicals (W_{it}, P_{jlt}, S_{jlt})." The equivalent lot sizing reformulation to the preceding problem becomes apparent with the following substitutions.

Let

$$SQ_{it} = Q_{it} - W_{it} \qquad\qquad i = 1, NP \quad t = 1, NT \qquad\qquad (1.26)$$

$$d_{it} = \{W_{it} - \max_{T=0,\,t-1} W_{iT}\}^+ \qquad i = 1, NP \quad t = 1, NT \qquad\qquad (1.27)$$

where $W_{i0} = Q_{i0}$. The lot sizing reformulation of **P** for each process i is then:

Model LSR-i

$$\min \sum_{t=1}^{NT} (\alpha_{it} E_{it} + \beta_{it} y_{it})$$

subject to

$$
\begin{aligned}
E_{it} &\le U y_{it} & t &= 1, NT \\
SQ_{it-1} + E_{it} &= d_{it} + SQ_{it} & t &= 1, NT \\
SQ_{i0} &= 0 & & \\
E_{it}, SQ_{it} &\ge 0 & t &= 1, NT \\
y_{it} &\in \{0, 1\} & t &= 1, NT
\end{aligned}
$$

where U is a large positive quantity.

In model **LSR-i**, SQ_{it} can be viewed as "inventory" of capacity for process i, that is, excess of capacity installed at early times in order to serve demand during subsequent periods. Accordingly, E_{it} can be regarded as "production" of capacity to satisfy "demand" for capacity as determined by the operating levels W_{it} in (1.27).

Model **LSR-i** is a MILP, which can be further reformulated so that its relaxed LP solutions produce 0–1 values for the y_{it} [12]. For this, we introduce $\phi_{it\tau}$ to denote the capacity expansion of process i made in period t to serve production requirements up to period $\tau(\tau \geq t)$. Model **LSR-i** then becomes:

Model MLSR-i

$$\min \sum_{t=1}^{NT} (\alpha_{it} E_{it} + \beta_{it} y_{it})$$

subject to

$$E_{it} \geq \phi_{it\tau} \qquad t = 1, NT \quad \tau \geq t$$
$$\phi_{it\tau} \leq C_{it\tau} y_{it} \quad t = 1, NT \quad \tau \geq t \tag{1.28}$$

$$\sum_{\tau=1}^{t} \phi_{i1t} \geq C_{ilt} \qquad t = 1, NT \tag{1.29}$$

$$E_{it}, \phi_{it\tau} \geq 0 \qquad t = 1, NT \quad \tau \geq t$$
$$0 \leq y_{it} \leq 1 \qquad t = 1, NT \quad \tau \geq t$$

where $C_{it\tau}$ are upper bounds for the required capacity expansions and can be obtained by

$$C_{it\tau} = \sum_{T=t}^{\tau} d_{iT}$$

The evaluation of d_{it} by (1.27) results in nonconvexities in the model. The alternative suggested in reference [13] is to *a priori* estimate the bounds $C_{it\tau}$ as follows. Constraint (1.29) can be relaxed as

$$\sum_{\tau=1}^{t} \phi_{i\tau t} \geq W_{it} - Q_{i0} \qquad i = 1, NT \quad t = 1, NT$$

Furthermore, $C_{it\tau}$ in constraint (1.28) can be estimated by $\hat{C}_{it\tau}$, which is given by

$$\hat{C}_{it\tau} = [\min\{E_{it}^U, \max_{T=t,\tau} \Omega_{iT}\} - (Q_{i0} + \{\max_{T=1,t-1} \omega_{iT} - Q_{i0}\}^+)]^+$$

where Ω_{it} and ω_{it} for each process i and period t are obtained by maximizing and minimizing the operating level W_{it} subject to the mass balance constraints in the network, respectively.

The reformulation of **P** by inclusion of the previously mentioned variables and constraints is as follows.

Model RP

$$\max NPV \quad \text{as given in } (1.9)$$

subject to

$$E_{it} \geq \phi_{it\tau} \qquad\qquad i = 1, NP \quad t = 1, NT \quad \tau \geq t \tag{1.30}$$

$$\phi_{it\tau} \leq \hat{C}_{it\tau} y_{it} \qquad\qquad i = 1, NP \quad t = 1, NT \quad \tau \geq t \tag{1.31}$$

$$\sum_{\tau=1}^{t} \phi_{i\tau t} \geq W_{it} - Q_{i0} \qquad i = 1, NP \quad t = 1, NT \tag{1.32}$$

$$\phi_{it\tau} \geq 0 \qquad\qquad i = 1, NP \quad t = 1, NT \quad \tau \geq t$$

and constraints (1.2), (1.3), (1.7) to (1.14).

The preceding reformulation results in a tighter LP relaxation as stated in the following theorem.

Theorem [13]: *The optimal NPV of the linear programming relaxation of* **RP** *is not greater than the optimal NPV of the linear programming relaxation of* **P**, *and it may be strictly less.*

With this formulation, while the relaxation becomes more accurate, the number of new variables and constraints introduced in the model are increased. In fact, we are adding $NP \times NT \times (NT + 1)/2$ variables and $NP \times NT^2$ constraints. This increase in the number of constraints can be handled by a constraint generation scheme prior to the branch and bound algorithm. This scheme is described next.

0 Add constraints (1.32) to **P** and solve the LP relaxation of the resulting formulation. If the solution is integral, STOP. Else, set $NPV' = NPV$ (the optimum from the relaxed LP).

1 For each nonzero $\phi_{it\tau}$, find one or more constraints from among (1.30) and (1.31) that are violated by the solution to the current LP relaxation. If such inequalities are found, append them to the current LP and go to Step 2. Else, go to Step 3.

2 Solve the new LP relaxation. If $NPV' - NPV > \epsilon$ (where NPV is the current LP solution and ϵ is a prescribed tolerance), then set $NPV' = NPV$ and repeat Steps 1 and 2. Otherwise, go to Step 3.

3 Start a branch and bound procedure or any other exact algorithm to find the optimum to the current LP formulation.

In the preceding algorithm, the violated constraints are added on an as needed basis. Because the number of ϕ variables is quite high, a large number of constraints may need to be added in each step.

Projection Approach

The increased number of variables in **RP** can be reduced by projecting the problem from the (E,W,P,S,y,ϕ) space onto the (E,W,P,S,y) space. Let $x = (E,W,P,S,y)$ and P be the polyhedron defining the feasible space of **RP**: $P = \{(x, \phi) : Ax + G\phi \leq b, \phi \geq 0, x \in X\}$. Also, let $C = \{v : vG \geq 0, v \geq 0\}$. If $v^1, v^2,..., v^H$ are the extreme rays of C, then the projection of P onto the space of x is given by: $\text{proj}_x(P) = \{x : v^h Ax \leq v^h b, h = 1,...,H; x \in X\}$. This expression requires finding the extreme rays of the polyhedral cone C [2], which calls for the complete characterization of the basic solutions of $vG = 0$. In reference [6], the authors show that the projected model is as follows.

Model PP

$$\max NPV \qquad \text{as given in (1.9)}$$

subject to

$$W_{it} - \sum_{\tau=1}^{t} \{\sigma_{i\tau t} E_{i\tau} + (1 - \sigma_{i\tau t}) U_{i\tau t} y_{i\tau}\} \leq Q_{i0},$$

$$\sigma_{i\tau t} = 0 \quad \text{or} \quad 1 \qquad i = 1, NP \quad t = 1, NT \tag{1.33}$$

and constraints (1.7), (1.10) to (1.14).

Model **PP** can be solved using a cutting plane method similar to the constraint generation technique described in the previous section. In this method, Step 1 involves the solution of a separation problem to determine (i^*, t^*) corresponding to the maximum violation of constraint (1.33) for a given LP relaxation solution. If such an inequality is found, it is appended to the current LP and the method progresses as before. The procedure for solving the separation problem for given solution is as follows.

1. Given (E^*, W^*, y^*), the current LP relaxation solution, calculate for each i and t.

$$\Theta_{it}^* = W_{it}^* - Q_{i0} - \sum_{\tau=1}^{t} U_{i\tau t} y_{it}^*$$

$$\theta_{it}^* = E_{it}^* - U_{i\tau t} y_{it}^* \qquad \tau \le t$$

$$\sigma_{i\tau t}^* = \begin{cases} 1 & \text{if} \quad \theta_{i\tau t}^* \le 0 \\ 0 & \text{if} \quad \theta_{i\tau t}^* > 0 \end{cases} \qquad \tau \le t$$

$$\zeta_{it}^* = \sum_{\tau=1}^{t} \sigma_{i\tau t}^* \, \theta_{i\tau t}^*$$

2. Let $(i^*, t^*) = \arg \max_{i,t} (\Theta_{it}^* - \zeta_{it}^*)^+$.

The lot sizing reformulation **RP** and the projection **PP** of **P** result in much larger problems than **P** because of additional variables and constraints, while providing tighter LP relaxations. The number of constraints in **PP** increases exponentially with the number of time periods. Computational results in [6] using branch and bound, indicate that for up to four time periods, **RP** and **PP** require smaller CPU times and fewer search nodes than **P**. However, for more than four time periods, **P** performs best and **PP** worst in terms of CPU times. This happens in spite of the fact that **RP** and **PP** require fewer search nodes than **P**. The increased CPU time for **RP** and **PP** can be attributed to the large number of additional constraints involved. However, when a constraint generation scheme is coupled with branch and bound for solving **RP** and **PP**, formulation **RP** outperforms **P** whereas **PP** outperforms both **P** and **RP** as the problem size increases. The success of the projection cutting plane approach is owing to the fact that only a small fraction of the projection constraints (1.33) is sufficient to significantly reduce the relaxation gap.

Remarks
Previously we discussed some of the solution strategies for model **P**. Bounding and integer cuts, strong cutting planes, and Benders decomposition were described as methods of direct solution of **P**. Two nonconventional reformulations of the same model, **RP** and **PP**, and their solution strategies were also discussed. In summary, we can make the following remarks.

1. Straightforward branch and bound strategy is computationally expensive for **P** because of the large number of feasible nodes that need to be examined.
2. For large-scale models of the form of **P**, the LP relaxation gap can be reduced by generating lower bounds, integer cuts, and strong cutting planes. A branch bound strategy with the relaxation gap reduced performs much better that Benders decomposition for these problems.
3. The lot sizing substructure embedded in the planning problem can be used to obtain nonconventional formulations. Model **RP** contains more variables and constraints than **P**. Yet, it possesses a tighter linear programming relaxation.
4. The large number of constraints in **RP** can be handled through a constraint generation scheme.
5. Projection of **RP** onto a lower dimensional space produces **PP**, a model with fewer variables but many more constraints.
6. The large number of constraints in **PP** can be handled through a cutting plane strategy along with branch and bound, in which only violated constraints are added.
7. Computational results indicate that, for small models, reformulation and projection models do not provide appreciable gains.

8. Straightforward branch and bound becomes extremely computationally expensive for **RP** and **PP** as the problem size increases.
9. For large problems, the constraint generation scheme for **RP** and cutting plane method for **PP** coupled with branch and bound are very efficient.
10. The best method for large problems is to project it in the form of **PP** and apply the cutting plane method.

Extensions of the MILP Model

Model **P**, in the previous section, was developed under several simplifying assumptions. In this section, we shall show how some of these assumptions can be relaxed to obtain more realistic formulations.

A Concave Programming Model

In model **P**, economies of scale in the investment cost functions were modeled by the introduction of a set of binary decision variables (y_{it}) to impose a fixed charge on the decision to expand, and a linear cost function for variable costs. A drawback of this formulation is that, in reality, variable costs are not directly proportional to expansion quantity. Rather, the investment cost is a concave function because of the presence of quantity discounts. Thus, a more realistic model for the investment cost would be

$$INVT_{it}(E_{it}) = \begin{cases} 0 & \text{when} \quad E_{it} = 0 \\ \beta_{it} + a_{it}E_{it}^{b_{it}} & \text{when} \quad E_{it} > 0 \end{cases} \tag{1.34}$$

where $a_{it} > 0$ and $0 < b_{it} < 1$. In this formulation, the integer variables have been discarded and the linear variable cost function has been replaced by a concave function in E_{it} with coefficient a_{it} and exponent b_{it}. Note that, this function is discontinuous at $E_{it} = 0$.

In reference [10], the authors present two formulations with simplifications of the general investment function (1.34). In the fixed charge concave programming model (FCP), the linear cost relation is retained but the discrete variables are eliminated by using the following concave function.

$$INVT_{it}(E_{it}) = \begin{cases} 0 & \text{when} \quad E_{it} = 0 \\ \beta_{it} + \alpha_{it}E_{it} & \text{when} \quad E_{it} > 0 \end{cases}$$

In the continuous concave programming model (CCP), the discontinuity at $E_{it} = 0$ is avoided by using the following function.

$$INVT_{it}(E_{it}) = a_{it}E_{it}^{b_{it}}$$

Both FCP and CCP are problems with concave objective functions subject to a set of linear constraints. These can be solved by a concave programming method based on the branch and bound procedure discussed earlier. Here, *branching* of the original problem into small subproblems is done by partitioning the feasible space by a hyperplane normal to the axis of the selected branching variable. The subproblems are maintained in a list. In each iteration, the procedure selects one of these subproblems for *bounding*, that is, generation of a numerical interval consisting of an upper and a lower bound between which the optimal value of the subproblem must lie. The algorithm can then utilize this information in its search for the global minimum. Because the global minimum must lie between the largest of the lower bounds and the smallest of the upper bounds, the algorithm may delete any subproblem which has an associated bound that is larger than or equal to the least upper bound. Details of the method for both FCP and CCP formulations are presented in reference [10]. Computational results presented in this paper suggest that the solution of FCP by the proposed algorithm is superior to the performance of straightforward branch and bound for formulation **P**. The authors also present means for approximating the FCP formulation with the CCP formulation, and vice versa, by minimizing the least squared error of the approximation. Comparison between the solution times of the two alternative formulations indicate that FCP models are much easier to solve.

Processing Networks with Dedicated and Flexible Plants

Model **P** was developed for a network in which each of the processing plants produced a set of products in fixed proportions at all times. Such plants are known as *dedicated* facilities. Moreover, the plants were assumed to be operating in continuous production mode. This is usually the case in the manufacturing of high volume chemicals. However, for low volume manufacture of chemicals, *flexible* processing plants are frequently employed. These plants are capable of manufacturing different sets of products at different times. In general, a processing establishment may consist of either or both dedicated and flexible plants, which can operate either *continuously* or in *batch* mode. Most industrial production facilities involve continuous dedicated units. For example, paper mills that operate continuously and can produce several different types of paper are examples of flexible continuous plants. Dedicated batch processes can be found in the food industry and flexible batch processes are common in the production of polymers and pharmaceuticals.

The choice among competing dedicated and flexible technologies, and the sizing of each type of facility to be used, is an important concern at the level of planning the capacity expansion policy of a processing network. In [15], the authors present an extension of the general model **P** to take into account the presence of both dedicated and flexible processing units, which can operate either continuously or in batch mode. This model is described next.

The chemical processing network is now assumed to consist of two types of processing nodes. Let B denote processes operating in batch mode and C denote those operating continuously. Let us now consider the following cases.

Continuous Flexible Processes

Each flexible process $i \in C$ can operate at a number of alternative production schemes, each of which is characterized by a main product. It is assumed that the production rate (r_{ijt}) of the main product j of each such scheme is proportional to the capacity of the plant.

$$r_{ijt} = \rho_{ij}Q_{it} \qquad i \in C \quad j \in M_i \quad t = 1, NT$$

where M_i denotes the index set of main products of the alternative production schemes of flexible process i, and ρ_{ij} represent relative production rates for main product $j \in M_i$.

Let a variable T_{ijt} be defined as the time interval in period t during which the flexible process i is allocated to the production scheme defined by main product j. The amount of each of these main products produced is then given by

$$W_{ijt} = (\rho_{ij}Q_{it})T_{ijt} \qquad i \in C \quad j \in M_i \quad t = 1, NY \qquad (1.35)$$

Let H_{it} be the time for which plant i is available for operation in period t. Then, the total allocation of production times T_{ijt} cannot exceed the total available time,

$$\sum_{j \in M_i} T_{ijt} \le H_{it} \qquad i \in C \quad t = 1, NT \qquad (1.36)$$

Let L_i denote the index set of chemicals that are inputs and outputs to process i, and $\mu_{ijj'}$ be the proportionality factor between the amounts of associated products $j \in L_i$ and the main product j',

$$W_{ijt} = \sum_{j' \in M_i} \mu_{ijj'} W_{ij't} \qquad i \in C \quad j \in L_i \quad t = 1, NT \qquad (1.37)$$

The nonlinear constraint (1.35) can be avoided by introducing the variable

$$\theta_{ijt} = Q_{it}T_{ijt} \qquad i \in C \quad j \in M_i \quad t = 1, NT$$

Then, Eqs. (1.35) and (1.36), respectively, become

$$W_{ijt} = \rho_{ij}\theta_{ijt} \qquad i \in C \quad j \in M_i \quad t = 1, NT \tag{1.38}$$

$$\sum_{j \in M_i} \theta_{ijt} \le Q_{it}H_{it} \qquad i \in C \quad t = 1, NT \tag{1.39}$$

Continuous Dedicated Processes
This is just a special case of the previous one, where we have only one production scheme and, hence, one main product m_i. Thus, here we have $\rho_{ij} = 1$ and Eqs. (1.37) to (1.39) are simplified to

$$W_{im_it} \le Q_{it}H_{it} \qquad i \in C \quad t = 1, NT$$
$$W_{ijt} = \mu_{ijm_i}W_{im_it} \qquad i \in C \quad j \in L_i \quad t = 1, NT$$

Batch Processes
Here, each process ($i \in B$) is considered separately with no consideration for in-process inventory and setup costs. Furthermore, no distinction is made between continuous and dedicated units as their modeling is identical. If W_{ijt} is the total amount of chemical j produced by process i in period t and σ_{ij} is the conversion factor for the units of Q_{it} and W_{ijt}, then the number of batches that can be processed for product j in plant i in period t is given by $(W_{ijt}\,\sigma_{ij})/Q_{it}$. Let τ_{ij} denote the batch time for the production of product j in process i. The constraint that the total available time in unit i cannot be exceeded during period t is then expressed as follows.

$$\sum_{j \in M_i} W_{ijt}\sigma_{ij}\tau_{ij} \le Q_{it}H_{it} \qquad i \in B \quad t = 1, NT \tag{1.40}$$

The production amounts of the secondary products in the batch plants are calculated similarly to the continuous case,

$$W_{ijt} = \sum_{j' \in M_i} \mu_{ijj'}W_{ij't} \qquad i \in B \quad j \in L_i \quad t = 1, NT$$

Note that, by letting $\rho_{ij} = 1/(\sigma_{ij}\tau_{ij})$ for all batch processes and defining the variables $\theta_{ijt} = W_{ijt}/\rho_{ij}$ as in Eq. (1.38), constraint (1.40) reduces to constraint (1.39).

Thus, we can use the same constraint representation for both the continuous and the batch processes provided the rate coefficients ρ_{ij} and the corresponding capacities Q_{it} are defined accordingly. Model **P** can be extended to include dedicated and flexible plants operating continuously or in batch mode by the inclusion of the following constraints and variables:

$$W_{ijt} = \rho_{ij}\theta_{ijt} \qquad\qquad i = 1, NP \quad j \in M_i \quad t = 1, NT$$

$$\sum_{j \in M_i} \theta_{ijt} \le Q_{it}H_{it} \qquad\qquad i = 1, NP \quad t = 1, NT$$

$$W_{ijt} = \sum_{j' \in M_i} \mu_{ijj'}W_{ij't} \qquad i = 1, NP \quad j \in L_i \quad t = 1, NT$$

$$\theta_{ijt} \ge 0 \qquad\qquad i = 1, NP \quad j = 1, NC \quad t = 1, NT$$

where

$$\rho_{ij} = \begin{cases} 1 \text{ for continuous dedicated plant } i \\ C \text{ for a continuous flexible plant } i \\ 1/(\sigma_{ij}\tau_{ij}) \text{ for a (dedicated or flexible) batch process } i \end{cases}$$

and C is a specified constant.

The extended model described is also a mixed integer linear program, and the solution strategies discussed previously can be applied here.

1.4   Hedging against Uncertainty

As mentioned earlier, many of the parameters of the long-range planning model may be uncertain. Previously, we assumed that the uncertainties had been sufficiently characterized in the prediction of these parameters. This section presents ways to deal with the parameter uncertainties more explicitly. There are two distinct philosophies regarding problems of this type: fuzzy programming and stochastic programming. Uncertain parameters in fuzzy programming are modeled as fuzzy numbers, while in stochastic programming they are modeled as random variables with an underlying probability distribution. Both these philosophies will be discussed and contrasted.

In what follows, first a discussion on the sources and consequences of uncertainties in the model parameters is presented. Next, the fuzzy programming and stochastic programming approaches are discussed and a comparison of these two techniques is presented. Finally, we discuss an extension of the stochastic programming approach known as robust optimization.

Sources and Consequences of Uncertainty

Refer to the original model **P**. In a general framework, all of the parameters of this model could be uncertain. However, costs associated with capacity expansions has smaller variability as compared to those associated with the production, purchase, and sales of chemicals. Also, bounds on the demand and availabilities of chemicals are very much uncertain, while bounds on the capacity expansions are more or less fixed. Thus, some of the parameters must be treated as uncertain while others can be considered deterministic.

Once uncertainties are included in **P**, several problems arise. The planner's goal is to determine the capacity expansion plan *before* the realizations of the uncertain parameters are revealed. If a plan is obtained by solving the model for one set of realizations (a scenario) of the uncertain parameters, it may be infeasible or suboptimal for some other scenario. Another approach is to replace all of the uncertain parameters with their extremal values, thus creating the "worst-case" scenario, and solve the model for this scenario. Although the resulting plan will be feasible for all other scenarios, it is most often suboptimal because the probability of such extreme scenarios is usually very small. Such a plan, known as a "fat solution," reflects the planner's complete risk aversion. Also, it is not always straightforward to determine the worst-case scenario. One could alternatively solve the problem for all possible scenarios and determine which of the resulting plans is feasible as well as optimal when applied to all other scenarios. Such an approach is not practical as the number of possible scenarios may run into hundreds of thousands. Moreover, none of the plans produced from solving for individual scenarios may be feasible or optimal for all other scenarios. Thus, a different approach has to be adopted where one considers all scenarios in an aggregate sense, instead of looking at them individually.

Fuzzy Programming

According to fuzzy programming, uncertain parameters in a model are considered to be fuzzy numbers with a known range of values, and constraints are treated as "soft," that is, some violation is allowed.

Consider the following linear program where uncertainties appear only in the constraint right-hand sides

$$\max z = c^t x$$
$$\text{subject to } a_i^t \le \tilde{b}_i \quad \text{for} \quad i = 1,\ldots,m$$
$$x \ge 0$$

Each of the right-hand side parameters $\tilde{b}_i$ is a fuzzy number in the range $[b_i, b + \Delta b_i]$. Here, ideally, we want the decision variables x to satisfy the constraints $a_i^t x \le b_i$. However, a violation of up to $a_i^t x \le b_i + \Delta b_i$ is allowed. The degree of satisfaction of the constraints is then measured in terms of *membership functions* which are defined as follows.

$$u_i(x) := \begin{cases} 1 & \text{if} \quad a_i^t x \le b_i \\ 1 - \dfrac{a_i^t x - b_i}{\Delta b_i} & \text{if} \quad b_i < a_i^t x \le b_i + \Delta b_i \quad \text{for } i = 1,\ldots,m \\ 0 & \text{if} \quad b_i + \Delta b_i < a_i^t x \end{cases}$$

A similar membership function can also be considered for the objective function if the *aspiration levels* are known. For example, if the decision maker wants the objective function z to be above some level v^+ but will be satisfied with a smaller degree of satisfaction up to a level v^-, then the membership function for the objective is given by

$$u_0(z) := \begin{cases} 1 & \text{if} \quad v^+ \le z \\ 1 - \dfrac{v^+ - z}{v^+ - v^-} & \text{if} \quad v^- < z \le v^+ \\ 0 & \text{if} \quad z < v^- \end{cases}$$

Ideally, one would like to maximize the objective while at the same time maximize the degree of satisfaction of all constraints. However, as these objectives are conflicting, the approach in fuzzy programming is to maximize the smallest membership function of all constraints. Thus, the fuzzy programming formulation of the preceding linear program with uncertain right-hand side parameters is

$$\max \lambda$$

subject to

$$\lambda \le 1 - \frac{v^+ - z}{v^+ - v^-}$$
$$\lambda \le 1 - \frac{a_i^t x - b_i}{\Delta b_i} \quad \forall i$$
$$0 \le \lambda \le 1$$
$$x \ge 0$$

By considering uncertainties in upper bounds of the purchase and sales variables, in [Reference 9], the authors present the following fuzzy programming formulation of **P**

Model FP

$$\max \lambda$$

subject to

$$NPV - (v^+ - v^-)\lambda \geq v^-$$
$$a_{jlt}^L \leq P_{jlt} + \Delta a_{jlt}^U \lambda \leq a_{jlt}^U + \Delta a_{jlt}^U \qquad \forall j, l, t$$
$$d_{jlt}^L \leq S_{jlt} + \Delta d_{jlt}^U \lambda \leq d_{jlt}^U + \Delta d_{jlt}^U \qquad \forall j, l, t$$
$$0 \leq \lambda \leq 1$$

and constraints (1.2), (1.3), (1.7), and (1.10) to (1.12); and *NPV* as defined in (1.9).

The aspiration level $[v^-, v^+]$ can be determined by solving **P** with the smallest and the largest upper bounds, respectively, on the purchase and sales variables. Thus, the fuzzy programming formulation requires the solution of three mixed-integer linear programs.

In reference [9], the previous formulation is also extended to include uncertainties in the objective function coefficients. The resulting formulation is a mixed-integer nonlinear program and was solved in reference [9] by global optimization algorithms.

Stochastic Programming

The approach taken for dealing with uncertain model parameters whose probability distribution is known or can be approximated somehow is known as stochastic programming. Two different formulations have been developed in this area. *Recourse* models involve making some of the decisions before realizations of the uncertain parameters occur and, subsequently, taking corrective measures. On the other hand, *chance constrained* models include probabilistic constraints which might be violated owing to uncertainty. Formulations combining both these philosophies have also been considered. For modeling the long-range planning problem under uncertainty, recourse formulations are ideal. This will become clear as we describe a two-stage recourse formulation for the process planning problem.

A standard formulation of the two-stage stochastic linear program is as follows [see Reference 5].

Model 2S-SLP

$$z = \min_x \{cx + E_{\omega \in \Omega}[Q(x, \omega)] | Ax \leq b, x \geq 0\} \qquad (1.41)$$

where

$$Q(x, \omega) = \min_y \{f(\omega)y | D(\omega)y \geq h(\omega) - T(\omega)x, y \geq 0\} \qquad (1.42)$$

Here, $x \in R^{n_1}$, $A \in R^{m_1 \times n_1}$, $c^T \in R^{n_1}$, $b \in R^{m_1}$, $y \in R^{n_2}$, $D(\omega) \in R^{m_2 \times n_2}$, $T(\omega) \in R^{m_2 \times n_1}$, $f(\omega)^T \in R^{n_2}$, $h(\omega) \in R^{m_2}$, ω is a random vector and Ω is the set of all possible realizations of the random vector ω. In most cases, the matrix D is assumed to be "fixed," that is, independent of ω, in which case Eq. (1.42) becomes

$$Q(x, \omega) = \min_y \{f(\omega)y | Dy \geq h(\omega) - T(\omega)x, y \geq 0\} \qquad (1.43)$$

In the preceding formulation, problem (1.41) is the "first stage" problem and problem (1.43) is the "second stage" problem. The interpretation of 2S-SLP is as follows. The decision maker must select the activity levels of the design variables (x) "here and now," that is, before the uncertainties are realized.

Depending upon the realizations of ω, an appropriate choice of the operating variables (y) can then be made. As the second stage cost $Q(x,\omega)$ is a function of the random vector, an appropriate objective is to minimize the expectation of this cost function.

The problem of process planning under uncertainty has been formulated as a two stage stochastic linear program in reference [8], and with a similar formulation in reference [4], as follows.

Model SP

$$\max NPV = -\sum_t \sum_i (\alpha_{it} E_{it} + \beta_{it} y_{it})$$

$$+ E_\omega \left[\sum_t \left\{ -\sum_i \delta_{it}(\omega) W_{it} + \sum_j \sum_l (\gamma_{jlt}(\omega) S_{jlt} - \Gamma_{jlt}(\omega) P_{jlt}) \right\} \right]$$

subject to

$$a_{jlt}^L(\omega) \le P_{jlt} \le a_{jlt}^U(\omega) \quad \forall j, \forall l, \forall t \tag{1.44}$$

$$d_{jlt}^L(\omega) \le S_{jlt} \le d_{jlt}^U(\omega) \quad \forall j, \forall l, \forall t \tag{1.45}$$

and constraints (1.2), (1.3), (1.7), and (1.10) to (1.12).

The relationship between formulations **2S-SLP** and **SP** becomes obvious by recasting **SP** as a minimization problem and observing that the first stage problem is defined by variables $x = (E_{it}, Q_{it}, y_{it})$ and constraints (1.2), (1.7), (1.11), and (1.12). The second stage problem is then defined by variables $y = (W_{it}, P_{jlt}, S_{jlt})$ and constraints (1.3), (1.7), (1.10), (1.44), and (1.45).

The main difficulty with solving **SP** is the evaluation of the expectation functional. Several approximate methods have been developed. The deterministic approach is to *a priori* discretize the probability distribution of the uncertain parameters and construct scenarios or sets of realizations of these parameters. A probability weight is associated with each scenario, and a deterministic equivalent of the problem can be constructed. For example, the deterministic equivalent of **2S-SLP** with $s \in S$ scenarios with probability p^s is given as follows.

Model Det-2SSLP

$$z = \min cx + \sum_{s \in S} p^s f^s y^s$$

subject to

$$Ax \le b \tag{1.46}$$

$$-T^s x + D^s y^s \ge h^s \qquad \forall s \in S \tag{1.47}$$

$$x \ge 0 \quad y^s \ge 0 \qquad \forall s \in S \tag{1.48}$$

The size of the deterministic equivalent grows rapidly as the number of scenarios increase. The special block diagonal structure of the problem can be exploited using decomposition methods. One such method is the L-shaped decomposition method of Van Slyke and Wets [16]. The method is the same as Benders decomposition and is based on the observation that, once the first stage variables are fixed, the subproblems corresponding to each of the scenarios are independent and separable.

Another approach is to use a Gaussian quadrature approximation for the integration to evaluate the expectation function. Ierapetritou and Pistikopoulos [4] proposed such a method in conjunction with Benders decomposition for the planning problem. However, such a method is very expensive computationally.

Stochastic approaches have also been proposed in which, instead of using deterministic approximations of the expectation function, sampled estimates are employed. Liu and Sahinidis [8] used one such approach, where Monte Carlo sampling was incorporated into a decomposition algorithm, and lower bounds were estimated by solving subproblems for a sampled set of scenarios. In theory, these methods converge asymptotically, that is, as the sample size approaches infinity. However, for the planning problem, computational experiments of Liu and Sahinidis show good convergence behavior even for small sample sizes.

Fuzzy (FP) vs. Stochastic Programming (SP)

By comparing the two formulations, the following remarks can be made.

1. Formulation **SP** requires information about the probability distribution of the uncertain parameters, while model **FP** only requires their tolerances.
2. Model **FP** is easier to solve because it requires solving three small sized MILPs. On the other hand, model **SP** is a large-scale MILP and requires special methods that exploit structure.
3. The constraints for each scenario are strictly enforced in **SP** by associating a cost of satisfying these constraints. **FP** allows some violation of constraints.

A natural question to ask is then, "If there is no information regarding the distribution of the uncertain parameters, is the fuzzy programming approach better than using stochastic programming with an arbitrary distribution assumption?" Liu and Sahinidis [8] investigated this question in the context of the planning problem by computationally comparing the fuzzy formulation with a stochastic programming approach where a triangular distribution was constructed from ranges of uncertain variables. They found that, although **FP** provides a computational advantage, the **SP** solutions are superior because they provide feasibility over the entire range of the uncertain parameters.

Robust Optimization

The stochastic programming approach deals with minimizing expected future operational costs without any conditions on the variability of these costs. It is likely that, under high-risk situations, a design with low expected second stage cost might run into a specific realization of the uncertain parameters for which the second stage cost is unacceptably large. The importance of controlling variability of the costs as opposed to just optimizing its first moment is well recognized in portfolio management applications.

To handle the trade-off associated with the expected cost and its variability in stochastic programs, Mulvey et al. [11] introduced the framework of robust optimization (RO). RO is a goal programming approach where the objective function of the stochastic program is extended by weighted variability contribution. Let $v(y^1, y^2, \ldots, y^{|S|})$ be a measure of the variability of the recourse costs. Then, the deterministic equivalent of the stochastic programming formulation is modified as follows.

Model RO

$$\min cx + \sum_{s \in S} p^s f^s y^s + \rho v(y^1, y^2, \ldots, y^{|S|}) \quad \text{subject to constraints (1.46) to (1.48)} \quad (1.49)$$

where ρ is a goal programming weight. By changing the weight ρ, the relative importance of the expectation and variability of the second stage costs in the objective can be controlled to construct an "efficient frontier" of solutions.

Another approach for enforcing robustness is the restricted recourse framework of Vladimirou and Zenios [17]. This method differs from robust optimization in that, here, the variability of the second stage solution is reduced by adjusting the bounds on the second stage variables. The original formulation in reference [17] has been developed by the inclusion of a constraint on the mean Euclidean deviation of the second stage solution vector from the expected solution vector. In this way, the dispersion of the second stage solutions is restricted to a prescribed level. Instead of restricting the dispersion of the solution vectors, the approach to restrict the dispersion of the random recourse costs can also be considered. Because the variability of the recourse costs is a measure of its dispersion, the restricted recourse formulation can then be stated as follows.

Model RR

$$\min cx + \sum_{s \in S} p^s f^s y^s$$

subject to constraints (1.46) to (1.48) and

$$\nu (y^1, y^2, \dots, y^S) \le \epsilon \tag{1.50}$$

where ϵ is a prescribed tolerance level for the variability of the second stage costs.

An important issue in enforcing recourse robustness in both the robust optimization and restricted recourse approaches is the choice of the variability metric $\nu(y^1, y^2, \dots, y^{|S|})$. In most applications, the natural choice is variance. Variance is a symmetric risk measure, penalizing costs both above and below the expected recourse cost equally. As is the case for the chemical process planning problem we are interested in solving and, in general, whenever cost minimization is involved, it is more appealing to use an asymmetric risk measure that would penalize only costs above the expected value. Moreover, the use of variance results in the inclusion of nonlinearities in the formulation. For capacity expansion problems that involve integer decisions, inclusion of the variance would transform the original mixed integer linear program into a mixed integer nonlinear one making it computationally very difficult to solve. To overcome these difficulties, Ahmed and Sahinidis [1] propose to use the upper partial mean as the measure of variability of the recourse costs. The upper partial mean (UPM) of the recourse costs is defined as follows.

$$\bar{\Delta}(y^1, y^2, \dots, y^{|S|}) = \sum_{s \in S} p^s \Delta^s(y^1, y^2, \dots y^{|S|}) \tag{1.51}$$

where

$$\Delta^s(y^1, y^2, \dots, y^{|S|}) = \max \left\{ 0, f^s y^s - \sum_{s' \in S} p^{s'} f^{s'} y^{s'} \right\} \tag{1.52}$$

The preceding variability measure has been used to develop robust optimization and restricted recourse formulations of the process planning problem in reference [1]. These models can be used to generate a spectrum of solutions with varying degrees of recourse robustness and provide the decision maker with a tool to analyze the trade-off associated with the expected profit and its variability.

1.5 Conclusions

The purpose of this chapter has been to review the various optimization techniques developed in recent years for long-range planning of chemical processes. This discussion is summarized next.

1. First, a general model was developed, which was subsequently reduced to a mixed-integer linear program under some simplifying assumptions.
2. Various solution strategies for the MILP model were then discussed. Important approaches in this regard are the various types of cutting planes which help to reduce the integrality gap.

3. Extensions of the MILP model to account for power economies of scale and differing plant types were also presented.
4. Uncertainty in the problem data was approached through fuzzy and stochastic programming formulations of the same problem. Solution strategies developed for these models make possible the solution of large-scale problems.

There are several avenues that could be further explored. Some important research directions are identified next.

1. Most of the bounding and cutting plane generation techniques could be used in the context of capacity planning problems from other industrial sectors.
2. A complete complexity classification of the problem would be interesting.
3. The problem, being an integer program, is inherently difficult. Thus, there is considerable motivation for the development of heuristics or approximation schemes. Worst- and average-case performance measures of these heuristics for the process planning problem could be an important contribution. Liu and Sahinidis [7] recently initiated some work in this area, such as high variability settings, that could also be explored.

In conclusion, the problem of long-range planning in the chemical industry is a very intriguing one. The complexity of the problem holds considerable challenge for researchers, while its application potential is attractive to practitioners.

Acknowledgments

The authors are grateful for partial financial support from the National Science Foundation under CAREER award DMII 95-02722 to N.V.S.

References

1. Ahmed, S. and Sahinidis, N. V., Robust process planning under uncertainty, *Ind. Eng. Chem. Res.,* 37, 1883, 1998.
2. Balas, E. and Pulleyblank, W., The perfectly matchable subgraph polytope of a bipartite graph, *Networks,* 13, 495, 1983.
3. Benders, J. F., Partitioning procedures for solving mixed variables programming, *Num. Math.,* 4, 238, 1962.
4. Ierapetritou, M. G. and Pistikopoulos, E. N., Novel optimization approach of stochastic planning models, *Ind. Eng. Chem. Res.,* 33, 1930, 1994.
5. Kall, P. and Wallace, S. W., *Stochastic Programming,* John Wiley & Sons, Chichester, U.K., 1994.
6. Liu, M. L. and Sahinidis, N. V., Long range planning in the process industries: a projection approach, *Comput. Oper. Res.,* 23, 237, 1995.
7. Liu, M. L. and Sahinidis, N. V., Optimization in process planning under uncertainty, *Ind. Eng. Chem. Res.,* 35, 4154, 1996.
8. Liu, M. L. and Sahinidis, N. V., Process planning in a fuzzy environment, *Eur. J. Oper. Res.,* 100, 142, 1996.
9. Liu, M. L. and Sahinidis, N. V., Bridging the gap between heuristics and optimization: the capacity expansion case, *AIChE J.,* 43, 2289, 1997.
10. Liu, M. L., Sahinidis, N. V., and Shectman, J. P., Planning of chemical processes via global concave minimization, in *Global Optimization in Engineering Design,* I. E. Grossmann, Ed., Kluwer Academic, Boston, MA, 1996.
11. Mulvey, J. M., Vanderbei, R. J., and Zenios, S. A., Robust optimization of large-scale systems, *Oper. Res.,* 43, 264, 1995.
12. Nemhauser, G. L. and Wolsey, L. A., *Integer and Combinatorial Optimization,* John Wiley & Sons, New York, 1988.

13. Sahinidis, N. V. and Grossmann, I. E., Multiperiod investment model for processing networks with dedicated and flexible plants, *Ind. Eng. Chem. Res.*, 30, 1165, 1991.
14. Sahinidis, N. V. and Grossmann, I. E., Reformulation of the multiperiod milp model for capacity expansion of chemical processes, *Oper. Res.*, 40, Suppl. 1, S127, 1992.
15. Sahinidis, N. V. Grossmann, I. E., Fornari, R. E., and Chathrathi, M., Optimization model for long range planning in the chemical industry, *Comput. Chem. Eng.*, 13, 1049, 1989.
16. Van Slyke, R. and Wets, R., L-shaped linear programs with applications to optimal control and stochastic programming, *SIAM J. Appl. Math.*, 17, 638, 1969.
17. Vladimirou, H. and Zenios, S. A., Stochastic linear programs with restricted recourse, *Eur. J. Oper. Res.*, 101, 177, 1997.

2

Feature-Based Design in Integrated Manufacturing

Venkat Allada
University of Missouri-Rolla

2.1 Introduction

The sequential engineering approach to product design and development typically treats design and manufacturing as isolated activities. In this approach, the design department designs an artifact and throws it "over the wall" to the manufacturing department without taking into consideration the manufacturing capabilities and limitations of the shop floor. The manufacturing department, in turn, studies the design from a manufacturability viewpoint and throws it back "over the wall" to the design department with a list of manufacturing concerns. Typically, the artifact drawings go back and forth between the two departments until, eventually, the drawings are approved for production. Obviously, this situation prolongs the product realization time. Also, the cost of making design changes increases sharply with time. Owing to global competition, many manufacturing industries are under intense pressure to compress the product realization time and cost.

These industries have realized that the sequential engineering approach should be discarded in favor of the Concurrent Engineering (CE) approach. The CE approach assumes that design and manufacturing activities are highly interdependent. It emphasizes that crucial manufacturing issues should be considered at the design stage in order to decrease the number of design iterations. Within the CE context, major research effort is being devoted in the development of seamless integrated engineering design and manufacturing systems. These integrated systems should emphasize both the syntactic-level and semantic-level sharing of information.

One of the major bottlenecks in building integrated design-manufacturing systems is the incomprehension of the language of Computer-Aided Manufacturing (CAM) systems by the Computer-Aided Design (CAD) systems. The CAD systems were initially envisioned to serve as drafting systems. Currently, the CAD systems are being continuously enhanced to conduct Computer-Aided Engineering (CAE) analysis at various levels of sophistication. The design information provided by the CAD system is implicit and in terms of low level primitives which has limited use in conducting a comprehensive manufacturing analysis. The design information provided by the CAD system needs to be translated into explicit manufacturing information such as part features in order to be understood by various CAM application systems. Thus, features serve as a link between the CAD and CAM systems. This link would be beneficial to many manufacturing applications such as process planning, Group Technology (GT) coding, Numerical Control (NC) code generation, inspection, and assembly.

The rest of the chapter is organized as follows: Section 2.2 presents the various x-refs definitions of the term "feature" and feature taxonomies; Section 2.3 discusses the various feature-based design approaches; Section 2.4 presents the relation between CAD modeling and automatic feature recognition systems; Section 2.5 presents feature-based design applications; Section 2.6 presents the major research issues in the area of feature-based manufacturing; and, finally, Section 2.7 presents the chapter summary.

2.2   Definition of Features and Feature Taxonomies

CAM-I (1981) defined a form feature as, "A specific geometric configuration formed on the surface, edge, or corner of a work-piece intended to modify outward appearance or to aid in achieving a given function." There seems to be no consensus amongst researchers regarding the definition of the term "feature." For example, the manufacturing, design, and analysis features for a given part may not be the same. This means that the definition of the feature is context dependent [Woodwark, 1988; Shah, 1991a]. Also, within the realm of manufacturing features, features can be categorized as prismatic part features, rotational part features, sheet metal features, welding features, casting features, forging features, die casting features, and so on. Typically, manufacturing features are manufacturing process dependent.

The different definitions of features put forward at the NSF-sponsored workshop on *Features in Design and Manufacturing* [NSF, 1988] include, "a syntactic means to group data that defines a relationship to other elements of design," "a computer representable data relating to functional requirements, manufacturing process or physical properties of design," "attributes of work pieces whose presence or absence affects any part of the manufacturing process starting from process planning to final packaging," "regions of a part with some manufacturing significance," and so on. Pratt and Wilson [1985] defined a form feature as a "region of interest on the surface of a part." Shah [1991a] defined features as "elements used in generating, analyzing, or evaluating design." Pratt [1991] defined a form feature as, "A related set of elements of a product model, conforming to characteristic rules enabling its recognition and classification, which, regarded as an entity in its own right, has some significance during the life cycle of the product."

The feature-based product definition is a high-level semantic description of shape characteristics of a product model. Though the number of features are infinite, it is possible to form a finite categorization of the form features. Several research studies have been conducted to develop feature taxonomy. Pratt and Wilson [1985] have developed a scheme for CAM-I, which has been adopted by the form features information model (FFIM) of the Product Data Exchange Specification (PDES). In PDES [1988], features are classified as follows.

- Passages that define negative volumes that intersect the part model at both ends.
- Depressions that define negative volumes that intersect the part model at one end.
- Protrusions that are positive volumes that intersect the part model at one end.
- Transitions that are regions present in the smoothing of intersection regions.

- Area features that are 2-D elements defined on the faces of the part model.
- Deformations that define shape changing operations such as bending, stretching, and so on.

Cunningham and Dixon [1988] classified form features based on the role they play in the product design activity. Form features are classified as kinetic features and static features. Kinetic features are defined as elements that encompass energy or motion transfer. Static features are further classified as follows.

- Primitives that define the major shape of the part model.
- Add-ons that describe local changes] on the part model.
- Intersections that define the type of interaction between primitives and add-ons.
- Whole forms that describe the attributes of the entire part model.
- Macros which are essentially combinations of primitives.

Pratt [1991] classified features as manufacturing features, design features, analysis features, tolerance and inspection features, assembly features, robotics features, and overall shape features. A good review of feature taxonomy for rotational parts is given by Kim et al. [1991]. Shah and Mäntylä [1995] distinguished various geometric features using a classification of features such as the following.

- Form features that describe portions of nominal geometry.
- Tolerance features that describe deviations from nominal form/size/location.
- Assembly features that describe assembly relations, mating conditions, fits, and kinematic relations.
- Functional features that describe feature sets related to specific function such as design intent, performance, and so on.
- Material features that describe material composition, treatment, and so on.

2.3 Feature-Based Design Approaches

A review of the literature on feature-based design systems has been provided by many researchers [Joshi, 1990; Chang, 1990; Shah, 1991a; Shah et al., 1991b; Singh and Qit, 1992; Salomons et al., 1993; Allada, 1994; Shah et al., 1994; Allada and Anand, 1995; Shah and Mäntylä, 1995]. The three popular feature-based design approaches are as follows:

- Human-assisted feature recognition.
- Automatic feature recognition.
- Design by features approach.

In the human-assisted feature recognition systems, the designer interacts with the CAD model to define a feature by picking up the entities from the part drawing that constitutes a particular feature. Examples of such systems are the TIPPS system by Chang and Wysk [1983] and the KAPPS system by Iwata and Fukuda [1987a]. These systems generally do not have feature validation procedures to verify user actions.

Automatic feature recognition systems recognize the features after a part is modeled using a CAD system. Typically, these automatic feature recognition systems use geometric and/or topological information to infer the presence of a particular type of feature. The approach of extracting manufacturing features seems very logical given the fact that these features can be mapped onto a limited number of manufacturing processes. For example, the possible manufacturing processes that can be employed for making a feature "hole" are drilling, boring, or reaming. While a number of robust methodologies have been devised to recognize primitive features (noninteracting), devising algorithms/methodologies to recognize interacting features is still an open-ended research problem that needs deeper investigation. To date, there exists no general automatic feature recognition methodology that would recognize all types of features interactions. One of the drawbacks of automatic feature recognition systems is that they tend to be fairly complex and computationally intensive.

In the design by features approach, the designer creates a part model using boolean operations and by instantiating the primitive features (from the feature library) at a desired location. While this approach eliminates the need for feature recognition from a part model, it can run into major problems when features interact with each other. Feature validation needs to be performed every time a new feature is added. This is to ensure that the new feature is placed in the correct position or if the new feature distorts the validity of the existing features. Another issue which comes up in the design by features approach is the determination of what features must be present in the feature library. A feature library with too many predefined features may be cumbersome for the designer. One solution to this problem is to have a limited set of features in the feature library (hopefully the commonly used features) and provide the designer with an option to create user defined features (UDFs). Furthermore, the design by features approach assumes that the designer is capable of choosing the best set of features to model a given artifact that has complex interacting features. The notion of capturing only one set of features (in other words, a single interpretation of features) for defining a part model may impose serious limitations while performing the manufacturing analysis. Dixon et al. [1990] identified the following unresolved issues in the development of design by features systems.

- Need for formal definition of the term "feature."
- System architecture issues.
- Developing methods to handle interacting features.
- Nature and scope of the feature library.
- Provision for user-defined features.
- Use of features in conceptual assembly design systems that enable design at various levels of abstraction and in multiple functional viewpoints.
- Mechanism to capture the design intent for its use in managing the propagation of design changes.

Based on the discussion so far it is clear that neither the design by features approach nor the automatic feature recognition approach is problem free. This has lead to a consensus amongst researchers that a hybrid approach incorporating both the approaches is best suited for feature-based design systems. The development of such a hybrid system is still in its infancy. The feature validation requirement by design by features approach reinforces the belief that automatic feature recognition is closely linked to it and would play a dominant role in the feature-based product modeling systems of the future [Meeran and Pratt, 1993].

2.4 Automated Feature Recognition and CAD Representation

Most automatic feature recognition systems proposed by researchers are dependent on the type of solid modeling representational scheme. Table 2.1 depicts the classification of automated feature recognition systems based on the CAD representational scheme employed.

TABLE 2.1 Automated Feature Recognition Systems and CAD Representation Scheme Used

CAD Representation Scheme	Representative Automated Feature Recognition Work
1. Constructive Solid Geometry (CSG)	Woo [1984], Lee and Fu [1987], Woodwark [1988], Perng et al. [1990], and Kim and Roe [1992]
2. Boundary Representation (B-Rep)	Kyprianou [1980], Jared [1984], Falcidieno and Giannini [1989], Sakurai and Gossard [1988], Joshi and Chang [1990], Prabhakar and Henderson [1992], Marefat and Kashyap [1992], Laakko and Mäntylä [1993], and Allada and Anand [1996]
3. Cellular Decomposition	Grayer [1977], Armstrong et al. [1984], Yamaguchi et al. [1984], and Yuen et al. [1987]
4. Wireframe	Meeran and Pratt [1993], Li et al. [1993], and Agarwal and Waggenspack [1992]

At first, the CSG representation seems to be ideally suited for developing automated feature recognition systems. However, a CSG tree poses numerous problems in feature recognition. It forces the designer to understand the manufacturing processes in order to select the appropriate primitives. The CSG tree contains information in an "unevaluated" form wherein the geometry and topology of the part is not readily available. Furthermore, the CSG tree representation is "nonunique." For these reasons, very few researchers have used a CSG scheme for developing feature recognition systems. Woodwark [1988] proposed three ways of simplifying CSG models for their potential use in feature recognition.

- Restrict the domain of the model by restricting the range of primitives and/or of the orientations that they may assume.
- Restrict the allowable ways in which the primitives may interact spatially.
- Restrict the set-theoretic expressions defining the part model.

The automated feature recognition systems reported in the literature can also be classified into two types as volumetric feature recognition systems or surface feature recognition systems. These two types of systems can be further classified based on the feature recognition approach that is employed (such as graph-theoretic, neural net, or rule-based approaches). Readers are referred to Allada [1994] and Allada and Anand [1995] for further details.

2.5 Feature-Based Design Applications

Feature-based technology has been widely used for a variety applications. Some of the applications of the feature-based design approach are listed in Table 2.2.

TABLE 2.2   Some Feature-Based Design Applications

Application Domain	Representative Research Work
1. Group Technology (GT) Coding	Kyprianou [1980], Iwata et al. [1987b], and Srikantappa and Crawford [1992]
2. NC Code/Cutter Path Generation	Grayer [1977], Parkinson [1985], Woo [1984], Yamaguchi et al. [1984], Armstrong et al. [1984], Yuen et al [1987], and Lee and Chang [1992]
3. Generative Process Planning	Hummel and Brooks [1986], CAM-I [1986], Requicha et al. [1988], Joshi and Chang [1990], van Houten [1990], Vandenbrande and Requicha [1993], Han and Requicha [1997], and Regli et al. [1997]
4. Tolerance Representation	Requicha and Chan [1986], Gossard et al. [1988], Shah and Miller [1990], Martino [1992], and Roy and Liu [1988, 1993]
5. Automated Inspection	Henderson et al. [1987], Park and Mitchell [1988], Hoffman et al. [1989], and Pahk et al. [1993]
6. Automated Assembly	Rosario and Knight [1989], Nnaji and Lick [1990], Li and Huang [1992], Shah and Tadepalli [1992], Lin and Chang [1993], and Arai and Iwata [1993]
7. Automated Grasp Formulation	Huissoon and Cacambouras [1993]
8. Fixturability/Setup Planning	Wright et al. [1991], Fuh et al. [1992], and Kumar et al. [1992], Chang [1990], Delbressine et al. [1993], Chu and Gadh [1996]
9. Finite Element Method (FEM) Analysis	Henderson and Razdan [1990]
10. Mold Design	Irani et al. [1989], Hui [1997]
11. Manufacturability/Tooling Cost Evaluation	Luby et al. [1986], Gadh and Prinz [1995], Rosen et al. [1992], Yu et al. [1992], Terpenny and Nnaji [1992], Poli et al. [1992], Mahajan et al. [1993], Gupta et al. [1995], Raviwongse and Allada [1997a,b]

2.6   Research Issues in Feature-Based Manufacturing

Architecture of the Feature-Based Design System

As was mentioned earlier, the widely shared belief by experts in the field of features technology is that the feature-based system architecture should be a blend of the design by features approach and automatic feature recognition systems [CAM-I, 1990; Dixon et al., 1990; Falcidieno et al., 1992; Chamberlain et al., 1993; Laakko and Mäntylä, 1993]. Both approaches rely on special-purpose geometric reasoning/algorithms for identifying a nonprimitive (interacting/compound) 3-D manufacturing feature. Martino et al. [1993] have developed an integrated system of design by features and feature recognition approaches based on a unified model.

A common feature library and a unified model link the geometric modeler and the feature-based modeler. The unified model is expressed as a hierarchial graph with each node corresponding to a shape feature volume represented in a boundary form. The connecting arcs represent connections between volumes expressed by their overlapping faces. In the system architecture proposed by Martino et al. [1993], the user interacts with the CAD system in three ways — through the feature editor, the feature modeler, and the solid modeler. The authors view two important issues in a hybrid feature-based system — the development of intertwined data structures which associate the geometric model of a part with its feature-based description and the system flexibility for supporting user-defined features and procedures.

Feature Recognition Techniques for Complex Parts

Independent machining features, such as slot, step, pocket, boss, and so on, can be easily recognized by most automatic feature recognition systems or can be easily modeled using the design by features approach. The number of 3-D primitive features are finite, but the number of features resulting from the interactions of the primitive 3-D features are infinite. However, recognizing interacting features such as boss originating from a pocket, or a feature originating from more than one face, is relatively difficult.

Interacting features pose major problems in automatic feature recognition. These problems occur because interacting features[1] cause the destruction of topological relations in a part model. For example, interacting features may cause some of the faces to be completely deleted, partially missing, or fragmented in several regions [Vandenbrande and Requicha, 1993]. Thus, feature recognition systems based on a syntactic pattern approach may not be suitable for recognizing arbitrary feature interactions. Vandenbrande and Requicha [1993] favored the use of a CSG tree representation for accommodating arbitrary types of feature interactions. They concluded that the feature recognition techniques cited in the literature suffer from one or more of the following problems.

- Features identified by the feature recognition algorithms do not contain comprehensive information that is required by the process planning activity, such as the ability to perform volumetric tests to detect intrusions or feature interactions, tool collisions, feature precedence analysis, and so on.
- Feature recognition algorithms do not provide "multiple" interpretations of features necessary to generate alternative process plans.
- Feature recognition algorithms often employ a number of special case or enumerative approaches to detect feature interactions. These algorithms often cannot be generalized (or extended) to provide a broader coverage of arbitrary feature interaction cases.
- The full potential of solid modeling system is seldom used to perform geometric reasoning on features.

[1]In this context interacting features are assumed to be physically interacting where their volumes are adjacent or intersect with each other.

Tseng and Joshi [1994a,b] described a methodology for detecting interacting features for certain classes of features such as slots, steps, and pockets. However, their study is limited to detecting interacting "depression" features for prismatic parts. Gadh and Prinz [1995] used a high-level abstract entity called the "loop" to define feature classes and their boundaries. The concept of "bond-cycle" has been defined to determine on which side of the boundary the closed curve (loop) lies. The feature interactions problem in this paper has been essentially viewed as one that exists owing to interaction between feature boundaries. Feature interaction cases have been classified as follows.

1. Interacting features sharing edges.
2. Interacting features sharing vertices.
3. Interacting features sharing faces.

Narang [1996] proposed a feature recognition methodology that is application independent (independent of manufacturing process) and one that generates explicit representation of geometric feature interactions. The methodology has been tested for 21/2-D parts. Suh and Ahluwalia [1995] developed an approach for classifying various feature interactions. They classified feature interaction cases into the following categories.

1. An existing feature removed by a new primitive feature.
2. An existing feature remains without any interaction with the new primitive feature.
3. An existing feature is modified by the new primitive feature.

The third case where an existing feature is modified by the new primitive feature is classified into three cases.

1. A part of a feature set including its boundary edges is removed.
2. A part of a feature set excluding its boundary edges is removed.
3. The convexity of the feature boundary edges is changed.

Feature modification methods have been developed for each of these feature interaction cases. Suh and Ahluwalia [1996] concluded that additional investigations are needed to cover the general operations (other than Boolean operations) with primitives and for cases where more than two features are mutually interacting.

Regli and Pratt [1996] have raised many interesting research issues relating to feature interactions. While a number of research studies have been directed for recognizing interacting features, as yet no general approach has been devised. Addressing the issue of interacting features (irrespective of whether design by features or an automatic feature recognition approach is used for feature information) is certainly important for conducting manufacturing analysis.

Multiple Interpretation of Features

Physically interacting features may result in multiple interpretation of features. A given set of interacting features can have multiple interpretations. Multiple interpretation of features is especially useful to generate alternate process plans. The process planning system reported by Chang [1990] uses heuristic techniques for refining features (either combining features or splitting features for machining). However, heuristic systems may not produce alternative feature interpretations for some cases.

Karinthi and Nau [1992] described an algebraic approach for determination of alternative feature interpretations. However, the work described cannot be used directly for manufacturing planning purposes because it has some limitations such as generation of infeasible feature interpretations and the inability of the algebraic approach to generate all possible feature interpretations. The choice of the optimal process plan usually involves the deployment of search engines to investigate the performance characteristics of the feasible alternative process plans. For example, Gupta [1994] reported a methodology for the selection of process plans from a set of various alternatives based on adherence to specified design tolerances and a rating system. Generation of feasible process plans (through multiple interpretations of features under various constraints) and subsequent identification of an optimal process plan under multiattribute objective function is an area which needs to be researched further.

Han and Requicha [1997] developed an Integrated Incremental Feature Finder (IF^2) based upon the earlier work on Object Oriented Feature Finder (OOFF) system[2] developed at the Programmable Automation Laboratory, University of Southern California. The IF^2 system generates the part interpretation in terms of machining features by analyzing *hints* from nominal geometry, direct user input, tolerances, and design features. It uses heuristics to derive the part interpretation (but is capable of generating alternative interpretations only when the user requests for it) and emphasizes finding a satisfactory solution as opposed to finding an optimal solution.

Regli et al. [1997] presented multiprocessor algorithms (using the distributed computing approach) to recognize machining features from solid models. The feature recognition method uses a trace-based approach (hint-based) to reconstruct feature instances from the "partially destroyed" feature information present on the final part model.

Most of the feature recognition systems that recognize interacting features split the complex interacting feature into a set of simple independent features. However, this set of independent features may be nonunique. Methods need to be devised which split the complex feature into a set of features based on factors such as minimum machining time, precedence analysis, and process capability of the shop floor.

Incorporation of Tolerancing Information in the Feature Model

Incorporation of tolerancing information in a feature modeling system has been pursued by many researchers. The various tolerance representational schemes used by researchers are [Shah and Miller, 1990]:

- Evaluated entity structures (for example, the EDT model of Johnson, 1985; Ranyak and Fridshall, 1988; Shah and Miller, 1990).
- CSG-based structures (for example, the VGraph structure by Requicha and Chan, 1986; Elgabry, 1986).
- Constraint-based face adjacency graphs (for example, Faux, 1990; Gossard et al., 1988; Roy and Liu, 1988).
- Constructive variational geometry (CVG) approach by Turner and Wozny [1988].

Shah and Miller [1990] suggested that the tolerance modeler should not only store the tolerances but should be capable of storing the meaning of the tolerances in the data structure. The guidelines for developing a tolerance modeler as envisioned by them are listed next.

- Support all the information needed to define all ANSI tolerance classes.
- Flexible to incorporate special tolerances to certain company-specific products.
- Support data reference frames to be tagged and their precedence to be specified where applicable.
- Network tolerances with the geometric and feature elements.
- Provide validity checking of geometric elements.
- Support material modifiers (material condition or tolerance zone modifiers).
- Automatic checks on legality of tolerances.
- Apply default tolerances to untoleranced elements.
- Provide graphic display of all features, data, and tolerance frames to the designer.

Roy and Liu [1993] have developed a geometric tolerance representational scheme that has been interfaced with the TWIN solid modeling system. The user has the flexibility to input the tolerance information in either a CSG or a B-Rep database. The tolerance representation is based on two kinds of features, namely, low-level entities such as face, edge, point, and high-level features such as slot, hole,

---

[2]See Vandenbrande and Requicha [1993].

and so on. Guilford and Turner [1992] identified some of the deficiencies in the tolerancing models proposed by researchers prior to the year 1990. They reported that the committee on Shape Tolerance Resource Model within the Standard for the Exchange of Product Model Data (STEP part 47) is attempting to define an unambiguous representation of tolerances compatible with the ANSI Y14.5 and the ISO 1101 family of standards. They identified a problem that exists in identifying locations and directions while defining tolerances and data. In STEP, these are represented by Cartesian vectors, but the problem of locating the part in the co-ordinate system exists. Guilford and Turner [1992] modified the approach employed by STEP in order to overcome the problems. For example, STEP describes the direction along which the straightness tolerance is measured as a vector in the global co-ordinate system, while the authors described it by using some virtual geometry entities attached to the actual geometry of the part. The authors have discussed the representation which covers almost all of the ANSI Y14.5 tolerances except for items such as knurls, gears, and screws; equalizing data, free state variations, conical position tolerance zones, and position tolerances on elongated holes.

Feature Data Exchange Mechanisms

Standards for Exchange of Product Data (STEP) is an international standard (designated as ISO 10303) that deals with the computer interpretable representation and exchange of product model data. The intent is to provide a neutral interface which is capable of describing all the life-cycle properties of a given artifact independent of the CAD platform used for product modeling. This will also serve as a basis for implementing and sharing product databases and archives. The various parts of ISO 10303 are divided into the following categories: description methods, application protocols, abstract test suites, implementation methods, and conformance testing.

Application protocols (AP) provide a basis for developing implementations of STEP (ISO 10303) and abstract test suites for conformance testing of Application Protocol (AP) implementations. AP 224 (developed by TC184/SC4/WG3) is a part of the application protocol category that defines the context, scope, and information requirements of producing mechanical product definition for process planning application, and it directs the integrated resources necessary to satisfy these requirements. These requirements specify items such as part identification, shape, and material data, necessary for product definition. The basic premise of AP 224 is that the process planning function will be greatly assisted by identifying machining features present on the part model. Knowledge about the machining features will help in proper identification of machining equipment, tooling, and processes to manufacture a part. AP 224 provides a schema for representation and exchange of part feature information.

ISO 10303-AP 224 employs two ways to represent the shape of part features: implicit shape representation and explicit shape representation. The explicit shape representation is specified by using a B-Rep (boundary representation) scheme. The implicit shape representation is specified by defining parameters (attributes) associated with each type of feature. Currently, three basic types of features are employed in AP 224, namely, machining features (such as hole, groove, boss, thread, etc.), replicate features (such as circular pattern, rectangular pattern, etc.), and transition features (such as chamfer, fillet, and rounded edge). Compound features (user defined features) can be created by the union of one or more machining features. The technical content of AP 224 provides good coverage on part features and associated attributes. However, it can be extended in scope (though the actual approval process needs input from representatives of several countries) to include some of the following issues.

- Multiple part mechanical parts as opposed to single piece mechanical parts
- Inclusion of features produced by manufacturing processes other than turning and milling
- Interacting features and feature relations deemed critical from the process planning standpoint
- Multiple "viewpoints" of features
- Support other CAD representation schemes than just the B-Rep scheme
- Support "redesign" product development by providing part retrieval mechanisms
- Provide definition of commonly used catalog parts such as nuts, bolts, gear, etc.

Shah and Mathew [1991] expressed the view that it would be necessary to develop feature data-exchange mechanisms as companies might use feature modelers from more than one vendor. They feel that establishing standards will enable feature data exchange between two feature modelers and allow the transfer of feature information from a feature modeler to an application. The FFIM (Form Feature Information Model) developed by the PDES committee has the capability of exchanging geometric and topological information. FFIM is not well suited for exchanging semantic information such as design intent and product type. Shah and Mathew [1991] identified some of the problems in FFIM, such as the lack of relational positioning/location information, multiple representations of a single feature, representation of some commonly used profiles using complex data structures, loss of semantic information, minor numerical inaccuracies, and nonunique mapping of features. The extensiveness of the feature library is an issue of concern for the feature standardization problem. The possible trade-offs suggested by Shah and Mathew [1991] are listed next.

- Standardization of "common" shapes and semantic information.
- Standardization of general classes of shapes only.
- Standardization of "common" shapes and facility to convey nonstandard information to the application programs.
- Standardization of low-level shape and semantic information.

Feature Mapping

Feature mapping refers to transformation of product information from one application (or viewpoint) to another. A feature space is dependent to a large extent upon the product type, application domain, and level of abstraction [Shah, 1988; Shah and Rogers, 1988]. Readers are referred to the work reported by Shah [1988] for a complete treatment on the concepts involved in feature mapping. Shah et al. [1990] developed a conjugate mapping shell for mapping between conjugate features. The concept of feature mapping was employed by Duan et al. [1993] to provide two kinds of application mappings — mapping from design definition into FEM analysis application and mapping into process planning and NC programming application. Rosen and Peters [1992] developed a mathematical foundation for the conversion of a design representation into a manufacturing representation. As an example, they converted the design representation into a tooling cost representation for conducting tooling cost evaluations for molding and casting operations. They found that small design changes may have a large influence on manufacturability and, hence, the conversion function is discontinuous. They suggested the use of topological spaces to represent both design and manufacturing. However, many possible choices of topology could exist and the issue as to whether topology can be used to get a continuous conversion function needs to be resolved. Shah and Mäntylä [1995] summarized various methods used in literature to avoid the problem of feature mapping as follows.

- The designer assumes a known machining stock primitive and uses Destructive Solid Geometry (DSG) to instantiate the various machining volumes (primitives).
- Use geometric models to communicate between various applications by developing domain-specific feature recognition techniques.

Shah and Mäntylä [1995] mentioned that the general methods for implementation of feature mapping are still not discovered. However, they presented some of the evolving methods used for feature mapping which include heuristic methods, mapping with intermediate structures, cell-decomposition mapping methods, and graph-based methods.

Feature Relations Taxonomy

While performing engineering analysis, features cannot be treated as isolated entities. Many researchers have attempted to identify application-specific feature relations. Feature relations such as "contained_side," "contained_bottom," and so on, were used to determine the sequence of setups [Kanumury and Chang, 1991].

Joshi [1990] used the "open_into" relation for performing a machining precedence analysis. The concept of "handles" was used by Turner and Anderson [1988] to establish the positional relationship between two or more features. Feature relations such as the "branch_connect" relation were used for determining the machining precedence [Inui and Kimura, 1991]. Anderson and Chang [1990] used the nesting and intersection relations to aid the process planning activity. However, noncontact type feature relations were not considered. ElMaraghy [1991] reported the development of a feature-based design language that could be used with a feature-based modeler to allow the designer to specify part names and attributes such as surface finish and relations with other features in a textual form. Chen et al. [1992] used relations such as "Is_in" and "adjacent_to" to support the manufacturability assessment (specifically, castability and moldability). Shah [1991b] listed three possible situations with regard to feature relations.

1. Features related by a geometric constraint that can be parameterized; for example, bolt holes, or gear teeth.
2. Features related by a geometric constraint (such as adjacency, tangency, edge sharing, etc.) but cannot be parameterized; for example, stepped holes or complex pockets.
3. Features with no geometric relationships but grouped together for reference or convenience.

Allada [1994] discussed various feature relations from a machining perspective for the following purposes.

1. Identification of design violations.
2. Identification of avenues for performing gang operations and, thus, help in the creation of efficient process plans.
3. Mapping design features into manufacturing features by either collecting design features and forming one or more manufacturing (feature gathering) or decomposing a design feature into two or more manufacturing features (feature decomposition).

Work on feature relations is primarily devoted to contact-type feature relations. Research in the area of noncontact-type feature relations is still evolving. Much of the existing literature in feature relations is in fragments and, to the best of the author's knowledge, a systematization and organization of the work in feature relations (from various application perspectives such as machining, casting, welding, etc.) has not been attempted. The research issues involved in a feature relations study could include

- Formalization of feature relations.
- Development of application-independent feature relations taxonomy as well as an application-specific feature relations taxonomy.

Manufacturability Evaluation

Typically, the designer designs a product with minimal knowledge about the capabilities and limitations of the manufacturing technology. The designer then tries to find out if the design can be manufactured. An alternate way of designing products could be first to find the answer to the question: What can be manufactured? This helps the designer to understand the limitations and capabilities of manufacturing technology concurrently while designing a product. While the latter seems to be an ideal approach to reach the goal of manufacturing it right the first time, there are a number of serious research issues that need to be addressed.

- How to represent the knowledge base and data base regarding the limitations and capabilities of the manufacturing technology in a computer interpretable format? How to build incremental manufacturing process models? Most of the manufacturability systems cited in literature are not current with the rapid advances in machine tool technology. Fast-paced technological advances may make some of the current manufacturing knowledge base outdated.
- Many companies use in-house manufacturing technology domains to determine the product manufacturability. However, this will preclude the designer from experimenting with his/her creative boundaries; a product may be identified as not being *manufacturable* or *manufacturable*

but at a high cost if the knowledge base is limited to just the in-house manufacturing technology capabilities. In many such situations, outsourcing (buy decision) may be a viable option. The design creativity of the designer should not be limited to company-specific manufacturing practices. Rather, a much broader manufacturing technology should reside in the system's knowledge base. Primarily, the designer will use the in-house technology knowledge base to perform manufacturability checks. If the evaluation is not satisfactory, the designer can perform a manufacturability check using an expanded knowledge-base (which should include the manufacturing capabilities of vendor companies).

A vast amount of literature exists in the area of feature-based manufacturability evaluation. Readers are referred to the paper by Gupta et al. [1995] for a comprehensive review of the work on automated manufacturability analysis. Some of the more recent work is presented in this section.

Das et al. [1996] developed a methodology for automatically generating redesign suggestions for machined parts by using the setup cost as the criterion. Their methodology allows for the designer's restrictions on the redesign solutions such as the type and extent of modifications allowed on certain faces and volumes. Dissinger and Magrab [1996] proposed and implemented a manufacturability evaluation approach for a powder metallurgy manufacturing process. The geometric model of the part consisted of a set of basic arbitrarily shaped manufacturable entities such as plates, blind cavities, and through cavities.

Geometric reasoning is employed to

1. Determine the part's orientation and tooling requirements.
2. Identify the features such as sharp corners, feathered edges, and thin walls which affect the die fill, tooling cost, part integrity, and density control.
3. Automatically "redesign" the part to desirable features such as fillets, edge rounds, tapers, and axial flats.

Chu and Gadh [1996] developed a feature-based approach to minimize the number of set-ups in process planning a machined part. In the first step, the machining features are classified into two classes — single tool approach direction (STAD) features and multiple tool approach direction (MTAD) features. The MTAD features are further classified into two groups — MTAD features such as a double-ended open slot with finite tool approach directions and MTAD features such as a flat surface with infinite tool approach directions. The second step consists of the following substeps.

1. Setup determination for all the STAD features.
2. Setup determination for all the MTAD features.
3. Determination of fixturing features for each setup (assumed to be a pair of parallel surfaces on the part that will be held by a standard vice).
4. Determination of machining sequencing within each setup using knowledge-based rules, feature attributes, and feature relations.
5. Determination of setup ordering based on knowledge-based rules, feature parameters, and other constraints for setup ordering.

Chu and Gadh [1996] have noted the limitations of their approach which include the following.

1. Limited to three-axis vertical milling machining center.
2. Cannot handle sculptured surfaces.
3. Raw stock is always assumed to be a rectangular bar.
4. The tool approach directions and knowledge-based rules for a given feature must be defined in advance.
5. The process plan yielded by this approach is based on the minimization of set-ups criterion and may not necessarily be the optimal one.

An automated design-to-cost (DTC) system using feature-based design, Group Technology (GT), and activity-based costing concepts was developed by Geiger and Dilts [1996]. In this prototype system, the designer creates a part model in a feature-based environment after which the system classifies the part using Optitz GT code. Hui [1997] developed three algorithms to study the relation between the parting direction of an injection molded or die-cast component and its external and internal undercuts, and their influence on part moldability. The first algorithm is concerned with the search for main parting direction, the second one with the search for side cores, and the third one with the search for split cores. Gupta et al. [1995] identified the following issues that are important for developing future automated manufacturability systems.

- Ability to handle multiple manufacturing processes such as casting, machining, injection molding, and so on.
- Ability to generate alternative manufacturing plans to produce a product.
- Ability of the system to work in a virtual enterprise and distributed manufacturing mode where multiple vendors with varying capabilities exist.
- Development of a manufacturing knowledge base based on process models and manufacturing simulations.
- Development of appropriate measures of manufacturability.
- Accounting for design tolerances by the manufacturability system.
- Automatic generation of redesign suggestions for the design violations detected by the system.
- Product-life cycle considerations such as manufacturability, assembly, and so on, and associated trade-offs.
- Use of emerging information technologies such as the World Wide Web to build manufacturability systems.
- Manufacturability system validation studies in industrial settings.
- Effective Human–Computer Interaction (HCI) so that the designer can easily interact with the system.

Ranking of Redesign Alternatives

Another area requiring research attention in the context of generation of redesign solutions (design advisor) is the ranking of the generated redesign alternatives [Allada, 1997]. One of the important tasks of an automated manufacturability evaluator is to check for any design violations and provide redesign alternatives to the designer. Most manufacturability evaluation systems cited in the literature provide redesign advice by enumerating all possible alternative solutions for a given design violation. This is of little use to the designer who may be constrained from implementing many of those redesign solutions. Allada [1997] presented a preliminary framework for the generation of intelligent "contextual" redesign solutions based on the concepts of functional representation of features, feature flexibility, and ranking of redesign solutions. Features present on the part model are closely tied to the design intent. The designer may have a variety of reasons (functional, weight reduction, safety, aesthetics, etc.) for having the feature on the part model. Each of these reasons have a different degree of importance depending on the context of the product. The concepts of feature flexibility and feature importance can be used to reduce the redesign solution search space. The problem of ranking of the redesign solutions (on the reduced redesign solution set) can then be formulated using a goal programming method [Allada, 1997].

Product Design Optimization

Another important issue that calls for further research is the development of a feature modeling system capable of performing product design optimization [Allada and Anand, 1995]. The implications of different types of features of a product model on product-life cycle issues such as safety, ergonomic issues, aesthetics, manufacturing, assembly, fixturing, recycling, and so on, should be considered. Each of these product-life cycle issues may have a distinct set of features representing a particular viewpoint. These product-life cycle issues often conflict with each other. The feature modeling system should have the capability to resolve these conflicts while performing the design optimization process. This could be augmented by sensitivity analysis procedures, which would perform "what if" analysis by tightening and relaxing the user assigned weightage points to the different product design implication factors. Alternative ways of redesigning a product should be indicated to the designer by the feature modeling system.

Dimension-Driven Geometric Approach

Dimensions are represented explicitly in both the CSG and B-Rep data structures [Sheu and Lin, 1993]. It is well-known that dimensions play a crucial role in product modeling as well as serve to relate part features. In the design-by-features approach, the features are predefined by a set of parameters. This means that only the feature size is variant but the feature form is invariant. Li et al. [1993] used the concept of composite feature in which the feature parameters can edit the parameters and several different shapes (forms) can be created. A composite feature is defined as a virtual feature which consists of several design and process attributes. Each design and process attribute is defined by a set of parameters associated with the feature. In the traditional dimension-driven geometric (DDG) approach, the design changes (but not the shape changes) are made through dimensional changes. Typically, the part geometry can be rescaled by first editing the annotated dimensions instead of first changing the geometric primitives such as lines, arcs, and surfaces.

Gossard et al. [1988] employed an object graph based on a hybrid B-Rep/CSG scheme for explicitly representing dimensioning, tolerances, and features on the part model. Dimensions are explicitly represented by using the concept of Relative Position Operator (RPO). RPO is a scalar quantity equal to the nominal dimension value by which a particular face is to be moved with respect to the other face so that its position is appropriate in the object space. In addition, the RPO has upper and lower bounds representing the tolerances.

Li et al. [1993] addressed the issue of incorporating composite feature and variational design into a feature-based design system for 2-D rotational parts. They used a dimensional operation unit (DOU), a modification of RPO, to include both position and feature changes. A part hierarchial graph is used to compute and evaluate the changes in the geometric shape.

Sheu and Lin [1993] proposed a representation scheme suitable for defining and operating form features. The five basic elements used to represent a form feature are as follows.

1. B-Rep data structure (similar to the half-edge data structure).
2. Measure entities, which attach dimensions to the solid model.
3. Size dimension, which is a high level abstraction of a specific dimension controlling the intrinsic size of the form feature.
4. Location dimension, which represents the relative positional relationship between child and parent features.
5. Constraints, which restrict the special behavior of the form feature.

The part is then represented using a feature dependency graph (FDG). In the FDG structure, the dimensions are used to determine the location and size of form features.

Effects of Using Parallel Numerical Control (NC) Machines

Most research studies in feature-based manufacturing are limited to conventional machining technology. Today, it is not uncommon for even a small-scale industry to have NC machines. Levin and Dutta [1992] identified the characteristics of an NC machine as follows.

1. Capability to perform simultaneous operations.
2. Can have functional combinations (for example, capability of a lathe and a milling machine).
3. Possess secondary spindles which can machine rear face of the part (portion not machinable from the main spindle).

Levin and Dutta [1992] defined a parallel NC machine as one having multiple tool-holding devices and possessing multiple work-holding devices. This opens up an entire spectrum of research avenues for determining the best possible machining strategy, including the following [Levin and Dutta, 1992].

- *Sharing of machining parameters* — the sequence in which machinable features are removed has greater effect on total machining time because operations performed on parallel machines may share machining parameters.
- *Modes of parallel machines* — the various modes of parallel machines are part rotating as in turning, part stationary as in drilling, and both part and tool in motion as in contour milling. The machine mode has direct implications on the process plan of the part model. For instance, a radial drilling operation and a turning operation cannot be performed simultaneously.
- *Dynamic collision avoidance* — in parallel machining, two or more tools may occupy the same spatial location at a given time which is unacceptable. Hence, the swept tool path volume as a function of time has to be reconfigured.
- *Fixturing and setup planning* — these issues need to be addressed in case of parallel machines.
- *Batch machining* — this situation arises when multiple parts are machined on the same machine.
- *Optimization of machining time* — minimizing the machining time is a major goal of parallel machines as opposed to minimizing transit time in conventional machines.

In addition to the research directions provided by the Levin and Dutta [1992], some of the other research directions from the perspective of feature-based design are worth mentioning. Most previous research studies are based on a single tool accessibility direction (for a given setup). However, for parallel NC machines the precedence analysis and tool accessibility analysis from multiple-viewing directions need to be considered. Also, interfeature relations (how a feature on the rear face of the part accessible to the secondary spindle but inaccessible to the main spindle would be related to the features accessible by the main spindle) need to be considered from the multi-view perspective. Additionally, the feature-based design concepts could be extended further in the actual configuration design of Special Purpose Machine tools (SPMs). For instance, if a company identifies a family of parts to be machined beforehand, then the best configuration of the spindles for the parallel machine tool can be deduced. The best configuration of the spindles could be based on the types and number of features present on the part models, feature accessibility direction, and interfeature relations.

2.7 Summary

For the realization of an effective integrated manufacturing environment, the features technology is probably the best known approach. Features provide semantic information of the part model that is useful in automating many of the downstream manufacturing applications. In this chapter, a general understanding of the feature-based design systems and its usefulness in building integrated manufacturing systems has been presented. Major research issues in the area of feature-based manufacturing systems are identified and discussed at length. It is clear that the current research in feature-based design is quite diverse but, in many instances, applicable only to narrow domains. However, with the rapid pace of research advances in this area, we hope that many of the research questions will be unraveled.

References

Agarwal, S. C. and Waggenspack, W. N., Jr., Decomposition method for extracting face topologies from wireframe models, *Comput. Aid. Des.*, 24(3), 123, 1992.

Allada, V., Feature-Based Modeling for Concurrent Engineering, Ph.D. dissertation, University of Cincinnati, 1994.

Allada, V., Generation of Intelligent Redesign Solutions, presented at the International Conference for Industrial Engineering, San Diego, CA, November 1997.

Allada, V. and Anand, S., Feature-based modeling for integrated manufacturing: the state-of-the art survey and future research directions, *Int. J. Comput. Integr. Manuf.,* 6(8), 411, 1995.

Allada, V. and Anand, S., Machine understanding of manufacturing features, *Int. J. Prod. Res.,* 34(7), 1791, 1996.

Anderson, D. C. and Chang, T. C., Automated process planning using object-oriented feature based design, in *Advanced Geometric Modeling for Engineering Applications,* Krause, F. L. and Jansen, H., Eds., Elsevier Science, Amsterdam, 1990, 247.

Arai, E. and Iwata, K., CAD system with product assembly/disassembly planning function, *Robot. Comput. Integr. Manuf.,* 10(1/2), 41, 1993.

Armstrong, G. T., Carey, G. C., and De Pennington, A., Numerical code generation form a solid modeling system, in *Solid Modeling by Computers,* Picket, M. S. and Boyce, J. W., Eds., Plenum Press, New York, 1984, 139.

CAM-I, *CAM-I's Illustrated Glossary of Workpiece Form Features,* Report R-80-PPP-02.1 (Revised), May 1981.

CAM-I, *Feature Extraction and Process Planning Specific Study,* Report R-86-GM/PP-01, January 1986.

CAM-I, *Proceedings Features Symposium,* Report No. P-90-PM-02, Woburn (Boston), MA, August 9–10, 1990.

Chamberlain, M. A., Joneja, A., and Chang, T. C., Protrusion-features handling in design and manufacturing planning, *Comput. Aid. Des.,* 25(1), 19, 1993.

Chang, T. C., *Expert Process Planning for Manufacturing,* Addison-Wesley, Reading, MA, 1990.

Chang, T. C. and Wysk, R. A., CAD/generative process planning with TIPPS, *J. Manuf. Syst.,* 2(2), 127, 1983.

Chen, Y. M., Miller, R. A., and Lu, S. C., Spatial reasoning on form feature interactions for manufacturability assessment, *ASME Comput. Eng. Conf.,* 1, 29, 1992.

Chu, C. P. and Gadh, R., Feature-based approach for set-up minimization of process design from product design, *Comput. Aid. Des.,* 28(5), 321, 1996.

Cunningham, J. and Dixon, J., Designing with features: the origin of features, *Proceedings of ASME Computers in Engineering Conference,* San Francisco, 1988, 237.

Das, D., Gupta, S. K., and Nau, D. S., Generating redesign suggestions to reduce setup cost: a step towards automated redesign, *Comput. Aid. Des.,* 28(10), 763, 1996.

Delbressine, F. L. M., De Groot, R., and Van Der Wolf, A. C. H., On the automatic generation of set-ups given a feature-based design representation, *Ann. CIRP,* 42(1), 527, 1993.

Dissingeer, E. T. and Magrab, B. E., Geometric reasoning for manufacturability evaluation — application to powder metallurgy, *Comput. Aid. Des.,* 28(10), 783, 1996.

Dixon, J. R., Libardi, E. C., and Nielsen, E. H., Unresolved research issues in the development of design-with-features systems, in *Geometric Modeling for Product Engineering,* Wozny, M. J., Turner, J. U., and Preiss, K., Eds., Elsevier Science, Amsterdam, 1990, 183.

Duan, W., Zhou, J., and Lai, K., FMST: a feature solid-modeling tool for feature-based design and manufacture, *Comput. Aid. Des.,* 25(1), 29, 1993.

Elgabry, A. K., A framework for a solid-based tolerance analysis, *ASME Computers in Engineering Conference,* July 1986.

Elmaraghy, H. A., Intelligent product design and manufacture, in *Artificial Intelligence in Design,* Pham, D. T., Ed., Springer-Verlag, London, 147, 1991.

Falcidieno, B. and Giannini, F., Automatic recognition and representation of shape-based features in a geometric modeling system, *J. Comput. Vis. Graph. Imag. Process.,* 48, 93, 1989.

Falcidieno, B., Giannini, F., Porzia, C., and Spagnuolo, M., A uniform approach to represent features in different application contexts, *Comput. Ind.,* 19, 175, 1992.

Faux, I. D., Modeling of components and assemblies in terms of shape primitives based on standard dimensioning and tolerancing surface features, in *Geometric Modeling for Product Engineering,* Wozny, M. J., Turner, J. U., and Preiss, K., Eds., Elsevier Science, Amsterdam, 1990, 259.

Fuh, J. Y. H., Chang, C. H., and Melkanoff, M. A., A logic-based integrated manufacturing planning system, *ASME Comput. Eng. Conf.*, 1, 391, 1992.

Gadh, R. and Prinz, F. B., Automatic determination of feature interactions in design-for-manufacturing analysis, *Trans. ASME*, 117, 2, 1995.

Geiger, T. S. and Dilts, D. M., Automated design-to-cost: integrating costing into the design decision, *Comput. Aid. Des.*, 28(6/7), 423, 1996.

Gossard, D. C., Zuffante, R. P., and Sakurai, H., Representing dimensions, tolerances, and features in MCAE systems, *IEEE Comput. Graph. Appl.*, March, 51, 1988.

Grayer, A. R., The automatic production of machined components starting from a stored geometric description, in *Advances in Computer Aided Manufacturing*, McPherson, D., Ed., North Holland, Amsterdam, 137, 1977.

Guilford, J. and Turner, J., Representing geometric tolerances in solid models, *ASME Comput. Eng. Conf.*, 1, 319, 1992.

Gupta, S. K., Automated Manufacturability Analysis of Machined Parts, Ph.D. thesis, University of Maryland, College Park, 1994.

Gupta, S. K., Regli, W., Das, D., and Nau, D., Automated Manufacturability Analysis: a Survey, University of Maryland, Technical Report UMIACS-TR-95-08, 1995.

Han, J. H. and Requicha, A. A. G., Integration of feature based design and feature recognition, *Comput. Aid. Des.*, 29(5), 392, 1997.

Henderson, M. R., Extraction of feature information from three-dimensional CAD data, Ph.D. thesis, Purdue University, Indiana, 1984.

Henderson, M. R. and Razdan, A., Feature based neighborhood isolation techniques for automated finite element meshing, *Geometric Modeling for Product Engineering*, Wozny, M. J., Turner, J. U., and Preiss, K. (Eds.), Elsevier Science Publishers: North Holland, 301, 1990.

Hoffman, R., Keshavan, H. R., and Towfiq, F., CAD-driven machine vision, *IEEE Trans. Syst. Man Cybern.*, 19(6), 1477, 1989.

Hui, K. C., Geometric aspects of the mouldability of parts, *Comput. Aid. Des.*, 29(3), 197, 1997.

Huissoon, J. P. and Cacambouras, M., Feature model-based grasp formulation, *Int. J. Prod. Res.*, 31(2), 351, 1993.

Hummel, K. E. and Brooks, S. L., Symbolic representation of manufacturing features for an automated process planning system, *Proceedings of the ASME Winter Annual Meeting*, Anaheim, CA, 233, 1986.

Inui, M. and Kimura, F., Design of machining processes with dynamic manipulation of product models, in *Artificial Intelligence in Design*, Pham, D. T., Ed., Springer-Verlag, London, 195, 1991.

Irani, R. K., Kim, B. H., and Dixon, J. R., Integrating CAE, features, and iterative redesign to automate the design of injection molds, *ASME Comput. Eng. Conf.*, 1, 27, 1989.

ISO 10303, Part 47, Shape variation tolerances, draft, ISO TC184/SC4/WG3/P3, Version 2B, December 1989.

ISO, International Organization for Standardization, Product data representation and exchange, Part AP 224, Mechanical product definition for process planning using machining features, TC 184/SC4/SC4, July, 1995.

Iwata, K. and Fukuda, Y., KAPPS: Know-how and knowledge assisted production planning system in the machining shop, *Proceedings of the 19th CIRP International Seminar on Manufacturing Systems*, June 1–2, 1987a.

Iwata, K., Sugimura, N., and Fukuda, Y., Knowledge-based flexible part classification system for CAD/CAM, *Ann. CIRP*, 36(1), 317, 1987b.

Jared, G. E., Shape features in geometric modeling, in *Solid Modeling by Computers: from Theory to Applications*, Plenum Press, New York, 1984.

Johnson, R. J., Study of Dimensioning and Tolerancing Geometric Models, CAM-I Report R-84-GM-02.2, May 1985.

Joshi, S., Feature recognition and geometric reasoning for some process planning activities, in *Geometric Modeling for Product Engineering*, Wozny, M. J., Turner, J. U., and Preiss, K., Eds., Elsevier Science, Amsterdam, 1990, 363.

Joshi, S. and Chang, T. C., Feature extraction and feature based design approaches in the development of design interface for process planning, *J. Intell. Manuf.*, 1, 1990.

Kanumury, M. and Chang, T. C., Process planning in an automated manufacturing environment, *J. Manuf. Syst.*, 10(1), 67, 1991.

Karinthi, R. R. and Nau, D., An algebraic approach to feature interactions, *IEEE Trans. Patt. Anal. Mach. Intell.*, 14(4), 469, 1992.

Kim, J. G., O'Grady, P., and Young, R. E., Feature taxonomies for rotational parts: a review and proposed taxonomies, *Int. J. Comput. Integr. Manuf.*, 4(6), 341, 1991.

Kim, Y. S. and Roe, K. D., Conversions in form feature recognition using convex decomposition, *ASME Comput. Eng. Conf.*, 1, 233, 1992.

Kumar, A. S. K., Nee, A. Y. C., and Prombanpong, S., Expert fixture-design system for an automated manufacturing environment, *Comput. Aid. Des.*, 24(6), 316, 1992.

Kyprianou, L. K., Shape Classification in Computer Aided Design, Ph.D. thesis, Christ College, University of Cambridge, Cambridge, U.K., 1980.

Laakko, T. and Mäntylä, M., Feature modeling by incremental feature recognition, *Comput. Aid. Des.*, 25(8), 479, 1993.

Lee, Y. and Fu, K., Machine understanding of CSG: extraction and unification of manufacturing features, *IEEE Comput. Graph. Appl.*, 7(1), 20, 1987.

Lee, Y. S. and Chang, T. C., Cutter selection and cutter path generation for free form protrusion feature using virtual boundary approach, *Proceedings of the 2nd Industrial Engineering Research Conference*, Los Angeles, CA, 1992, 375.

Levin, J. B. and Dutta, D., On the effects of parallelism on computer-aided process planning, *ASME Comput. Eng. Conf.*, 1, 363, 1992.

Li, R. K. and Huang, C. L., Assembly code generation from a feature-based geometric model, *Int. J. Prod. Res.*, 30(3), 627, 1992.

Li, R. K., Tu, Y. M., and Yang, T. H., Composite feature and variational design concepts in a feature-based design system, *Int. J. Prod. Res.*, 31(7), 1521, 1993.

Lin, A. C. and Chang, T. C., An integrated approach to automated assembly planning for three-dimensional mechanical parts, *Int. J. Prod. Res.*, 31(5), 1201, 1993.

Luby, S. C., Dixon, J. R., and Simmons, M. K., Designing with features: creating and using a features database for evaluation of manufacturability of castings, *ASME Comput. Eng. Conf.*, 1, 285, 1986.

Mahajan, P. V., Poli, C., Rosen, D., and Wozny, M., Features and algorithms for tooling cost evaluation for stamping, *Towards World Class Manufacturing*, Wozny, M. and Olling, G., Eds., Elsevier Science, Amsterdam, 37, 1993.

Marefat, M. and Kashyap, R. L., Automatic construction of process plans from solid model representations, *IEEE Trans. Syst. Man Cybern.*, 22(5), 1097, 1992.

Martino, P., Simplification of feature based models for tolerance analysis, *ASME Comput. Eng. Conf.*, 1, 329, 1992.

Martino de, T., Falcidieno, B., Giannini, F., Haßinger, S., and Ovtcharova, J., Integration of design by features and feature recognition approaches through a unified model, in *Modeling in Computer Graphics*, Falcidieno, B. and Kunii, T. L., Eds., Springer-Verlag, Berlin, 423, 1993.

Meeran, S. and Pratt, M. J., Automated feature recognition from 2-D drawings, *Comput. Aid. Des.*, 25(1), 7, 1993.

Narang, R. V., An application-independent methodology of feature recognition with explicit representation of feature interaction, *J. Intell. Manuf.*, 7, 479, 1996.

National Science Foundation (NSF), *NSF Workshop on Features in Design and Manufacturing*, UCLA, February 1988.

Nnaji, B. O. and Liu, H. S., Feature reasoning for automatic robotic assembly and machining in polyhedral representation, *Int. J. Prod. Res.*, 28(3), 517, 1990.

Pahk, H. J., Kim, Y. H., Hong, Y. S., and Kim, S. G., Development of computer-aided inspection system with CMM for integrated mold manufacturing, *Ann. CIRP*, 42(1), 557, 1993.

Park, H. D. and Mitchell, O. R., CAD-based planning and execution of inspection, *Proceedings Computer Vision Pattern Recognition Conference,* Ann Arbor, MI, June 1988, 858.

Parkinson, A., The use of solid models in BUILD as a data base for NC machining, *Proceedings PROLO-MAT,* Paris, France, June 1985, 293.

PDES, PDES Form Feature Information Model (FFIM), PDES Form Features Group, Dunn, M., Co-ord., June 1988.

Perng, D. B., Chen, Z., and Li, R. K., Automatic 3-D machining feature extraction from 3-D CSG solid input, *Computer Aided Design,* 22, 285, 1990.

Poli, C., Dastidar, P., and Graves, R., Design knowledge acquisition for DFM methodologies, *Res. Eng. Des.,* 4(3), 131, 1992.

Prabhakar, S. and Henderson, M. R., Automatic form-feature recognition using neural-network based techniques on boundary representation of solid models, *Computer Aided Design,* 24, 381, 1992.

Pratt, M. J., Aspects of form feature modeling, *Geometric Modeling: Methods and Applications,* Hagen, H. and Roller, D., Eds., Springer-Verlag, Berlin, 1991, 227.

Pratt, M. J. and Wilson, P. R., Requirements for support of form features in a solid modeling system, Final Report, CAM-I Report R-85-ASPP-01, 1985.

Ranyak, P. and Fridshall, R., Features for tolerancing a solid model, *ASME Computers in Engineering Conference,* San Francisco, July 31–August 4, 1988, 275.

Raviwongse, R. and Allada, V., Injection mold complexity evaluation model using a Back propagation network implemented on a parallel computer, *Int. J. Adv. Manuf. Technol.,* 1997a.

Raviwongse, R. and Allada, V., A fuzzy-logic approach for manufacturability evaluation of injection molded parts, *J. Eng. Eval. Cost Anal.,* 1997b.

Regli, W. C. and Pratt, M. J., What are feature interactions? *ASME Computers in Engineering Conference,* August 18–22, Irvine, CA, 1996.

Regli, W. C., Gupta, S. K., and NAU, D. S., Towards multiprocessor feature recognition, *Comput. Aid. Des.,* 29(1), 37, 1997.

Requicha, A. A. G. and Chan, S. C., Representation of geometric features, tolerances and attributes in solid modelers based on constructive geometry, *IEEE J. Robot. Autom.,* 2(3), 156, 1986.

Requicha, A. A. G. and Vandenbrande, J., Automated systems for process planning and part programming, *Artificial Intelligence in Industry: Artificial Intelligence Implications for CIM,* Kusiak, A. (Ed.), IFS Publications Ltd., UK, 301, 1988.

Rosario, L. M. and Knight, W. A., Design for assembly analysis: extraction of geometric features from a CAD data base, *Ann. CIRP,* 38(1), 13, 1989.

Rosen, D. W. and Peters, T. J., Topological properties that model feature-based representation conversions within concurrent engineering, *Res. Eng. Des.,* 4, 147, 1992.

Rosen, D. W., Dixon, J. R., Poli, C., and Dong, X., Features and algorithms for tooling cost evaluation in injection molding and die casting, *ASME Comput. Eng. Conf.,* 1, 45, 1992.

Roy, U. and Liu, C., Feature based representational scheme of a solid modeler for providing dimensioning and tolerancing information, *Robot. Comput. Integr. Manuf.,* 4(3–4), 333, 1988.

Roy, U. and Liu, C., Integrated CAD frameworks: tolerance representation scheme in a solid model, *Comput. Ind. Eng.,* 24(3), 495, 1993.

Salomons, O. W., Van Houten, F. J. A. M., and Kals, H. J. J., Review of research in feature-based design, *J. Manuf. Syst.,* 12(2), 113, 1993.

Sakurai, H. and Gossard, D. C., Shape feature recognition from 3-D solid models, *Proceedings of the ASME International Computers in Engineering Conference,* San Diego, CA, 1988, 515.

Shah, J. J., Feature-transformations between application-specific feature spaces, *Comput. Aid. Eng. J.,* December, 247, 1988.

Shah, J. J., Assessment of features technology, *Comput. Aid. Des.,* 23(5), 331, 1991a.

Shah, J. J., Conceptual development of form features and feature modelers, *Res. Eng. Des.,* 2, 93, 1991b.

Shah, J. J. and Mathew, A., Experimental investigation of STEP form-feature information model, *Comput. Aid. Des.,* 23, 282, 1991.

Shah, J. J. and Mäntylä, M., *Parametric and Feature-Based CAD/CAM*, John Wiley & Sons, New York, 1995.

Shah, J. J. and Miller, D. W., A structure for supporting geometric tolerances in product definition systems for CIM, *Manuf. Rev.*, 3(1), 23, 1990.

Shah, J. J., and Rogers, M. T., Functional requirements and conceptual design of the feature-based modeling system, *Comput. Aid. Eng. J.*, February, 9, 1988.

Shah, J. J., and Tadepalli, R., Feature based assembly modeling, *ASME Comput. Eng. Conf.*, 1, 253, 1992.

Shah, J. J., Mäntylä, M., and Nau, D., Introduction to feature based manufacturing, in *Advances in Feature Based Manufacturing*, Shah, J. J., Mäntylä, M., and Nau, D., Eds., Elsevier Science, Amsterdam, 1994, 1.

Shah, J. J., Rogers, M. T., Sreevalsan, P. C., Hsiso, D. W., Mathew, A., Bhatnagar, A., Liou, B. B., and Miller, D. W., The A.S.U. features testbed: an overview, CAM-I Proccedings: Features Symposium, Report P-90-PM-02, Woburn, MA, August 1990, 129.

Shah, J., Sreevalsan, P., and Mathew, A., Survey of CAD/feature-based process planning and NC programming techniques, *Comput. Aid. Eng. J.*, vol. 8, 25, 1991.

Sheu, L. C. and Lin, J. T., Representation scheme for defining and operating form features, *Comput. Aid. Des.*, 25(6), 333, 1993.

Singh, N. and Qi, D., A structural framework for part feature recognition: a link between computer-aided design and process planning, *Integr. Manuf. Syst.*, 3(1), 4, 1992.

Srikantappa, A. B. and Crawford, R. H., Intermediate geometric and interfeature relationships for automatic group technology part coding, *ASME Comput. Eng. Conf.*, 1, 245, 1992.

Suh, H. and Ahluwalia, R. S., Feature modification in incremental feature generation, *Comput. Aid. Des.*, 27(8), 627, 1995.

Terpenny, J. P. and Nnaji, B. O., Feature based design evaluation for machine/tool selection for sheet metal, *Proceedings of the 2nd Industrial Engineering Research Conference*, Los Angeles, CA, 1992, 26.

Tseng, Y. J. and Joshi, S. B., Recognizing multiple interpretations of 21/2 D machining of pockets, *Int. J. Prod. Res.*, 32(5), 1063, 1994a.

Tseng, Y. J. and Joshi, S. B., Recognizing multiple interpretations of interacting machining features, *Comput. Aid. Des.*, 26(9), 667, 1994b.

Turner, G. P. and Anderson, D. C., An object-oriented approach to interactive feature-based design for Quick Turnaround Manufacturing, *Computers in Engineering Conference*, 1988, 551.

Turner, J. and Wozny, M., A framework of tolerances utilizing solid models, *Third International Conference on Computer Aided Production Engineering*, Ann Arbor, MI, June 1988.

Vandenbrande, J. H. and Requicha, A. A. G., Spatial reasoning for automatic recognition of machinable features in solid models, *IEEE Trans. Pat. Anal. Mach. Intell.*, 15(12), 1269, 1993.

van Houten, F. J. A. M., PART: A Computer Aided Process Planning System, Ph.D. dissertation, University of Twente, 1990.

Woo, T. C., Interfacing solid modeling to CAD and CAM: Data structures and algorithms for decomposing a solid, *IEEE Comput.*, vol. 17, 44, 1984.

Woodwark, J. R., Some speculations on feature recognition, *Comput. Aid. Des.*, 20(4), 483, 1988.

Wright, P. K., Englert, P. J., and Hayes, C. C., Applications of artificial intelligence to part setup and workholding in automated manufacturing, in *Artificial Intelligence in Industry: Artificial Intelligence in Design*, Pham, D. T., Ed., Springer Verlag, London, 1991, 295.

Yamaguchi, K., Kunii, T. L., Rogers, D. F., Satterfield, S. G., and Rodriguez, F. A., Computer integrated manufacturing of surfaces using octree encoding, *IEEE Comput. Graph. Appl.*, January, 60, 1984.

Yu, T. T., Chen, Y. M., Miller, R. A., Kinzel, G. L., and Altan, T., A framework for sheet metal part design and manufacturing assessment, *ASME Comput. Eng. Conf.*, 1, 53, 1992.

Yuen, M. F., Tan, S. T., Sze, W. S., and Wong, W. Y., An octree approach to rough machining, *Proc. Inst. Mech. Eng.*, 201(B3), 157, 1987.

3

Flexible Factory Layouts: Issues in Design, Modeling, and Analysis

Saifallah Benjaafar
University of Minnesota

In this chapter, we address several issues related to design, modeling, and analysis of flexible factory layouts. We present a framework for defining and identifying sources of layout flexibility and for mapping different dimensions of flexibility to specific layout configurations. We use this framework to develop several potential measures of layout flexibility and compare the usefulness and limitations of each. We also examine the relationship between factory layout and material handling and show that the realization of layout flexibility largely depends on the configuration of the material handling system. Finally, we present an integrated procedure for flexible layout design in stochastic environments. We use the procedure to highlight the desirability of disaggregating functional departments into smaller subdepartments and distribute them throughout the factory floor. We show that increased disaggregation and distribution can be, indeed, effective in enhancing a layout's ability to cope with variability.

3.1 Introduction

In today's volatile and competitive environment, manufacturing facilities must be designed with enough flexibility to withstand significant changes in their operating requirements. The shortening of product life cycles and the increased variety in product offerings require that facilities remain useful over many product generations and support the manufacturing of a large number of products. Because the proliferation of products makes it exceedingly difficult to produce accurate forecasts of demand volumes and demand distribution, facilities must be able to rapidly reallocate capacity among different products without major retooling, resource reconfiguration, or replacement of equipment. Furthermore, increased emphasis on customer satisfaction places simultaneous requirements on shorter manufacturing lead times, on time deliveries, and greater product customization. All of these competing constraints require that manufacturing facilities be designed to exhibit high levels of flexibility, responsiveness, and performance despite the high volatility and unpredictability of the environment.

Although a significant effort has been made in recent years to re-engineer manufacturing enterprises to meet the challenges of flexibility and agility, much of this effort has centered around either introducing new technologies (e.g., automation, computer integrated manufacturing, and advanced information systems) or new production management techniques (e.g., JIT production, TQM, and lean manufacturing) [2, 16, 41]. As a result, little attention has been devoted to the physical design and organization of facilities (i.e., factory layout and material handling). This is despite the fact that this activity is often critical in determining the amount of operational flexibility and agility that a facility would eventually be able to deliver [7, 26, 33]. Because the selection of a plant layout and a material handling system are often irreversible, it is important that these decisions explicitly incorporate the need for flexibility and agility. In a variable environment, this would minimize the need for costly relayouts and assure a consistent level of performance despite potential changes in production requirements.

Unfortunately, existing layout design procedures have been, for the most part, based on a deterministic paradigm where design parameters, such as product mix, product demands, and product routings, are assumed to be all known with certainty [9, 20, 22]. The design criterion used in selecting layouts is often a static measure of material handling cost which does not capture the need for flexibility in a stochastic environment [4, 7, 8]. In fact, the relationship between layout flexibility and layout performance remains poorly understood and analytical models for its evaluation are still lacking. The structural properties of layouts that make them more or less flexible also are not well understood. Indeed, there exists little consensus as to what makes one layout more flexible than another or as to how layout flexibility should be measured [10, 15, 33, 45]. In turn, this has led to difficulty in devising systematic procedures for the design and implementation of flexible layouts.

It has been conventionally accepted that, when product variety is high and/or production volumes are small, a functional layout, where all resources of the same type share the same location, offers the greatest flexibility. However, a functional layout is notoriously known for its material handling inefficiency and scheduling complexity [14, 27, 32–34, 43]. In turn, this often results in long lead times, poor resource utilizations and limited throughput rates. While grouping resources based on their functionality allows for some economies of scale and simplicity in workload allocation, it makes the layout vulnerable to changes in the product mix and/or routings. When they occur, these changes often result in a costly relayout of the plant and/or an expensive redesign of the material handling system [10].

Clearly, there is a need, in dynamic and stochastic environments, for an alternative class of layouts that is more flexible and responsive than a traditional functional layout. There is also a need for an alternative design criterion for the simple measure of material handling cost that is routinely used in most existing layout design procedures. More importantly, there is a need for systematic methods for designing and implementing layouts in stochastic environments that explicitly account for the value of flexibility. In this chapter, we review recent progress in these three areas. In Section 3.2, we provide an extensive review of existing literature on the subject. In Section 3.3, we present a framework for defining and identifying sources of layout flexibility in manufacturing plants and mapping these to specific layout configurations. In Section 3.4, we present several potential measures of layout flexibility and compare their usefulness and limitations. In Section 3.5, we present a procedure for flexible layout design. We use the procedure to illustrate various issues related to flexibility modeling, design, and implementation. Material for this chapter draws on results we have recently published in references [7] and [8].

3.2   Literature Review

Although there exists an abundant literature on manufacturing flexibility as it relates, for example, to machines, product mix, and part routing and sequencing [6, 36, 47], very little of this literature deals with layout flexibility. Webster and Tyberghein 45 were among the first to recognize the importance of flexibility in layouts. They defined layout flexibility as the ability of a layout to respond to known and future product mixes. They considered the most flexible layout to be the one with the lowest material handling cost over a number of demand scenarios. Bullington and Webster [10] extended this definition to the multi-period case and presented a method for evaluating layout flexibility based on estimating the costs of future relayouts. They recommended that these costs be used as an additional criterion in determining the most flexible layout.

Gupta [15] presented a simulation approach for measuring layout flexibility. Using a pairwise exchange heuristic, layouts are obtained for several randomly generated flow matrices. For each such layout, the distance between all pairs of departments is computed. These distances were then used to compute the average distance between departments over the set of all generated layouts. A penalty function measuring the sum of absolute deviations from their distance mean of all pairs of departments for a given layout is calculated. A layout with the smallest penalty is considered to be the most flexible layout. Shore and Tompkins [35] also proposed a penalty function as a criterion for choosing the most flexible layout. Their penalty function measures the expected material handling inefficiency of each layout over all possible production demand scenarios. Assuming the probability of each demand scenario is known and the number of scenarios is finite, the layout with the least expected inefficiency can be identified. This layout is considered to be the most flexible layout.

Rosenblatt and Lee [31] presented a robustness approach to the stochastic plant layout problem. They considered an uncertain environment in which the exact values of the probabilities of the different possible scenarios are unknown. For such an environment, layout flexibility is defined in terms of the robustness of the layout's performance under different scenarios. Thus, the most flexible (robust) layout is the one whose cost performance remains close to the optimal layout for the largest number of scenarios. A robustness approach to the single and multiple period layout problem was also proposed by Kouvelis et al. [18]. Rosenblatt and Kropp [25] presented an optimal solution procedure for the single period stochastic plant layout problem. They showed that their procedure only requires solving a deterministic from to flow matrix, where the deterministic matrix is a weighted average of all possible flow matrices. They compared their results to the flexible layout measure developed by Shore and Tompkins [35] and showed that their approach will always result in the most flexible layout.

The preceding approaches are primarily applicable to job shops. However, as discussed earlier, job shops suffer from many serious limitations, including complex material flows and inefficient material handling. An alternative to the functional organization of job shops is a cellular configuration, where the factory is partitioned into cells, as shown in Figure 3.1b, each dedicated to a family of products with similar processing requirements [2]. Although cellular factories can be quite effective in simplifying workflow and reducing material handling, they can be highly inflexible because they are generally designed with a fixed set of part families in mind whose demand levels are assumed to be stable and their life cycles are considered to be sufficiently long. In fact, once a cell is formed, it is usually dedicated to a single part family with limited allowance for intercell flows. While such organization may be adequate when part families are clearly identifiable and demand volumes are stable, they become inefficient in the presence of significant fluctuations in the demand of existing products or with the frequent introduction of new ones. A discussion of the limitations of cellular manufacturing systems can be found in references [5], [12], [17], [26], [27], [37], and [41]. These limitations resulted in recent calls for alternative and more flexible cellular structures, such as overlapping cells [1], cells with machine sharing [5, 37], and virtual cellular manufacturing systems [12, 17]. For highly volatile environments, certain authors have even suggested completely distributed layouts, where copies of a given machine type are dispersed as much as possible throughout the shop floor [26].

For computer integrated manufacturing systems, Drolet [12] proposed virtual cellular manufacturing systems (VCMS) as a more flexible alternative to conventional cellular configurations. Instead of configuring a manufacturing facility into cells, each dedicated to a specific part family, machines of various types are distributed throughout the shop floor and reconfigured in real time in virtual cells in response to actual job orders. Upon completing the job order, the virtual cell is disbanded and the associated machines again are made available to the system. Figure 3.2 contrasts the differences between virtual cellular layouts and conventional cellular layouts. The author does not, however, provide a procedure for generating these layouts.

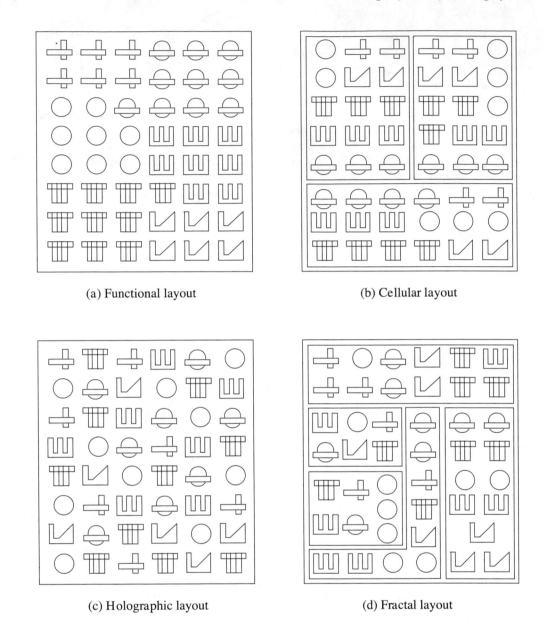

(a) Functional layout   (b) Cellular layout

(c) Holographic layout   (d) Fractal layout

FIGURE 3.1   Functional, cellular, holographic, and fractal layouts.

To support flexible system configurations, such as a VCMS, Montreuil et al. [26] introduced the concept of holographic layouts as an alternative to process layouts for systems operating in highly volatile environments. A holographic layout, as illustrated in Figure 3.1c, spreads the machines of each type throughout the manufacturing facility. For each machine of a particular type, an attempt is made to ensure its proximity to machines of every other type so that routings that are flow efficient can be created in real time by an intelligent shop floor control system. A heuristic design procedure was proposed where the objective is to generate a layout such that each machine is as centrally located, with respect to other machines of different type, as possible. The procedure assumes, however, that no distinguishable flow patterns exist and that a maximally dispersed layout is always desirable.

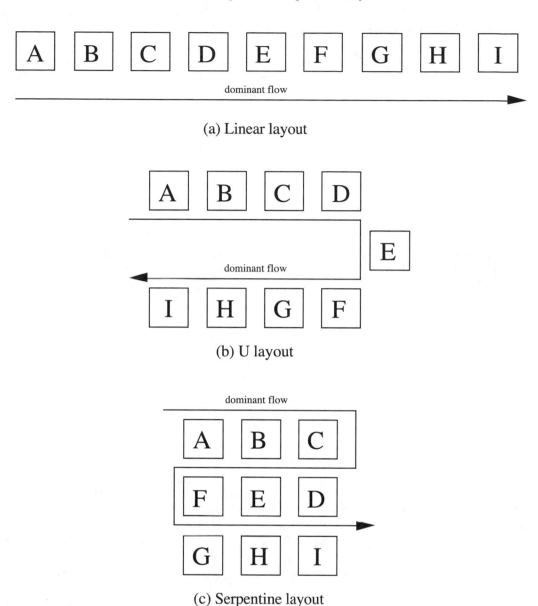

(a) Linear layout

(b) U layout

(c) Serpentine layout

FIGURE 3.2 Linear, U, and Serpentine layouts.

More recently, the same authors [27, 42] proposed the *fractal* layout as yet another alternative for job shop environments. In a fractal layout, as shown in Figure 3.1d, several almost identical cells are created, with each cell being allocated, whenever possible, an equal number of machines of each type. This results in a set of flexible cells, where every product can be produced in almost every cell. The authors propose a multistage design methodology for cell creation, cell layout, and cell flow assignment, in which cell layouts are generated in an iterative fashion with flow assignment decisions. Askin et al. [3] compare functional, fractal, and holographic layouts when part routings, part interarrival times, and processing times are random. Using a queueing model, they showed that the functional layout is dominated by both the fractal and holographic layouts. Although the holographic layout provides the best performance under a number of conditions, the authors find it not to be always robust.

3.3   Flexible Layout Configurations

In this section, we examine desirable properties of flexible layouts and identify structural features that make certain layouts more flexible than others. We define a flexible layout as one that is capable of coping effectively with variability and change. Sources of variability and change could include one or more of the following.

1. Fluctuations in product demand and product mix.
2. Introduction of new products or elimination of existing ones.
3. Changes in product design or manufacturing.
4. Facility expansion or contraction.
5. Introduction of new equipment or elimination of existing ones.
6. Changes in material handling configuration or replacement of material handling devices.

In general, sources of variability can be classified as being either product or process related. Product-related variability is due to changes in either the type or quantity of products produced in the facility, or to the way in which they are produced, which directly affect the volume of material flow between different locations in the layout. Process-related variability is due to physical changes made to the facility, through the addition/elimination of processing equipment or material handling and storage, which also affect the amount of material flow between different areas of the layout. A flexible layout should, therefore, accommodate changes in flow volumes resulting from either source of variability without significantly affecting performance.

If we measure layout performance by the amount of material travel between different areas, or departments, of the layout, a flexible layout can then be redefined as one that maintains short material travel despite variability in the amount of flow between different departments. Because material flow volumes are determined by the product routing sequences, a flexible layout should ideally provide equally efficient travel for all producible products. In other words, regardless of the sequence in which different departments must be visited by a product, material flow should remain efficient. This can be achieved by letting each department be easily accessible from any other department. For example, in Figure 3.2, we show a series of layouts with increasing degrees of routing flexibility. In the linear layout of Figure 3.2a, only products with routings similar to the dominant flow can be produced efficiently. In the U-layout of Figure 3.2b, efficiency is improved because the physical distances between most departments are substantially reduced. However, not all routings are yet equally efficient (e.g., the sequence $A \rightarrow F \rightarrow B \rightarrow E \rightarrow H \rightarrow D \rightarrow I$ has long travel distances). In the serpentine layout of Figure 3.2c, most of inefficiencies are eliminated, and almost all sequences have equal travel distances. Note that the gains in flexibility achieved in the U or the serpentine are not achieved at the expense of the efficiency of the dominant flow. In fact, in all three layouts, the efficiency of the dominant flow remains the same.

In addition to being affected by the relative location of different departments, layout flexibility is also affected by the number of copies of each department that can be independently and separately located on the plant floor. If multiple copies of the same department are available, layout flexibility can be enhanced by strategically locating these copies in different areas of the plant floor. The distribution of similar departments throughout the plant increases the accessibility to these departments from different regions of the layout. In turn, this improves the material travel distances of a larger number of product sequences. Note that department duplication does not necessarily require the acquisition of additional resources but can simply be achieved by disaggregating existing departments into smaller subdepartments. Figure 3.3 shows a series of layouts with increasing department duplication and distribution.

The degree to which layout flexibility is realized can be largely determined by the flexibility of the material handling devices and the material handling configuration. Departments may be in close physical proximity, but may not be accessible easily because of the configuration of the material handling system. In fact, this is often the case when material handling is assured by automated devices [e.g., automated conveyance systems, automated guided vehicles (AGV), and robots]. For instance, a unidirectional AGV could severely limit the flexibility of even a highly distributed/disaggregated layout. Similarly, the configuration of the

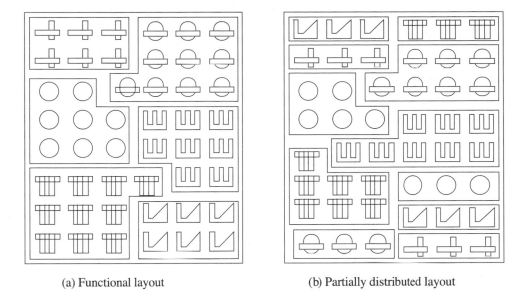

(a) Functional layout   (b) Partially distributed layout

(c) Fully distributed layout

FIGURE 3.3   Functional, partially distributed, and fully distributed layouts.

material handling paths, or traffic aisles, could directly determine how much flexibility is available. Figure 3.4 shows three material handling configurations with varying degrees of flexibility. The linear layout, where material flows are unidirectional, is clearly the least flexible. It is vulnerable to changes in the production operation sequence and cannot easily accommodate the introduction of new departments and/or the elimination of existing ones. The loop layout is more flexible because it provides for easier material flow between any pair of departments and can more easily accommodate changes in product routings. The star layout is, however, the most flexible of all three layouts because all departments are equidistant from each other. The introduction of new departments, or the elimination of existing ones, can be done with little impact on the rest of the plant or its overall efficiency.

In industries where production capacity must be added gradually, a layout should be able to expand gracefully without affecting efficiency. Similarly, in industries where processing equipment must be changed, replaced, or eliminated frequently, the ability to make small or local layout changes without affecting the

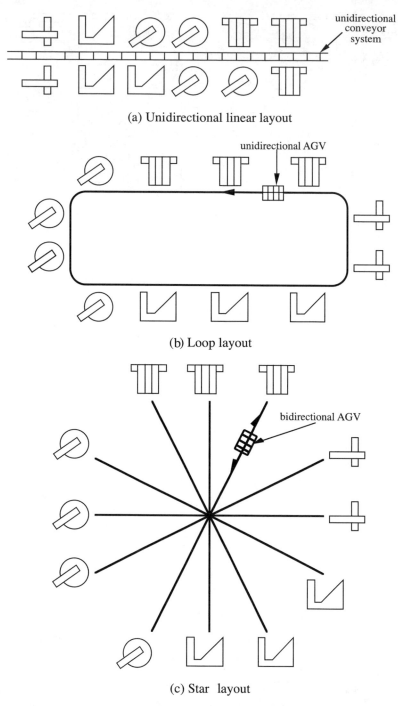

(a) Unidirectional linear layout

(b) Loop layout

(c) Star layout

FIGURE 3.4   Material handling system configuration.

overall efficiency is important. Layouts that are modular in nature or layouts where departments are either equidistant from each other or distributed throughout the plant certainly exhibit such capability. An example of a modular layout is the spine layout, as shown in Figure 3.5, in which different departments are organized around a central corridor for material travel [39, 40]. As illustrated in Figures 3.5a and b, department expansion and contraction can be realized without a need for reconfiguring the existing layout and with no effect on overall flow efficiency.

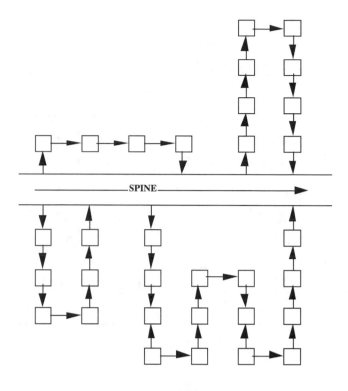

(a) Original spine layout

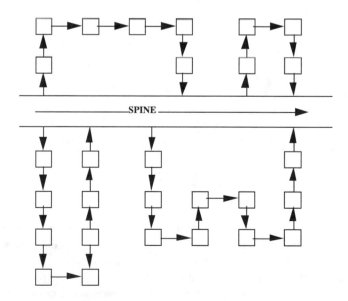

(b) Spine layout after expanding and contracting some departments

FIGURE 3.5 Spine layout.

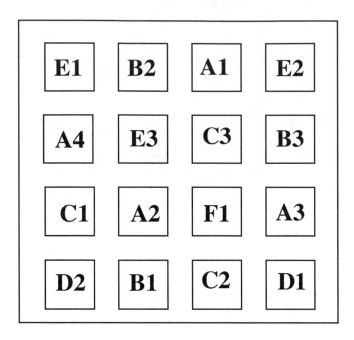

FIGURE 3.6   Selected flexible layout.

In view of the preceding discussion, four principles emerge as being critical in determining layout flexibility.

1. *Reducing variance in distances between different departments* — letting departments be as equidistant from each other as possible increases the number of product routings that can be accommodated.
2. *Improving material handling flexibility* — the realization of layout flexibility is contingent on a flexible material handling configuration and flexible material handling devices.
3. *Increasing department duplication and distribution* — disaggregating large functional departments and distributing the resulting duplicates throughout the plant floor significantly increases the ability of a layout to cope with changes in flow patterns.
4. *Increasing layout modularity* — designing layouts so that departments can be easily *plugged-in* and *plugged-out* improves the ability of a layout to expand and contract gracefully.

3.4   Measuring Layout Flexibility

Because of the many facets of flexibility, finding a single quantitative measure of layout flexibility can be challenging. Generally, two approaches can be taken: a geometry-based approach and a flow-based approach. In the first case, we attempt to relate flexibility to specific geometric characteristics of the layout (e.g., distances between departments). In the second case, we use, in addition to layout geometry, information regarding expected material flow patterns and material flow usage. Because flow patterns can be difficult to predict, especially in highly variable environments, the first approach is usually easier to implement. In the following sections, we provide example measures using each approach.

Geometry-Based Measures

Because a flexible layout should provide efficient material flow for all producible products, an obvious measure of flexibility is the average distance traveled by all producible products. The flexibility of a layout

L_i can, therefore, be measured as

$$\phi_1(L_i) = \sum_{J=1}^{P} \frac{d_j}{P} \qquad (3.1)$$

where d_j is the total distance traveled by a unit load of product j, and P is the total number of products. However, this measure becomes impractical when the set of all producible products is difficult to identify or when the number of products is very large. An alternative is to simply use the average distance between each pair of departments so that flexibility is measured as

$$\phi_2(L_i) = \sum_{j=1}^{N} \sum_{k=1(k\neq j)}^{N} \frac{d_{jk}}{N(N-1)} \qquad (3.2)$$

where d_{jk} is the distance between departments j and k and N is the total number of departments. Because the total travel distance associated with any product is simply the sum of a series of interdepartmental distances, this measure indirectly captures the distances traveled by different products.

Unfortunately, the preceding two measures do not account for performance variance. A layout with a small mean of interdepartment distances could have a high interdepartment distance variance. This means that while the layout is efficient for certain sequences, or products, it is highly inefficient for others. This goes counter to the spirit of our definition of a flexible layout as being equally effective for most producible products and sequences. This problem can be, in part, addressed by using distance variance as a secondary measure. A combined mean-variance criterion can then be used to rank different layouts. A popular alternative to variance is an entropic measure of the following form:

$$\phi_3(L_i) = \frac{\sum_{j=1}^{N} \gamma_j}{N} \qquad (3.3)$$

where γ_j is given by

$$\gamma_j = -\sum_{k\neq j}^{N} \rho_{jk} \log(\rho_{jk}) \qquad (3.4)$$

and

$$\rho_{jk} = \frac{1/d_{jk}}{\sum_{k\neq j}^{M} 1/d_{jk}} \qquad (3.5)$$

The parameter ρ_{jk} measures the relative ease with which department k can be accessed from department j. The function γ_j measures the amount of material handling flexibility available from department j. This measure of flexibility is particularly suitable for its many desirable properties. A few of these are listed next.

1. γ_j is minimum and equal to zero when $\rho_{jk} = 1$ (i.e., flexibility at department j is zero when there is only one feasible path from department j).
2. γ_j increases with decreases in the difference between the relative distance ratios and is maximum when $\rho_{j1} = \rho_{j2} = \cdots = \rho_{jM} = 1/(M-1)$ (i.e., flexibility at department j is maximum when all other departments are equidistant).
3. γ_j increases with increases in the number of accessible departments from department j when $\rho_{j1} = \rho_{j2} = \cdots = \rho_{jM}$.
4. γ_j does not change if an additional department $M+1$ is included, but where $\rho_{j(M+1)} = 0$ (i.e., when a department which is infinitely far from department j is added, flexibility at j does not increase).

A similar expression to Item 4 was initially introduced to study *entropy* in thermodynamics. Later, entropy was adopted as a measure of information uncertainty in information theory. Recently, Yao and Pei [47], Kumar [19], and Benjaafar and Ramakrishnan [6] proposed entropic measures for the study of routing flexibility in Flexible Manufacturing Systems. Despite its many useful properties, an entropic measure has several limitations. The most important of which is that it uses only relative distance information. This makes it difficult to rank layouts with similar relative distance structures but with different absolute distance values.

In each of the previous flexibility measures, we have weighed equally all possible sequences. In practice, certain sequences are simply not feasible (e.g., going from a shipping department to a receiving department). In such instances, the infeasible sequences should not be included. More generally, it is possible to modify the previous measures by associating a weight α_j, $0 \leq \alpha_j \leq 1$, with each interdepartment distance. The weights would reflect the likelihood, or frequency, with which a particular path between two departments would be taken. Unfortunately, determining appropriate weights, especially in highly volatile environments, can be difficult (see the next section).

We argued in the previous section that the degree to which duplicates of the same department type are distributed throughout the plant floor can significantly improve layout flexibility. In a highly distributed layout, copies of the same department would be uniformly placed throughout the plant floor to ensure easy access to all department types from any area of the layout. A possible measure of layout distribution is given by

$$\phi_4(L_i) = \sum_{j=1}^{N} \sum_{n_j=1}^{N_j} \frac{\delta_{n_j}}{NN_j} \tag{3.6}$$

where

$$\delta_{nj} = \sum_{k \neq j}^{N} d_{n_j k}^* \tag{3.7}$$

with $d_{n_j k}^*$ being the distance from the nth copy of department j to the closest copy of department k, n_j is the nth department of type j, and N_j is the number of copies of departments of type j. Because in a highly distributed layout, each department should be immediately adjacent to as many departments of other type as possible, an alternative measure of layout distribution is given by

$$\phi_5(L_i) = \sum_{j=1}^{N} \sum_{n_j=1}^{N_j} \frac{\beta_{n_j}}{NN_j} \tag{3.8}$$

where β_{n_j} is the number of departments immediately adjacent to department n_j that are of a different type. Other measures of layout distribution are described in Benjaafar and Seikhzadeh [7] and Montreuil et al. [26].

Flow-Based Measures

In contrast to geometry-based measures, flow-based measures account for expected flow patterns in measuring layout flexibility. In a deterministic environment where product demands and product mix are known, flow volumes between departments are given by the flow matrix $\mathbf{v} = [v_{ij}]$, where v_{ij} refers to the volume of flow between department i and j per period. Material handling efficiency, or cost, can then be measured as

$$z = \sum_{i=1}^{N} \sum_{j=1}^{N} \sum_{k=1}^{K} \sum_{l=1}^{K} v_{ij} d_{kl} x_{ik} x_{jl} \tag{3.9}$$

where $x_{ik} = 1$ if department i is assigned to location k and $x_{ik} = 0$, otherwise. In a stochastic environment, where product demands fluctuate and product mix changes, the amount of flow, v_{ij}, between departments i and j is variable and assumes different values for different product demand scenarios. Because a flexible layout should maintain efficient material flow over all demand scenarios, a possible measure of layout flexibility is the expected value of material handling cost

$$\phi_6(L_i) = E(z) = \sum_{i=1}^{N}\sum_{j=1}^{N}\sum_{k=1}^{K}\sum_{l=1}^{K}\sum_{s=1}^{S} \pi_s v_{ijs} d_{kl} x_{ik} x_{jl} \tag{3.10}$$

where π_s is the probability of occurrence of demand scenario s, v_{ijs} is the amount of flow between department i and j under demand scenario s, and S is the total number of scenarios (when product demand varies over a continuous range, the summation with respect to s can be substituted with integration). Unfortunately, this measure accounts only for average performance and ignores variance. This could be unacceptable if material handling cost is excessively high for one or more of these scenarios. Once again, variance can be accounted for by using a combined mean-variance criterion or an entropy-based measure, as discussed in the previous section.

In practice, the difficulty in implementing flow-based measures is in estimating the probability distribution of flow volumes. In a highly variable environment, where product demand and product mix vary widely, assigning equal probabilities to all scenarios is a reasonable option. Alternatively, a min–max criterion could be used where layouts are ranked based on their worst-case performance. However, this may lead to ranking highly a layout that underperforms under most scenarios. A more viable alternative is a robustness criterion where the robustness of a layout is measured by the number of times its material handling cost does not exceed the lowest cost layout by a certain percentage (see Rosenblatt and Lee [31]).

Finally, each of the preceding measures assumes that flow volumes, that is, the amount of flow between each pair of departments, remains constant from scenario to scenario. This is not necessarily the case when duplicates of the same department exist. In this case, it is usually beneficial to alter flow allocation among department duplicates from scenario to scenario. In the next section, we present a procedure for the design of flexible layouts that accounts for department duplication and flexible flow allocation.

It should be evident by now that the pursuit of a universal measure that captures the various dimensions of layout flexibility is unlikely to yield universal acceptance or usefulness. A less ambitious pursuit is the use of a portfolio of measures, each quantifying a particular aspect of flexibility and each used only as such. This makes the development and adoption of any measure of flexibility subordinate to its intended usage.

3.5 A Procedure for Flexible Layout Design

In this section, we present a procedure for design of flexible layouts in stochastic environments. A detailed description of the procedure has been published previously [7, 8]. Therefore, only key features will be presented here. Our primary objective is to provide insights into how a layout solution procedure may be developed to account for flexibility in a variable environment. The procedure departs from conventional methods of layout design by

1. Explicitly modeling the stochasticity in product demands and the resulting material flows between different processing departments.
2. Allowing for the possibility of multiple processing departments of the same type to exist in the same facility (e.g., not placing all machines of a given type in the same plant floor area).
3. Allowing material flows between pairs of individual departments be determined simultaneously with the layout and as a function of the demand scenarios. Accounting for the possibility of having multiple departments of the same type in the same facility is significant because duplicating departments or disaggregating existing ones can be an effective mechanism for enhancing layout flexibility. As we discussed earlier, by strategically locating duplicate departments in different areas of the plant floor, a facility can hedge against future fluctuations in job flow patterns and volumes. The fact that multiple copies of the same department can exist in the same facility allows us

additional flexibility in workload allocation. For a fixed layout, workload allocation can indeed be varied based on actual product demand realization, enabling us to further customize flows during operation. Because we place no restrictions on the number of department copies nor size of each department, our procedure can be used to evaluate layouts with varying degrees of department duplication and disaggregation.

Solution Procedure

In order to illustrate the procedure, we consider the case of a facility with multiple products, $p_1, p_2, \ldots, p_P$, whose demand varies according to a known distribution. For simplicity, we assume that the number of demand scenarios is finite — see Reference [7] for a discussion of the continuous case. Under each scenario s, $s = 1, \ldots, S$, demand for product i is d_{is}. Each scenario occurs with probability π_s, where $\pi_1 + \pi_2 + \cdots + \pi_S = 1$. Using the product demand distributions, the product process routings, and the product unit transfer loads, we determine for each possible demand scenario the amount of product flow, v_{ijp}, between each pair of department types i and j. This results in a multiproduct from–to flow matrix, $\boldsymbol{m}(s)$, for each demand scenario s. For each demand scenario, we then generate the corresponding optimal layout and the corresponding optimal flow allocation between copies of the same department. Once all layouts have been generated, we evaluate each layout over the entire set of possible demand scenarios and select the most flexible layout (e.g., we choose the layout with the lowest expected value of material handling cost over all demand scenarios).

For each demand scenario, the optimal layout and the corresponding optimal allocation of flow between departments can be obtained by solving the following model [7]

$$\min\ z = \sum_{P=1}^{P}\sum_{i=1}^{N}\sum_{j=1}^{N}\sum_{n_i=1}^{N_i}\sum_{m_j=1}^{N_j}\sum_{k=1}^{K}\sum_{l=1}^{K} v_{n_i m_j p}\, d_{kl}\, x_{n_i k}\, x_{m_j l}$$

subject to

$$\sum_{k=1}^{K} x_{n_i k} = 1 \qquad n_i = 1_i, 2_i, \ldots, N_i \quad i = 1, 2, \ldots, N \tag{3.11}$$

$$\sum_{i=1}^{N}\sum_{n_i=1}^{N_i} x_{n_i k} = 1 \qquad k = 1, 2, \ldots, K \tag{3.12}$$

$$\sum_{p=1}^{N}\sum_{i=0}^{N}\sum_{n_i=1}^{N_i} v_{n_i m_j p}\, t_{m_j p} \leq c_{m_j} \qquad m_j = 1_j, 2_j, \ldots, N_j \quad j = 1, 2, \ldots, N \tag{3.13}$$

$$\sum_{n_i=1}^{N_i}\sum_{m_j=1}^{N_j} v_{n_i m_j p} = v_{ijp} \qquad i, j = 0, 1, \ldots, N \quad p = 1, 2, \ldots, P \tag{3.14}$$

$$\sum_{i=0}^{N}\sum_{n_i=1}^{N_i} v_{n_i m_j p} = \sum_{q=0}^{N}\sum_{r_q=1}^{N_q} v_{m_j r_q p} \qquad m_i = 1_j, \ldots 2_j, \ldots, N_j \quad j = 1, 2, \ldots, N \quad p = 1, 2, \ldots, P \tag{3.15}$$

$$x_{n_i k} = 0, 1 \qquad n_i = 1_i, 2_j, \ldots, N_i \quad k = 1, 2, \ldots, K \tag{3.16}$$

where

$$x_{n_ik} = \begin{cases} 1, & \text{if } n\text{th department of type } i \text{ is assigned to location } k \\ 0, & \text{otherwise} \end{cases}$$

$v_{n_im_jp}$ = flow volume between the nth department of type i and the mth department of type j owing to product type p

v_{ijp} = flow volume between departments of type i and departments of type j owing to product type p

d_{kl} = distance between location k and location l (known parameter)

t_{m_jp} = processing time per unit load of product type p at department m_j

c_{ni} = capacity of department n_i (available operation time)

N_i = number of departments of type i

N = total number of department types

K = total number of locations

P = total number of product types

The preceding model solves simultaneously for department location and volume of flow allocation between individual departments, so that material handling costs are minimized. The decision variables are the x_{n_ik}'s and the $v_{n_im_jp}$'s. Constraints (3.11) and (3.12) ensure that each department is assigned to exactly one location and each location is assigned to one department. When the number of locations exceeds the number of departments, dummy departments with zero flows may, without loss of optimality, be used to account for the difference. Constraint (3.13) ensures that the flow volume allocated to a department does not exceed the capacity of that department. Constraint (3.14) equates the amount of flow between multiple copies of departments of type i and j to the amount of flow between department type i and department type j, as dictated by the from–to flow matrix. Constraint (3.15) ensures that the amount of input and output flow (per product) to and from a department are the same. Note that the index $i = 0$ is used to denote input–output departments. This is necessary in order to capture both entering and exiting flows. However, this formulation is used only for modeling convenience because we assume that a product can enter or exit at any department.

The preceding model can be viewed as a variation on the classical quadratic assignment problem (QAP) [13]. The differences between the two models are, however, important. In our model, the objective function is polynomial because department location and flow volume between departments are both decision variables. This difference is owing to the availability of multiple departments of the same type. Obviously, when there is only a single copy of each department, the model reduces to the classical QAP. As we discuss later on, the fact that flow volumes between departments are not predetermined allows us to optimize material handling costs even if the layout is fixed.

The model assumes that all department copies are of the same size. In practice, this may not always hold, especially if we consider duplicates of the same department not containing the same numbers of machines. This problem could be addressed by disaggregating departments into small grids with equal area and assigning artificially large flows between grids of the same department so that they are always placed in adjoining locations. A more detailed discussion of the general merits and limitations of this approach can be found in references [7] and [8], along with a discussion of alternative approaches.

A Heuristic Approach

Because the quadratic assignment problem has been shown elsewhere to be NP complete [28], the model proposed here is also NP complete. This means that obtaining an optimal solution for most problems in practice would require an excessive amount of computational effort. Therefore, a heuristic method is provided next. The method is an extension of existing layout heuristics, such as CRAFT [44], that use an iterative pairwise, or multi-step, exchange procedure in generating a final layout. Our approach differs from these heuristics in that we are not only solving for the layout but also for the flow volume allocation between departments. Consequently, at each iteration step and for each new layout considered, the flow volume allocation problem needs to be solved before the layout cost can be calculated. Fortunately, the problem can be formulated as a linear program and solved optimally in a reasonable amount of time. The steps of the heuristic are described next.

Step 1. Set $J = 1$.

Step 2. Generate an initial layout.

Step 3. Solve optimally for flow volume allocation (a linear program).

Step 4. Calculate $z(J)$, the resulting objective function value of the original layout problem.

Step 5. Set $J = J + 1$. If $J > J_{max}$, go to Step 9 (e.g., J_{max} is the maximum number of feasible pairwise interchanges).

Step 6. Generate the next layout (e.g., by a pairwise interchange).

Step 7. Solve optimally for flow volume allocation (a linear program).

Step 8. Calculate $z(J)$, the resulting value for the objective function. Go back to Step 5.

Step 9. Implement the minimum cost layout. If the minimum cost layout is the same as the previous one, then go to Step 10. Otherwise, set $J = 1$ and go back to Step 5.

Step 10. Stop.

The linear program that must be solved at each iteration (Steps 3 and 7) is

$$\min \ z = \sum_{P=1}^{P} \sum_{i=1}^{N} \sum_{j=1}^{N} \sum_{n_i=1}^{N_i} \sum_{m_j=1}^{N_j} \sum_{k=1}^{K} \sum_{l=1}^{K} v_{n_i m_j p} d_{kl} x_{n_i k} x_{m_j l}$$

subject to

$$\sum_{p=1}^{P} \sum_{i=0}^{N} \sum_{n_i=1}^{N_i} v_{n_i m_j p} t_{m_j p} \le c_{m_j} \qquad m_j = 1_j, 2, ..., N_j \quad j = 1, 2, ..., N \tag{3.17}$$

$$\sum_{n_i=1}^{N_i} \sum_{m_j=1}^{N_j} v_{n_i m_j p} = v_{ijp} \qquad i, j = 0, 1, ..., N \quad p = 1, 2, ..., p \tag{3.18}$$

$$\sum_{i=0}^{N} \sum_{n_i=1}^{N_i} v_{n_i m_j p} = \sum_{q=0}^{N} \sum_{r_q=0}^{N_i} v_{m_j r_q p} \qquad m_j = 1_j, 2_j, ..., N_j \quad j = 1, 2, ..., N \quad p = 1, 2, ..., P \tag{3.19}$$

Note that, in this case, the values of the variables $x_{n_i k}$ and $x_{m_j l}$ in the objective function are already known. The only decision variables are, therefore, the flow volumes $v_{n_i m_j p}$. Once the flow volumes between departments are known, the cost of the layout under consideration can be calculated (Steps 4 and 8).

Flexible Layout Selection

The result of either the optimization model or the heuristic is a layout of the various departments and an allocation of flow between these departments. Once such a layout has been identified for every possible demand realization scenario, a decision must be made as to which of these layouts is the most flexible. By definition, a flexible layout is one that maintains low material handling costs despite changes in demand levels. Therefore, a possible measure of layout flexibility is the layout's expected material handling cost over the range of feasible demand scenarios. A procedure for identifying the most flexible layout, can then be obtained by constructing the matrix $z(L(I), J)$, where $z(L(I), J)$ is the material handling cost resulting from using a layout generated for demand scenario I when the actual demand corresponds to that of scenario J and selecting the most flexible layout by choosing, for example, the layout with the smallest expected material handling cost over all demand scenarios. Note that in obtaining the material handling cost of operating a layout that was designed for demand scenario I under scenario J, a linear program is solved in order to determine the optimal flow volume allocation. The expected material handling cost of each layout, $E[z(L(I)]$, is given by

$$E[z(L(I))] = \sum_{J=1}^{S} \pi_{Jz}(L(I), J)$$

where π_J is the probability of occurrence of demand scenario J.

The preceding definition of a flexible layout, as the one with the lowest average cost, is by no means unique. Alternative criteria for selecting the most flexible layout can be used. In fact, any of the flexibility measures we discussed in Section 3.3 can be substituted. Also, note that in identifying the most flexible layout, we have restricted our attention to the set of the best layouts selected for different scenarios. In doing so, we may have overlooked a layout with a lower overall expected material handling cost that is not necessarily optimal for any scenario. This limitation can be overcome by directly solving for the layout with the lowest expected material handling cost. A procedure that implements this approach is described in Reference [8]. Extensive experimentation with both procedures indicates that the second procedure yields layouts with slightly lower expected material handling costs. The difference between the two procedures tends to diminish with increases in the level of department duplication. The original procedure is, however, more general because it allows for alternative criteria (other than expected cost) for measuring layout flexibility to be used.

Software Implementation and Analysis

The preceding flexible layout design procedure was implemented in a computer software application, FLEX-LAYOUT (a copy of the software is available from the author upon request). Extensive experimentation with FLEX-LAYOUT is reported in References [7] and [8], in which several hundred problems of varying sizes were evaluated under varying operating assumptions. The input to FLEX-LAYOUT consists of product demand distributions, product routings, layout geometry, and the number and size of different departments, while the output consists of probability distribution for scenarios, a flow-volume matrix for each demand scenario, the layout selection for each scenario, the flow-volume allocation for each scenario, and a flexible layout selection. Output for a simple example is shown in Tables 3.1 to 3.3 and Figure 3.6.

Based on our experimentation with the procedure, the following observations can be made.

- Most of the layouts obtained through the procedure are highly distributed. Departments of the same type tend to be either dispersed throughout the plant floor or centrally located. This results in layouts where each department is closely located to departments from every other type. For departments with a single copy, the department is generally located close to the center of the plant floor.

TABLE 3.1 Number of Copies per Department Type

Department Type	Number of Department Copies
A	4 (A1, A2, A3, A4)
B	3 (B1, B2, B3)
C	3 (C1, C2, C3)
D	2 (D1, D2)
E	3 (E1, E2, E3)
F	1 (F1)

TABLE 3.2 Generated Layouts and Material Handling Costs per Scenario

Scenario (s)	Scenario Probability (π_s)	Flow Matrix ($m(s)$)	Selected Layout $L(s)$	Material Handling Cost $z(L(s))$
1	0.04	$V_{AB} = 80, V_{AE} = 60$ $V_{BD} = 60, V_{CA} = 60$ $V_{CF} = 80, V_{DC} = 80$ $V_{EB} = 60, V_{FA} = 80$	[E1, B2, A1, E2] [A4, E3, C3, B3] [C1, A2, F1, A3] [D2, B1, C2, D1]	620
2	0.10	$V_{AB} = 150, V_{AE} = 60$ $V_{BD} = 60, V_{CA} = 60$ $V_{CF} = 150, V_{DC} = 150$ $V_{EB} = 60, V_{FA} = 150$	[E1, B2, A1, E2] [A4, E3, A3, B3] [C1, A2, F1, C3] [D2, B1, C2, D1]	935
3	0.06	$V_{AB} = 180, V_{AE} = 60$ $V_{BD} = 60, V_{CA} = 60$ $V_{CF} = 180, V_{DC} = 180$ $V_{EB} = 60, V_{FA} = 180$	[E1, B2, A1, E2] [A4, E3, A3, B3] [C1, A2, F1, C3] [D2, B1, C2, D1]	1080
4	0.10	$V_{AB} = 80, V_{AE} = 90$ $V_{BD} = 90, V_{CA} = 60$ $V_{CF} = 80, V_{DC} = 80$ $V_{EB} = 80, V_{FA} = 80$	[E1, A2, B2, D1] [C3, F1, A3, C1] [B1, C2, D2, A1] [E2, A4, B3, E3]	733
5	0.25	$V_{AB} = 150, V_{AE} = 90$ $V_{BD} = 90, V_{CA} = 90$ $V_{CF} = 150, V_{DC} = 150$ $V_{EB} = 90, V_{FA} = 150$	[D2, C3, B2, D1] [A2, F1, A3, C1] [B1, C2, E1, A1] [E2, A4, B3, E3]	1050
6	0.15	$V_{AB} = 180, V_{AE} = 90$ $V_{BD} = 90, V_{CA} = 90$ $V_{CF} = 180, V_{DC} = 180$ $V_{EB} = 90, V_{FA} = 180$	[D2, C3, B2, D1] [A2, F1, A3, C1] [B1, C2, E1, A1] [E2, A4, B3, E3]	1185
7	0.06	$V_{AB} = 80, V_{AE} = 180$ $V_{BD} = 180, V_{CA} = 180$ $V_{CF} = 80, V_{DC} = 80$ $V_{EB} = 180, V_{FA} = 80$	[B2, E1, A3, C2] [E3, B3, E2, A2] [A1, D2, C3, F1] [C1, D1, B1, A4]	1140
8	0.15	$V_{AB} = 150, V_{AE} = 180$ $V_{BD} = 180, V_{CA} = 180$ $V_{CF} = 150, V_{DC} = 150$ $V_{EB} = 180, V_{FA} = 150$	[D1, B2, E2, C3] [C1, A3, A1, F1] [A4, E3, B1, A2] [E1, B3, D2, C2]	1420
9	0.09	$V_{AB} = 180, V_{AE} = 180$ $V_{BD} = 180, V_{CA} = 180$ $V_{CF} = 180, V_{DC} = 180$ $V_{EB} = 180, V_{FA} = 180$	[D1, B2, E2, C3] [C1, A3, A1, F1] [A4, E3, B1, A2] [E1, B3, D2, C2]	1560

TABLE 3.3 Expected Material Handling Costs for Selected Layouts

Scenarios	$z(L(I),J)$									$E(z(L(I)))$
	$J = 1$	$J = 2$	$J = 3$	$J = 4$	$J = 5$	$J = 6$	$J = 7$	$J = 8$	$J = 9$	
$I = 1$	620	955	1120	770	1050	1185	1290	1500	1620	1153
$I = 2$	637	935	1080	805	1070	1185	1343	1575	1680	1178
$I = 3$	637	935	1080	805	1070	1185	1343	1575	1680	1178
$I = 4$	627	1020	1200	733	1100	1260	1157	1420	1560	1159
$I = 5$	620	955	1120	770	1050	1185	1290	1500	1620	1153
$I = 6$	620	955	1120	770	1050	1185	1290	1500	1620	1153
$I = 7$	693	1190	1420	783	1240	1440	1140	1510	1680	1282
$I = 8$	627	1040	1240	733	1100	1260	1150	1420	1560	1163
$I = 9$	627	1040	1240	733	1100	1260	1150	1420	1560	1163

- An increase in the variability of part demands tends to induce a more distributed layout. Under high variability conditions, the difference between a distributed layout and a layout where machines are grouped functionally can be very significant. A functional layout appears to be more vulnerable to demand shifts than a distributed one.

- In addition to department dispersion, department disaggregation seems to improve significantly the robustness of a layout. Comparisons between functionally grouped layouts and disaggregated and distributed layouts show that significant performance gains can be achieved even with limited amounts of disaggregation. In fact, the relationship between performance and department disaggregation/duplication was found to be of the diminishing kind with most of the performance gains realized with only limited disaggregation.

- The robustness of a layout seem to be affected by its geometry. Symmetric and multi-aisle layouts, where copies of the same department can be evenly distributed throughout the plant floor, are found to be more robust in the face of demand variability than those with asymmetric geometries or those with one or two aisles.

Further experimentation is, however, needed to extend these results. In particular, the relationship between layout performance and the degree of department distribution and disaggregation in the layout must be further characterized. Although examples show that department disaggregation/duplication is desirable, it is not clear how much disaggregation/duplication is generally needed and/or can be afforded. Methodologies for determining optimal levels of disaggregation and duplication must be developed. The fact that multiple copies of the same department may exist poses a challenge for workload assignments from period to period. The flow volume allocation procedure described earlier can certainly be used. However, the procedure must be enriched with additional details regarding tooling, labor, and material handling. The implementation and operation costs of a distributed layout must also be documented better. More importantly, the enabling technologies (e.g., material handling, tooling, in-process storage, shop floor control, and production scheduling) must be clearly identified. Finally, the impact of a distributed layout on the operational performance of the manufacturing facility (e.g., production lead times, work-in-process inventories, and production rates) need to be examined.

3.6   Conclusion

In this chapter, we addressed the issue of flexibility in design and implementation of factory layouts. We reviewed existing literature and discussed its limitations. We identified key principles for enhancing layout flexibility and introduced measures for quantifying various flexibility dimensions. Finally, we presented an integrated procedure for flexible layout design in stochastic environments. We used the procedure to highlight the desirability of disaggregating functional departments into smaller subdepartments and distributing them throughout the factory floor. We showed that increased disaggregation and distribution can be, indeed, effective in enhancing a layout's ability to cope with variability.

References

1. Ang, C. L. and Willey, P. C. T., A comparative study of the performance of pure and hybrid group technology manufacturing systems using computer simulation techniques, *Int. J. Prod. Res.*, 22(2), 193, 1984.

2. Askin, R. G. and Standridge, C., *Modeling and Analysis of Manufacturing Systems,* John Wiley & Sons, New York, 1993.

3. Askin, R. G., Lundgren, N. H., and Ciarallo, F., A material flow based evaluation of layout alternatives for agile manufacturing, in *Progress in Material Handling Research,* Graves, R. J., McGinnis, L. F., Medeiros, D. J., Ward, R. E., and Wilhelm, M. R., Eds., Braun-Brumfield, Ann Arbor, MI, 1997, 71.

4. Benjaafar, S., Design of plant layouts with queueing effects, *Manage. Sci.,* in review.

5. Benjaafar, S., Machine sharing in cellular manufacturing systems, in *Planning, Design, and Analysis of Cellular Manufacturing Systems,* Kamrani, A. K., Parasei, H. R., and Liles, D. H., Eds., Elsevier Science, 1995.

6. Benjaafar, S. and Ramakrishnan, R., Modeling, measurement, and evaluation of sequencing flexibility in manufacturing systems, *Int. J. Prod. Res.,* 34(5), 1195, 1996.

7. Benjaafar, S. and Sheikhazadeh, M., Design of flexible plant layouts, *IIE Trans.,* Vol. 32(4), 309, 2000.

8. Benjaafar, S., Soewito, A., and Sheikhzadeh, M., Performance evaluation and analysis of distributed plant layouts, working paper, Department of Mechanical Engineering, University of Minnesota, Minneapolis, 1996.

9. Bozer, Y. A. and Meller, R. D., A reexamination of the distance-based facility layout problem, *IIE Trans.,* 29, 549, 1997.

10. Bullington, S. F. and Webster, D. B. Evaluating the flexibility of facilities layouts using estimated relayout costs, *Proceedings of the IXth International Conference on Production Research,* 1987, 2230.

11. Drolet, J. R., Scheduling Virtual Cellular Manufacturing Systems, Ph.D. thesis. School of Industrial Engineering, Purdue University, West Lafayette, IN, 1989.

12. Francis, R. L. and White, J. A., *Facility Layout and Location: An Analytical Approach,* 2nd ed., Prentice-Hall, Englewood Cliffs, NJ, 1993.

13. Flynn, B. B. and Jacobs, F. R., A simulation comparison of group technology with traditional job shop manufacturing, *Int. J. Prod. Res.,* 24(5), 1171, 1986.

14. Gupta, R. M., Flexibility in layouts: a simulation approach, *Mater. Flow,* 3, 243, 1986.

15. Hopp, W. and Spearman, M., *Factory Physics,* Irwin, Chicago, IL, 1995.

16. Irani, S. A., Cavalier, T. M., and Cohen, P. H., Virtual manufacturing cells: exploiting layout design and intercell flows for the machine sharing problem, *Int. J. Prod. Res.,* 31, 791, 1993.

17. Kouvelis, P., Kurawarwala, A. A., and Gutierrez, G. J., Algorithms for robust single and multiple period layout planning for manufacturing systems, *Eur. J. Oper. Res.,* 63, 287, 1992.

18. Kumar, V., Entropic measures of manufacturing flexibility, *Int. J. Prod. Res.,* 25, 957, 1987.

19. Kusiak, A. and Heragu, S. S., The facility layout problem, *Eur. J. Oper. Res.,* 27, 229, 1987.

20. Meller, R. and Gau, K. Y., The facility layout problem: Recent and emerging trends and perspectives, *J. Manuf. Syst.,* 15(5), 351, 1996.

21. Montreuil, B. and Laforge, A., Dynamic layout design given a scenario tree of probable futures, *Eur. J. Oper. Res.,* 63(2), 271, 1992.

22. Montreuil, B., Venkatadri, U., and Lefrançois, P., Holographic layout of manufacturing systems, Tech. Rep. No. 91-76, Faculty of Management, Laval University, Quebec, 1991.

23. Montreuil, B., Venkatadri, U., and Rardin, R., The fractal layout organization for job shop environments, Tech. Rep. No. 95-13, School of Industrial Engineering, Purdue University, West Lafayette, IN, 1995.

24. Pardalos, P. M. and Wolkowicz, H., Eds., *Quadratic Assignment and Related Problems,* DIMACS Series, Vol. 16, American Mathematical Society, 1994.

25. Rosenblatt, M. J. and Kropp, D. H., The single period stochastic plant layout problem, *IIE Trans.,* 24(2), 169, 1992.

26. Rosenblatt, M. J. and Lee, H., A robustness approach to facilities design, *Int. J. Prod. Res.,* 25(4), 479, 1987.

27. Sarper, H. and Greene, T. J., Comparison of equivalent pure cellular and functional production environments using simulation, *Int. J. Comput. Integr. Manuf.,* 6(4), 221, 1993.

28. Sethi, A. K. and Sethi, S. P., Flexibility in manufacturing: a survey, *Int. J. Flex. Manuf. Syst.,* 2, 289, 1990.

29. Shafer, S. M. and Charnes, J. M., Cellular versus functional layout under a variety of shop operating conditions, *Dec. Sci.,* 36(2), 333, 1988.

30. Shore, R. H. and Tompkins, J. A., Flexible facilities design, *AIIE Trans.,* 12(2), 200, 1980.

31. Stecke, K. E. and Raman, N., FMS planning decisions, operating flexibilities, and system performance, *IEEE Trans. Eng. Manage.,* 42(1), 82, 1995.

32. Suresh, S. C., Partitioning work centers for group technology: analytical extension and shop-level simulation investigation, *Dec. Sci.*, 23, 267, 1992.
33. Tompkins, J. A., Modularity and flexibility: dealing with future shock in facilities design, *Ind. Eng.*, Vol. 12, 78, 1980.
34. Tompkins, J. A., and Spain, J. D., Utilization of spine concept maximizes modularity in facilities planning, *Ind. Eng.*, Vol. 15, 34, 1983.
35. Tompkins, J. A., White, J. A., Bozer, Y. A., Frazelle, E. H., Tanchoco, J. M. A., and Trevino, J., *Facilities Planning*, 2nd ed., John Wiley & Sons, New York, 1996.
36. Upton, D. M., What really makes factories flexible? *Harv. Bus. Rev.*, July, 74, 1995.
37. Venkatadri, U., Rardin, R., and Montreuil, B., A design methodology for the fractal layout organization, *IIE Trans.*, 29, 911, 1997.
38. Venkatadri, U., Rardin, R. L., and Montreuil, B., Facility organization and layout design: an experimental comparison for job shops, Tech. Rep., No. 96–27, Faculty of Management, Laval University, Quebec, 1996.
39. Vollmann, T. E. and Buffa, E. S., The facilities layout problem in perspective, *Manage. Sci.*, 12(10), 450, 1966.
40. Webster, D. B. and Tyberghein, M. B., Measuring flexibility of job shop layouts, *Int. J. Prod. Res.*, 18(1), 21, 1980.
41. Wemmerlöv, U. and Hyer, L. N., Cellular manufacturing in the U. S. industry: a survey of users, *Int. J. Prod. Res.*, 27(9), 1511, 1989.
42. Yao, D. D. and Pei, F. F., Flexible parts routing in manufacturing systems, *IIE Trans.*, 22, 48, 1990.

4

Structural Control of Large-Scale Flexibly Automated Manufacturing Systems

Spyros A. Reveliotis
Georgia Institute of Technology

Mark A. Lawley
Purdue University

Placid M. Ferreira
University of Illinois at Urbana-Champaign

Current strategic and technological trends in discrete-part manufacturing require extensive and flexible automation of the underlying production systems. However, even though a great deal of work has been done to facilitate manufacturing automation at the hardware component level, currently there is no adequately developed control methodology for these environments. In particular, it has been realized recently by the manufacturing community that any successful attempt toward the extensive automation of these environments requires the establishment of logically correct and robust behavior from the underlying system. The resulting line of research is complementary to the most traditional performance-oriented control, and it has come to be known as the structural control of the contemporary flexibly automated production environment. In this chapter, we initially discuss the role of structural control in the broader context of real-time control of flexible manufacturing systems (FMS) and, in continuation, we present the major results of our research program on manufacturing system deadlock, currently the predominant structural control problem in the manufacturing research literature. In the concluding section, we also indicate how these results should be integrated in contemporary FMS controllers, and highlight directions for future research in the FMS structural control area.

4.1   Introduction

Current Trends in Discrete-Part Manufacturing

Current business and technology trends in discrete-part manufacturing require the effective deployment of large-scale flexibly automated production systems. To a large extent, this is the result of the shift of discrete-part manufacturing strategies from the previously sought economies of scale to what has come to be known as economies of scope [1, 3]. In this new environment, emphasis is placed on the ability of the system to provide customized products and services, while meeting tighter time constraints, and with highly competitive prices. Large-scale flexible automation is the enabler of the afore-mentioned economies of scope because, in principle, this technology allows a variety of parts to be produced through the system (product flexibility), with each of them being profitably produced over a wide range of production volumes (volume flexibility). The high speed, increased accuracy, repeatability, and reconfigurability which are ascribed to these systems, can lead to high resource utilization, while maintaining high responsiveness to any shift in the production objective.

Furthermore, in certain industries the sheer complexity and sensitivity of the manufacturing operations, and/or the increased volume–weight of the transferred material, pose additional requirements for extensive automation of the production environment. For example, in the semiconductor industry, the need for extremely clean environments combined with the increased size of processed wafers has turned the modern semiconductor fab into one of the most technologically sophisticated shop floors.

The Inadequacy of Existing Control Methodologies

The last 15 years have seen a lot of success in developing the hardware features — computerized process control and factory networking infrastructure — to support the concept of flexibly automated production systems. However, it is becoming increasingly clear to the manufacturing community that these systems have not yet fully realized their advertised advantages, primarily because of the lack of an adequate control methodology. Quoting from a recently published paper [4],

> Currently most control implementations of flexible manufacturing cells (FMCs) have been developed specially to a particular facility, and no generic format or tools exist for systematic creation and planning for control software. In general, these systems are developed as "turnkey" systems by personnel other than the manufacturing engineers responsible by their operation.

The authors go on to identify some of the problems resulting from this ad hoc approach to the control of current FMCs as

1. High implementation times and costs.
2. Lack of portability of the resulting control software.
3. Limited operational flexibility of the production system.

These findings are corroborated by a recent survey regarding manufacturing flexibility in the printed circuit-board industry, where it was found that

[P]roduction technology turned out to be significantly related to mix and new-product flexibility, but with a pattern opposite to what [the authors] expected given the capabilities of the technology.... This touches a very important point: The fact that automated and programmable equipment in [the authors'] sample tends to be used to run the largest production batches, instead of being used in a more flexible way [5].

This lack of a sufficient methodology for the design and operation of effective controllers for large-scale flexibly automated production environments is mainly due to the extreme complexity of the underlying control problem. The most prominent causes of this high complexity can be traced to

1. The large variety of events taking place in such a production system and the time scales associated with these events.
2. The stochastic nature of the system operation.
3. The nondeterminism inherent to flexible systems.
4. The discrete system nature that invalidates the application of most well-known classical analytical techniques.

As a result, the development of an analytical methodology providing "globally optimal" solutions to the problem FMS control is currently deemed unlikely. In fact, practitioners typically resort to simulation for the synthesis of FMS controllers in real-life applications.

Simulation Approaches in FMS Production Planning and Control

The major advantage of simulation as an analysis and evaluation tool is extensive modeling flexibility. However, with existing computing technology, the amount of detail to be included in a simulation model is constrained by

1. The programming skills of the simulation developer and a thorough understanding of the modeled system.
2. The capabilities and constructs of the employed simulation platform.
3. The time available to set up and run the simulation experiments.

The third of these constraints becomes especially prominent in a flexibly automated system, where the system operations undergo frequent reconfigurations.

The limitations attributed to simulation-based approaches are partially addressed by the paradigm of object-oriented modeling and software development. The object-oriented methodology tries to systematically characterize the behavior of the underlying system components and their interaction, and to develop "standardized" libraries of object classes, from which customized models can be rapidly synthesized in a modular fashion (e.g., see Reference [6]). Furthermore, in their effort to systematically extract these standardized behaviors, object-oriented techniques resort to more formal analytical models and tools, such as Discrete Event Systems (DES) theory, and Generalized Semi-Markov Process models, thus moving simulation-based efforts closer to analytically oriented methodologies [7].

Analytical Approaches in FMS Production Planning and Control

The most prevalent analytical approach to real-time FMS control attempts to *hierarchically decompose* the problem into a number of more manageable subproblems. These subproblems are classified into number of layers, based on the time scales associated with the corresponding decision horizons. A typical classification scheme discerns

1. "Strategic-level" decisions, concerning the initial deployment and subsequent expansion(s) of the production environments.
2. "Tactical-level" decisions which determine the allocation patterns of the system production capacity to various products so that external demands are satisfied.
3. "Operational-level" decisions which coordinate the shop-floor production activity so that the higher-level tactical decisions are observed [8, 9].

Among the three (problem) layers of the hierarchical decomposition framework described previously, the ones relevant to this discussion are the tactical and the operational; strategic-level decisions are revised over long operational periods (typically measured in years) and, therefore, they fall beyond the scope of real-time system control. Hence, in the following, we briefly discuss the tactical planning and the operational control problem as defined in the hierarchical decomposition framework.

FMS Tactical Planning

Tactical planning attempts to efficiently track external product demand by exploiting the inherent flexibilities and reconfiguration capabilities of the system. The most prevalent framework for addressing this issue was proposed in Reference [8]. It further decomposes the production planning problem into the following subproblems:

1. Part-type selection for immediate and simultaneous processing.
2. Machine grouping into identically tooled machine pools.
3. Establishing production ratios according to which the different part types must be produced over time.
4. Resource (pallet and fixture) allocation to the different part types so that the determined production ratios are observed.
5. Loading of the machine tool magazines so that the resulting processing capacity is consistent with the aforementioned decisions.

The detailed formulation of each of these problems and their dependencies is determined by the operational details of the underlying environment. The resulting analytical models are typically Integer Programming (IP) [10] formulations, combined with some analytical or simulation-based performance evaluation tool that supports the "pricing" of the various system configurations examined by the IP solution process. Some representative work along these lines can be found in References [11–13] and the references cited therein.

FMS Scheduling

In hierarchical decomposition, the scheduling problem consists of part loading and dispatching so that target production objectives are met. Because this decision-making level addresses the real-time operation of the system, the relevant problem formulations assume a fixed system configuration. Unfortunately, the FMS scheduling problem is one of the hardest problems addressed by the manufacturing community. Many researchers from different methodological areas (e.g., "classical OR," control theory, AI, simulation, etc.) have attempted to address it. An interesting classification and general description of all the different approaches taken to the problem is presented in Reference [14]. The authors conclude that the most viable approach, in view of the problem complexity, is

> [T]o rely on the control paradigm to define the nature and objectives of scheduling problems, serving as a problem framework. Within this framework, combined solution heuristics from machine scheduling, resource-constrained project scheduling and AI search could be used to generate candidate schedules. The performance of candidate schedules could be verified using simulation models, where the simulations could be manipulated interactively by an expert scheduler. Such a synthesis would seemingly wed the rich modeling aspects of control and simulation with the solution-oriented techniques of heuristic algorithms.

Indeed, such a research program was developed in the years following the publication of these ideas and is manifested in the work presented in References [15–22]. This work recognizes that rather simple distributed scheduling schemes seem to be the most viable approach to the scheduling problem, and it resorts to control-theoretic techniques to formally analyze the system behavior, measure its performance, and eventually draw some definitive conclusions about the appropriateness of each of these policies in different operational settings.

FMS Structural Control

The hierarchical decomposition framework [8, 9] addresses issues related to the efficient operation and performance optimization of the controlled production system. However, recent attempts to extensively automate these environments indicate that robust and logically correct system behavior must be established before system performance can be addressed. In an analogy to the computing system environments, what needs to be developed is a "manufacturing operating system" which will facilitate the logically correct, efficient, and transparent allocation of the system resources to the various applications (i.e., production requirements). The entire suite of problems related to the logical components of such a control system software has been characterized as the structural control of flexibly automated production systems [23, 24]. The most typical problems currently identified in this area are

1. The resolution the *manufacturing-system deadlock*, that is, the establishment of undisturbed flow for all parts in the system.
2. The design of *protocols* according to which the system should react and recover from different disruptions occurring in it, such as machine breakdowns and the processing of expedited jobs.

Chapter Overview

Figure 4.1 presents the hierarchical decomposition control framework, extended to include the new area of structural control. Figure 4.2 indicates how the structural control problem is interrelated to the rest of manufacturing system control problems, identified in the previous discussion. The rest of this chapter focuses on our recent research results with respect to the manufacturing structural control problem and, in particular, the manufacturing deadlock. Hence, the next section introduces an FMS operational model, demonstrates how deadlock arises in these environments, and overviews the existing literature on the deadlock problem. Section 4.3 proceeds to the formal modeling and analysis of the problem in the context of Discrete Event Systems (DES) theory [25]. In that section, it is also shown that the optimal control of the manufacturing system with respect to deadlock-free operation is, in general, NP hard [26]. However, in Section 4.4, we establish that for a large class of manufacturing systems, optimal deadlock avoidance is attainable with polynomial computational cost. Section 4.5 develops a methodology for designing suboptimal, yet computationally efficient, deadlock avoidance policies for the remaining cases that do not admit polynomially optimal solutions. Section 4.6 considers the efficiency of these suboptimal policies and suggests a number of ways in which their efficiency can be enhanced. Section 4.7 concludes with a discussion of additional research topics related to this work. Finally, we note that the discussion in this chapter emphasizes methodological aspects and highlights the key results of the surveyed work. Technical details and formal proofs of the presented results are to be found in the cited publications.

4.2 The FMS Operational Model and the Manufacturing System Deadlock

The FMS Operational Model

The FMS model under consideration is depicted in Figure 4.3. It consists of a number of workstations, $W_1,...,W_N$, each capable of performing some fixed set of (elementary)-operations, a load and an unload station, and an interconnecting Material Handling System (MHS). Each workstation can be further described by a number of parallel processors/servers (i.e., identically configured machines) and a number of buffering positions holding the parts waiting for, or served by, the processors. The MHS consists of a number of carriers, each of which can transfer parts between any two workstations; an AGV system might be a good example. Jobs are loaded in the FMS through the load station and, in continuation, they go through a sequence of processing steps performed at the system workstations. Notice that a job can visit a workstation more than once in this sequence. On the other hand, the entire processing performed on

The Extended Hierarchical Decomposition Control Scheme

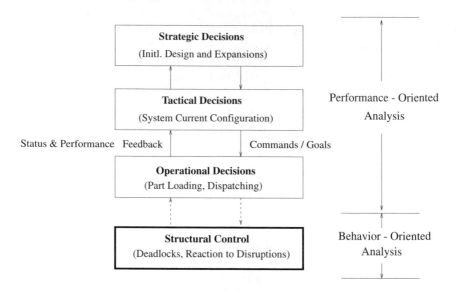

FIGURE 4.1   A hierarchical framework for the FMS design and control problem.

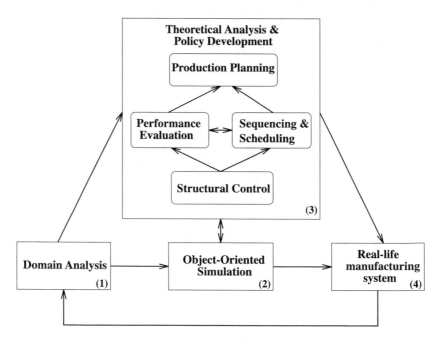

FIGURE 4.2   The major (discrete-part) manufacturing system control problems and their interaction.

a job during a single visit at a workstation is modeled as a single operation. The completed job exits the FMS through the unload station.

To see how deadlock arises in such an environment, consider the small FMS depicted in Figure 4.4. It consists of two workstations and an AGV, each able to accommodate one workpiece at a time. The jobs supported by the system can take only one of the two routes annotated in the figure, while it is

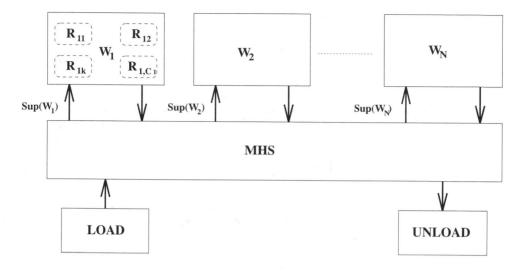

FIGURE 4.3   The FMS operational model.

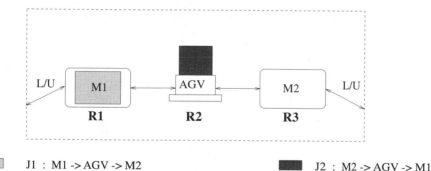

FIGURE 4.4   Example: the FMS model.

further assumed that all auxiliary equipment required for the processing of these jobs at their different stages is allocated to them upon their loading into the system. For this small FMS, it is not difficult to see that under the loading pattern depicted in Figure 4.4, the system is permanently stuck: the job loaded on workstation M_1 is waiting for the AGV to transport it to the next station M_2, while the AGV is currently occupied by another job which needs to advance to workstation M_1. This permanent blocking that can take place in such manufacturing environments is characterized as the *manufacturing system deadlock*.

Manufacturing deadlock is a major disruption for the system operation because its resolution requires external intervention. During its occurrence, the utilization of the resources involved is essentially zero. Furthermore, as it is demonstrated by the preceding example, this permanent blocking effect results from the fact that every part loaded in the FMS, being a physical entity has to hold its currently allocated operational space until it is transferred to a free unit of the resource type it requires next. Specifically, this implicit hold-and-wait behavior can give rise to a circular-waiting pattern, where every job involved requires a unit of equipment held by another job in this loop. It is exactly this behavioral pattern that is at the core of the deadlock problem.

Actually, deadlock has also been a problem in the operational environment of the more traditional (nonautomated) manufacturing shop floor. However, the manufacturing community has tended to ignore the deadlock problem in that context because the presence of human operators allowed for an easy solution, namely, the preemption of a number of deadlocked jobs to auxiliary buffers so that the remainder could proceed. At present, the increased levels of automation and autonomy sought for the contemporary FMS necessitate the development of automated solutions to the deadlock problem. It has become clear by now that establishing nonblocking behavior for the automated FMS is a prerequisite to any performance optimizing policy.

Because the primary cause for the deadlocking effects in contemporary FMS is the ineffective allocation of its buffering capacity, in the subsequent analysis we shall focus on this capacity of the FMS workstations and MHS, ignoring the detailed operational content of the different processing stages. Hence, all different classes of FMS equipment able to stage a part will be collectively characterized as the FMS resources. However, an additional aspect of the FMS operation to be taken into consideration during the FMS structural analysis is the management of the system auxiliary equipment, that is, the available pallets, fixtures, and cutting tools. Specifically, if there are sufficient units of this kind of equipment and/or every job is allocated the required auxiliary resources when loaded into the system and it keeps it until exiting, the management of this set of items is not a cause of deadlock. On the other hand, if auxiliary equipment is scarce and it is allocated to the requesting jobs only prior to executing the corresponding processing step(s) in an exclusive, nonpreemptive manner, then it should be obvious that careless management of these resources can give rise to circular-wait situations and, therefore, to deadlocks. A third feature that is relevant to the character-ization of the FMS structural behavior is whether the routing of a certain part processed through the system is determined online (dynamic routing) or during the loading of the part into the system (static routing). The issue here is that the machine flexibility [27] implemented in modern manufacturing processors, as well as the operation flexibility [27] supported by the design of the various products run through the system, allows for the production of a certain part through a number of alternate sequences of processing steps (process plans or routes). The resulting routing flexibility [27] allows for better exploitation of the system production capacity because the system workload can be better balanced. Hence, from a performance viewpoint, a dynamic routing scheme is preferable to static one(s). However, dynamic routing introduces an unpredictable element to the structural behavior of the different parts running through the system and makes the development of structural control policies a more difficult problem.

The Underlying Resource Allocation System (RAS) and the RAS Taxonomy

In order to systematically address the problem of FMS deadlock, we model the FMS operation as a Resource Allocation System (RAS). In general, a resource allocation system consists of a set of concurrently executing processes which, at certain phases of their execution, require the exclusive use of a number of the system resources to successfully run to completion [28]. The resources are in limited supply and they are characterized as reusable because their allocation and deallocation to requesting processes affect neither their nature nor their quantity. Furthermore, in the FMS case, the resulting RAS can also be characterized as sequential because it is assumed that every process/job, during its execution, undergoes a predefined sequence of resource allocation and deallocation steps. Specifically, every process can be described by a sequence of stages, with every stage defined by the subset of resources required for its successful execution. The detailed structure of the resource requests posed by the executed jobs at their different stages depends on the FMS operational features discussed in the previous section, that is, the allocation of the auxiliary equipment and the employed degree of routing flexibility. In fact, it turns out that the tractability of developing effective structural control policies and the details of the resulting solutions strongly depend on the way that the FMS is configured with respect to these operational characteristics. This finding has led to the following classification of the FMS-modeling RAS based on the structure of the allowable resource requests defining the job processing stages ([29]–© 1997 IEEE):

1. Single-Unit (SU) RAS — Every process stage requires only one unit from a single resource for its successful execution. This model applies to situations where only the limited buffering capacity of the FMS equipment is a cause of deadlock.
2. Single-Type (ST) RAS — Every process stage requires an arbitrary number of units but all of the same resource type. Similar to the SU-RAS, ST-RAS model FMS in which the only cause of deadlock is the limited buffering capacity of the FMS equipment. However, the ST-RAS allows the modeling of the aggregation of parts running through the system into tightly connected batches of varying size.

3. Conjunctive (AND) RAS — Every process stage, to be successfully executed, requires the simultaneous exclusive service of an arbitrary number of units of different resource types. This RAS supports the modeling of the more general case in which scarce auxiliary equipment can also be a cause for deadlock.
4. Disjunctive/Conjunctive (AND/OR) RAS — Every stage is associated with a set of conjunctive requests, the implication being that satisfaction of any of these requests by the system is sufficient for the successful execution of the step. The AND/OR RAS is the model to be used in case that dynamic routing is allowed.

Notice that higher-numbered members of this taxonomy subsume the lower-numbered ones, but the models get more complicated. The results presented in this chapter primarily concern the first (single-unit) class of this taxonomy. Focusing the discussion on this RAS model allows the exposition of the most significant problem features without being overwhelmed by the complicating details arising from the increased complexity of higher-level RAS models. The interested reader can find extensions of the results discussed herein to higher level models in References [23] and [30].

The Nature of the RAS Deadlock and Generic Resolution Approaches

The problem of system deadlock was first studied in the context of Computer System Engineering, back in the late 1960s and early 1970s. The work presented in such papers as References [31] to [34] set the foundation for understanding the problem nature. Specifically, one of the early findings in the study of reusable RAS [33] was that for the development of a deadlock in these systems, the simultaneous satisfaction of the next four conditions is necessary and sufficient.

Mutual exclusion — that is, tasks must claim exclusive control of the resources that they require.
"Wait for" condition — that is, tasks hold resources already allocated to them while waiting for additional resources.
"No preemption" condition — that is, resources cannot be forcibly removed from the tasks holding them until the resources are used to completion.
"Circular wait" condition — that is, a circular chain of tasks exists subject to each task holds one or more resources that are being requested by the next task in the chain.

The identification of the preceding set of necessary and sufficient conditions for deadlock occurrence has then driven the efforts for its resolution. In general, all the proposed solutions can be classified into three major categories, namely, prevention, detection and recovery, and avoidance [28]. Prevention methods stipulate that the RAS is designed to be deadlock-free by ensuring that the set of necessary conditions for deadlock cannot be simultaneously satisfied at any point in the RAS operation. This requirement is met by imposing a number of restrictions on how a process can request the required resources. Detection and recovery methods allow the RAS to enter a deadlock state. A monitoring mechanism is used to detect a deadlock when it occurs and a resolution algorithm rolls the system back to a safe state by appropriately preempting resources from a number of deadlocked processes. Notice that a RAS state is safe if there exists a sequence of resource acquisition and release operations that allows all the processes in the system to run to completion. Detailed definitions of deadlock-related concepts are provided in Section 4.3. Avoidance methods address the problem of deadlocks by controlling how resources are granted to requesting processes. They use online feedback about the allocation of the system resources, that is, the RAS state. Given the current RAS state, a process request is granted only if the resulting state is safe. In principle, such a scheme leads to informed decisions about the safety of a resource allocation operation, allowing the RAS to achieve maximum flexibility.

Among the three methodologies, avoidance approaches are considered to be the most appropriate for dealing with the problem of deadlock as it arises in the FMS context [35–37]. The rollback required by the detection and recovery approaches when the system is deadlocked is deemed to be too costly in the FMS setting because it implies the temporary "unloading" (preemption) of some currently processed jobs. On the other hand, prevention methods, by being designed offline, tend to be overly conservative, leading to significant underutilization of the system. The point is that *a priori* knowledge of the routes of the jobs run through the FMS, makes the information contained in the RAS state theoretically sufficient for the determination of its safety. Therefore, in principle, more efficient FMS deadlock resolution policies can be designed by using online feedback with respect to the system state and the available information on the anticipated job requirements during the policy decision making. As we shall see in the next section, however, this statement must be qualified by the findings presented in References [38] and [39] which indicate that, in the general case, the resolution of the state-safety decision problem — underlying the computation of the maximally permissive FMS DAP — is NP-complete.

In the next section, we report some results on FMS deadlock avoidance that can be found in the current literature in addition to the results discussed in the rest of this chapter.

Literature Review of the FMS Deadlock Problem

In one of the first papers to deal with deadlock in manufacturing systems [36], the authors remark that avoidance seems to be the most appropriate method for handling deadlocks in an FMS environment and, furthermore, they claim that avoidance methods developed by the computing community are not efficient for FMS applications because they ignore available information about the process structures and are thus "unduly conservative." A new deadlock avoidance policy, referred to as the (B-K) Deadlock Avoidance Algorithm (BKDAA), is proposed. BKDAA expands on the idea underlying Banker's algorithm[1] [40] — one of the seminal deadlock avoidance policies in computer systems — by exploiting the fact that, in some FMS, certain resources are used exclusively by a single processing step of a single job. These exclusive resources can function as buffers dedicated to the related job instances, thus allowing for the decomposition of the deadlock resolution problem over segments of process routes and the design of an efficient control algorithm. Using formal petri net (PN) modeling [41], the authors show that under BKDAA, the system will never enter a deadlock or a restricted deadlock.[2]

PN modeling for the analysis of FMS deadlocks is also used in Reference [35]. This work exploits the concept of PN model invariants. The proposed control policy performs a look-ahead search over the set of states reachable from the current state in a certain number of steps and selects as the next system step the one least likely to lead to deadlock. Under this policy, deadlocks are possible, even though the probability of their occurrence is reduced.

In fact, petri net related research has produced an extensive series of results that pertain to the FMS deadlock. To deal with the complexity of analyzing petri net structures modeling real-life FMSs, researchers in the field have pursued a modular approach. Specifically, the FMS operation is modeled as the interaction of a number of simpler petri net modules with well-defined properties. The study of these characteristic properties can lead to some conclusions about the boundness, liveness, and reversibility [41] of these modules and the combined networks. Typically, a set of sufficient conditions to be satisfied by the initial state (marking) of the system in order to ensure deadlock-free operation are eventually identified.[3] The work presented in References [42] to [46] follows this line of reasoning. In many of these cases, the undertaken analysis has provided significant theoretical insight to the deadlock-related properties and the broader structural behavior of the considered systems — that is, those that can be modeled and analyzed as a synthesis of the elementary modules. However in all cases known to us, the sufficiency conditions identified for deadlock-free operation require the enumeration of special structures in the net like siphons and traps [41] which is a task of nonpolynomial complexity. Hence, the applicability of these techniques to real-life systems is restricted by computational intractability.

A partial justification for the increased complexity of the petri net based approaches discussed previously, is that most of them evolved as specializations to the petri net framework of the broader problem of

---

[1]This algorithm will be discussed more extensively in a subsequent section.

[2]A restricted deadlock is defined as a state in which all feasible transitions are inhibited by the constraints imposed by the avoidance policy itself [36].

[3]Because these approaches resolve deadlock by selecting the initial configuration of the system, they should be classified as prevention methods.

supervisory control of discrete event systems (DES) [25]. The primary issue undertaken in this paradigm is the development of a formal framework for the structural analysis and control of DES based on the theory of automata and formal languages [47]. In the supervisory control context, the requirement of deadlock-free operation is equivalent to the establishment of nonblocking behavior [25] — a prerequisite to any further logical analysis of DES — while the deadlock avoidance problem can be considered as a special case of the forbidden state problem [48]. As it is observed in Reference [48]:

> The complexity of the proposed control designs depends mainly on the representation of the forbidden sets. Our methods are very efficient, provided that an efficient, parsimonious representation of forbidden state sets can be found.

In fact, this is the more general spirit of the supervisory control paradigm, as the emphasis is placed on the decidability of problems encountered References [47], [49], and [50] — that is, whether a given issue can be effectively/algorithmically resolved — rather than on the provision of computationally efficient algorithms. However, as it will become clear in the ensuing sections of this chapter, the most severe problem in designing practical solutions to the FMS deadlock is the lack of a "parsimonious representation" of the states to be forbidden.

A different approach to the problem is taken in References [37] and [51]. Here, the authors suggest that a deadlock detection algorithm, presented in Reference [52], be applied to the state that results if a considered request is granted. The request can be granted only if the resulting state is not a deadlock state. This approach is, however, susceptible to restricted deadlocks. Furthermore, the volume of information processed by the deadlock detection algorithm in Reference [52] grows exponentially with the size of the FMS configuration because the algorithm searches for all possible cycles in a generalized version of the resource allocation graph [34].

The resolution of the FMS deadlock problem is treated in a radically different way in Reference [53]. Recognizing the increased complexity — in fact, NP — completeness — of the exhaustive search for safe allocation sequences, the authors suggest that this search should be restricted along the path that is defined by the applied scheduling (dispatching) policy. Given a resource allocation request, the FMS operation is simulated by first ordering the executed jobs according to a given dispatching rule and, then, trying to advance them in that order as much as possible. If the state (marking) resulting from this simulation covers the minimal marking required to keep the system live [41], then the considered allocation is deemed to be safe, and it is granted to the requesting process.

A thorough analysis of the deadlock problem in single-unit RAS where every resource possesses unit capacity, is provided in Reference [54]. In that work, the authors use the concept of the (extended) resource allocation graph — an analytical structure expressing the interdependencies of the running jobs resulting from the current resource allocation and the job posed requests — to characterize topologically the concepts of deadlock and unsafety in such a system. From this characterization, they proceed to the definition of deadlock avoidance policies because a number of conditions that will guarantee that these problematic structures will never be realized during the system operation. Observance of these conditions during the system operation ensures that it remains deadlock and/or restricted deadlock free. Similar results over the entire class of single-unit RAS are derived in Reference [55].

Finally, notice that in this survey we have tried to focus on the relatively structured approaches to the problem of FMS deadlock. Given the current urgency of the problem, a number of more ad hoc solutions, some of them developed specially for certain system/cell configurations, have also appeared in the literature. We have opted not to include them here because of either the specialized or the opportunistic aspect of their nature.

Developing Correct and Scalable Deadlock Avoidance Policies for Contemporary FMS

With the exception of References [36] and [53], and some of the policies discussed in Reference [54], the approaches discussed previously suffer from problems arising from the NP-complete nature of the state-safety decision problem. Therefore, with these solutions either there is no guarantee that the system will never enter deadlock or restricted deadlock or if such a guarantee is provided, it is obtained by (potentially) excessive computational cost (required computation is exponential in the size of the FMS configuration). In the first case, we say that the avoidance policy is not provably correct and in the second case that it is not scalable. It is our position that *deadlock avoidance policies for future technological systems must be both provably correct and scalable*. Note that correctness in the context of this discussion implies only that states characterized as safe are indeed safe. There might exist a subset of safe states which will not be recognized as such by a provably correct and scalable avoidance policy. In fact, this is the price one must pay (the compromise one must make) to obtain scalable policies in spite of the NP-complete nature of the underlying decision problem. Provably correct policies that successfully recognize all safe states are, by definition, optimal.

In our research program, we have developed a series of FMS deadlock avoidance policies (DAPs) possessing the two properties of correctness and scalability. These results are discussed in Sections 4.3 to 4.5. As it has already been pointed out, however, the NP-hardness of the problem of obtaining optimal structural control policies implies that, in the general case, provably correct and scalable policies are going to be suboptimal, that is, preclude some safe states. Therefore, a naturally arising additional requirement is that correctness and scalability must not come at a high cost with respect to system performance. Furthermore, the proposed policies must be easily implementable in the context of the current FMS control practices. Efficiency considerations regarding our developed policies is the topic of Section 4.6.

4.3   The Single-Unit RAS and the Deadlock Avoidance Problem

As has already been pointed out, this chapter considers the deadlock avoidance problem primarily in the context of the single-unit RAS subclass of the taxonomy presented in an earlier section. Hence, here we provide a rigorous characterization of this RAS class and of the deadlock avoidance problem as it arises in this context. The results discussed in this section have previously appeared in Reference [57] (© 1996 IEEE). Generalization of these results to the broader conjunctive RAS class is quite straightforward, and it can be found in Reference [30].

The Single-Unit RAS

The single-unit RAS model is defined by a number of resource types, denoted by $R = \{R_i, i = 1,\ldots,m\}$ and a number of process (job) types, denoted by $J = \{JT_j, j = 1,\ldots,n\}$. Every resource R_i is further characterized by its capacity C_i, that is, a finite positive integer indicating how many units of resource type R_i the RAS possesses. Process type JT_j is defined by a sequence $\langle JT_{jk}, k = 1,\ldots,l(j)\rangle$, with element JT_{jk} corresponding to the resource R_q supporting the processing of the kth step of this type. Thus, JT_{jk} denotes the resource allocation request associated with the kth stage of the jth job type. The sequence of resource allocation requests defining process type $JT_j, j = 1,\ldots,n$, will be referred to as its route.[4]

The single-unit RAS state, $s(t)$, at a given time instance t, is defined as follows.

Definition 4.1 The RAS state $s(t)$ at time t is a vector of dimensionality $D = \sum_{j=1}^{n} l(j)$ — that is, equal to the total number of distinct route stages in the system — with component $s_i(t), i = 1,\ldots,D$, being equal to the number of processes executing step k of route JT_q at time t, where q is the largest integer subject to $i > \sum_{j=1}^{q-1} l(j)$ and $k = i - \sum_{j=1}^{q-1} l(j)$.

---

[4]The terms process type and process route are used interchangeably in this document. To model the effects of routing flexibility in a static routing context, a subtler distinction must be made between the concepts of job type and job route. A job type is associated with a distinct product type produced by the system, while a set of job routes characterizes all the possible ways in which this product can be produced through the system. Although this distinction is significant when considering performance aspects, like load balancing of the system, it does not affect the FMS structural behavior and, therefore, it has been downplayed in the proposed FMS modeling.

Notice that the information contained in the RAS state is sufficient for the determination of the distribution of the resource units to the various processes as well as the slack (or idle) capacity of the system. Furthermore, the finiteness of the resource capacities implies that the set of distinct RAS states is finite; let it be denoted by $S = \{s^i, i = 0, 1, \ldots, Z\}$.

The system state changes in one of the following three ways:

1. Loading a new process in the system.
2. Advancing an already loaded process to its next route stage.
3. Unloading a finished process.

During a single state transition only one process can proceed. The resulting step, however, is feasible only if the next required resource can be obtained from the system slack capacity. The FMS controller selects the transition to be executed next among the set of feasible transitions. The development of control schemes to guide this selection is the notorious problem of job-shop sequencing and scheduling (cf. Section 4.1). As such, it constitutes the major link between structural and performance-oriented control, and it is beyond the scope of this study. In any case, the selection of a feasible transition by the controller and its execution by the FMS will be called an event in the FMS operation. Furthermore, because the undertaken structural analysis is concerned only with the logical aspects (i.e., the deadlock-free operation) of the system behavior, we focus only on the sequencing of these events, ignoring the detailed event timing. Practically, we assume that the controller can defer making a decision until all transitions that are potentially feasible from the current state have been enabled. Therefore, in the following, we suppress explicit consideration of time.

When time is ignored in the FMS operation, the underlying single-unit RAS model can be corresponded to an FSA, where the event set, E, consists of all the possible route steps executed in the FMS, the set of states, S, corresponds to the set of underlying RAS states, and the state transition function, $f: S \times E \rightarrow S$, is defined by

$$f(s^i, e) = \begin{cases} \text{succ}(s^i, e) & \text{if } e \in F(s^i) \\ s^i, & \text{otherwise} \end{cases}$$

In the preceding equation, $F(s^i)$ denotes the set of feasible events in state s^i, and the function $\text{succ}(s^i, e)$ returns the state that results when event $e \in F(s^i)$ takes place with the FMS being in state s^i. The initial and final states of this automaton are identified by state s^0, the state in which the FMS is idle and empty of jobs. Hence, the language accepted by the FMS consists of those input (controller command) strings that, starting from the empty state, leave the FMS in the same idle condition. In a physical interpretation, these strings correspond to complete production runs.

Example We elucidate the previously defined concepts by returning to the example of Figure 4.4. The system resource set is $R = \{R_1, R_2, R_3\}$, with $C_i = 1, i = 1, 2, 3$. The job types supported by the system can be formally described as

$$JT_1 = \langle R_1, R_2, R_3 \rangle$$
$$JT_2 = \langle R_3, R_2, R_1 \rangle$$

The system state depicted in Figure 4.4 is denoted by $s(t) = \langle 1\ 0\ 0\ 0\ 1\ 0 \rangle$. Furthermore, it turns out[5] that the size of the entire FMS state space is $Z = 27$, with the state signatures running from 0 to 26. In particular, state $s^0 = \langle 0\ 0\ 0\ 0\ 0\ 0 \rangle$ denotes the initial empty state. Table 4.1 enumerates the FMS state transition function and Figure 4.5 provides the corresponding state transition diagram (STD).

---

[5]It can be shown that, for an SU-RAS, the number of all possible distinct allocations of the system resources is equal to $\Pi_{i=1}^m (c_i + Q_i)! \neq c_i! Q_i!$, where Q_i denotes the number of stages supported be resource R_i.

TABLE 4.1   Example: The FMS State Transition
Function

$i: s^i$	State Vector	Successor States
0	0 0 0 0 0 0	1, 2
1	1 0 0 0 0 0	3, 15
2	0 0 0 1 0 0	4, 15
3	0 1 0 0 0 0	5, 6, 16
4	0 0 0 0 1 0	7, 8, 17
5	1 1 0 0 0 0	9, 18
6	0 0 1 0 0 0	0, 9
7	0 0 0 0 0 1	0, 10
8	0 0 0 1 1 0	10, 19
9	1 0 1 0 0 0	1, 11
10	0 0 0 1 0 1	2, 12
11	0 1 1 0 0 0	3, 13
12	0 0 0 0 1 1	4, 14
13	1 1 1 0 0 0	5
14	0 0 0 1 1 1	8
15	1 0 0 1 0 0	16, 17
16	0 1 0 1 0 0	18
17	1 0 0 0 1 0	19
18	1 1 0 1 0 0	
19	1 0 0 1 1 0	
20	0 0 1 0 0 1	6, 7
21	0 1 0 1 0 1	16
22	1 0 1 0 1 0	17
23	0 1 0 0 0 1	20, 21
24	0 0 1 0 1 0	20, 22
25	0 1 1 0 0 1	11, 23
26	0 0 1 0 1 1	12, 24

Source: (c) 1996 IEEE. With permission.

Structural Analysis of the Single-Unit RAS

Next, we use the STD of the previous example to analyze the structural/deadlock properties of the single-unit RAS model underlying the FMS operation.

State Reachability and Safety

A directed path in the STD of Figure 4.5 represents a feasible sequence of events in the FMS. We are mainly interested in paths that start and finish in the empty state s^0. Notice that there is a subset of nodes for which there is no directed path from state s^0; these are shown as dashed nodes in the STD. This implies that when the system is started from empty state, (under normal[6] operation) the states (resource allocation) represented by the dashed nodes will never occur. These states will be referred to as unreachable. The remaining states are feasible states under normal operation and will be called reachable states. The set of reachable states will be denoted by $S_{\bar{r}}$ and the set of unreachable states will be denoted by $S_{\bar{r}}$. Notice that $S_{\bar{r}} = S \backslash S_r$. Formally,

Definition 4.2 State s^i is reachable from state s^j, denoted $s^i \leftarrow s^j$, iff there exists a sequence of events that can bring the system from state s^i to state s^i In the FSA notation,

$$\forall s^i, s^j \in S_s, \quad s^i \leftarrow s^j \Leftrightarrow \exists u \in E^* : f(s^j, u) = s^i$$

---

[6]By normal, it is meant that the FMS operation observes the assumptions regarding the feasibility of the state transitions discussed previously.

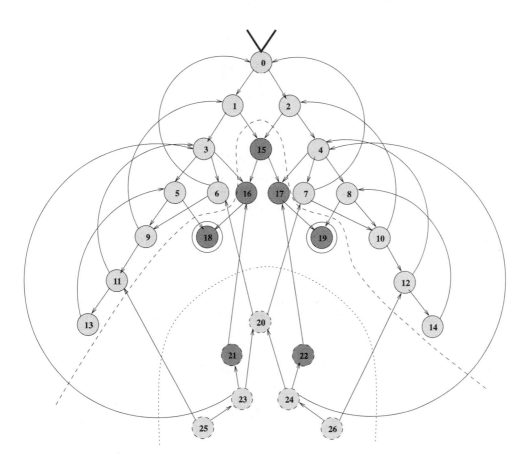

FIGURE 4.5   Example: the FMS state transition diagram. (From © 1996 IEEE. With permission.)

Furthermore, a state $s^2 \in S$ is a reachable state, iff $s^i \leftarrow s^0$.

Another important classification of the STD nodes/states results from the following observation. There are states from which the empty s^0 is reachable by following a directed path of the STD, and states for which this is not possible. In the STD of Figure 4.5, the former are lightly shaded while the latter are heavily shaded. If the FMS enters any of the heavily shaded states it will never be able, under normal operation, to complete all running jobs, that is, become idle and empty. For this reason, the heavily shaded states are characterized as unsafe, while the lightly shaded states, which provide accessibility to state s^0, are characterized as safe. The set of safe states is denoted by $S_s \subseteq S$ and the set of unsafe states is denoted by $S_{\bar{s}}$. Again, it holds that $S_{\bar{s}} = S \backslash S_s$.

Definition 4.3 State s^i is a safe state iff state s^0 is reachable from state s^i. A state that is not safe will be called an unsafe state. In the FSA notation,

$$\forall' s^i \in S, \qquad \text{safe}(s^i) \Leftrightarrow \exists u \in E^* : f(s^i, u) = (s^0 \Leftrightarrow s^0 \leftarrow s^i)$$

Furthermore, we extend the characterization of safety to FMS transitions emanating from safe states, by characterizing them as safe if they result in a safe state. Mathematically,

$$\forall' s^i \in S_s, \qquad \forall e \in F(s^i), \qquad \text{safe}(e/s^i) \Leftrightarrow f(s^i, e) \in S_s$$

Finally, we denote the intersection of any two classes resulting from the previous two classifications by S_{xy} where $x = r, \bar{r}$ and $y = s, \bar{s}$.

FMS Deadlock

It has already been stated in the previous section that an FMS state is a deadlock if there exists a set of jobs such that every job in the set is waiting for the release of some resources held by some other jobs in the set. This is now defined formally.

Definition 4.4 In a single-unit RAS, state s^i is a partial deadlock, if a subset of its resources, DR^i, is filled to capacity, and every process holding a unit of these resources requires transfer to another resource in DR^i for the execution of its next stage.

The class of FMS deadlock states for a given FMS configuration will be denoted by S_d. Deadlocks are the natural reason for the existence of the unsafe states in the FMS operation. This is established by the following two propositions, formally proven in Reference [23].

Proposition 4.1 *An FMS deadlock is an unsafe state.*

Proposition 4.2 *In the FMS-STD, every directed path that starts from an unsafe state, and does not involve the loading of a new job in the system, results in a deadlock.*

It should be noticed, however, that there can be unsafe states which are not deadlocks. As an example, consider state s^{15} in the STD of Figure 4.5. This state, although one step away from deadlock, it does not contain a deadlock itself because both of its loaded jobs can advance to their next requested resource (i.e., the AGV).

An Algebraic FMS State-Safety Characterization

The definition of the FMS deadlock, together with the two propositions linking the safety and the deadlock concepts, leads to the following algebraic characterization of the state-safety problem:

Consider the FMS state $s^i = s_0$ in which J_0 jobs, $\{j_1, j_2, \ldots, j_{J_0}\}$, are currently loaded in the system. Obviously, $J_0 \leq \sum_{k=1}^{m} C_k$, the total capacity of the FMS. For uniformity of presentation, we include an extra resource R_{m+1} with $C_{m+1} = \infty$. This resource accommodates all jobs exiting the system. Hence, all job routes are augmented by one step: $JR_{j,l'(j)} = R_{m+1}, j = 1,\ldots,r$, where $l'(j) = l(j) + 1$. Let $U_{it}, i = 1,\ldots, J_0$, denote the resource unit occupied by job j_i at step t. Then, assuming that state s_0 is safe, the total number of steps required for running all the currently loaded jobs to completion, is $T_t = \sum_{i=1}^{J_0} [l'(jr(U_{i0})) - st(U_{i0})]$, where $jr(U_{i0})$ and $st(U_{i0})$ are the functions returning the job type and the running stage of their argument. Finally, let $\{\delta_{ikt}\}$ denote a set of binary variables with

$$\delta_{ikt} = \begin{cases} 1 & \text{if job } j_i \text{ occupies a unit of resource } k \text{ at step} t \\ 0 & \text{otherwise} \end{cases}$$

where $i \in \{1,\ldots, J_0\}$, $k \in \{1,\ldots, m + 1\}$ and $t \in \{0,\ldots, T_t\}$.

State s_0 is safe iff the following system in variables δ_{ikt} is feasible.

$$\delta_{ik0} = \begin{cases} 1 & \text{if } \exists q \in \{1,\ldots,C_k\}: U_{i0} = R_{k_q} \\ 0 & \text{otherwise} \end{cases} \quad \forall i,k \tag{4.1}$$

$$\sum_{j=1}^{J_0} \delta_{ikt} \leq C_k \quad \forall k,t \tag{4.2}$$

$$\sum_{k=1}^{m+1} \delta_{ikt} = 1 \quad \forall i,t \tag{4.3}$$

$$\sum_{t=0}^{T_t} \delta_{i,l'(i),t} \geq 1 \qquad \forall i \tag{4.4}$$

$$\delta_{ikt} - (\delta_{ik(t+1)} + \delta_{i,\mathrm{sr}(i,k,p_i),(t+1)}) \leq 0 \qquad \forall i, k \ (\neq m+1), t \tag{4.5}$$

$$\delta_{ikt} + \delta_{i,\mathrm{sr}(i,k,p_i),(t+1)} + \sum_{j=1}^{J_0} \delta_{j\mathrm{sr}(i,k,p_i),t} \leq C_{\mathrm{sr}(i,k,p_i)} + 1 \qquad \forall i, k \ (\neq m+1), t \tag{4.6}$$

$$\delta_{ikt} \in \{0,1\} \qquad \forall i,k,t \tag{4.7}$$

In the preceding equations, $\mathrm{sr}(i, k, p_i)$ is a function returning the resource type required by job j_i for its next processing step, given that it is currently allocated one unit of resource type k for the execution of its p_ith processing step (s-uccessor r-esource). Eq. (4.1) introduces the description of the initial state s_0 into the program and it represents the set of initial conditions. Eq. (4.2) states that no resource type can hold more jobs than its capacity. Eq. (4.3) expresses the fact that every job instance always requires one complete unit of operational space. Eq. (4.4) simply states that for a RAS state to be safe, it is necessary that every job executes its last step at some point over the considered horizon. Given the convention introduced previously, this guarantees that the job has been run to completion and left the system. Eq. (4.5) represents the precedence constraints introduced by the job routes, that is, every movement of a job in the system must obey its routing sequence. Eq. (4.6) imposes the requirement that a job can proceed to its next step only if there is at least one unit of available capacity from the resource type required during that step. Finally, Eq. (4.7) states the binary nature of the δ variables.

This algebraic characterization of state safety has proven useful in developing a methodology for establishing the correctness of a class of deadlock avoidance policies discussed in Section 4.5.

Deadlock Avoidance Policies — General Definitions

The characterization of state safety and the FMS deadlock by the topological structure of the state transition diagram (STD) of the underlying FSA can also be used formally to define the avoidance approach in the resolution of the FMS deadlock. In the following discussion, it is assumed that the FMS always undergoes normal operation, so that only reachable states occur.

The subgraph consisting of the reachable states S_r and the arcs emanating from them is called the reachability graph of the FSA. A DAP must restrict the operation of a given FMS by limiting it to its reachable and safe subspace S_{rs}. Practically, we seek to identify an appropriate set of feasible transitions which, when removed from the STD (or equivalently, disabled by the DAP), render the unsafe subspace $S_{r\bar{s}}$ unreachable from state s^0. At the same time, it must be ensured that every state s^i in the remaining graph (i.e., reachable under the control policy) is still safe (i.e., there exists a directed path in the remaining graph leading from state s^i to s^0). States that are reachable under the DAP and from which progress is inhibited by the policy-imposed constraints and not by the RAS structure are characterized as restricted deadlocks in the deadlock literature [36].

An example of a DAP that gives rise to restricted deadlock is presented in Figure 4.6. This hypothetical policy, defined on the STD of the example FMS of Figure 4.4, admits (allows access to) only the states corresponding to the white-colored nodes in the depicted STD. Notice that the policy provides accessibility to state s^8, while it disables the only transition out of it by not admitting state s^{10}. As a result, whenever the system finds itself in state s^8, it is permanently blocked there by the policy logic itself.

A formal characterization of the preceding concepts is as follows:

Definition 4.5 An avoidance policy P for the FMS is a function

$$P : S \to 2^E \qquad P(s^i) = \{e \in F(s^i) : e \text{ is selected by the policy}\}$$

Events $e \in \cup_i P(s^i)$ are called the (policy-)enabled events.

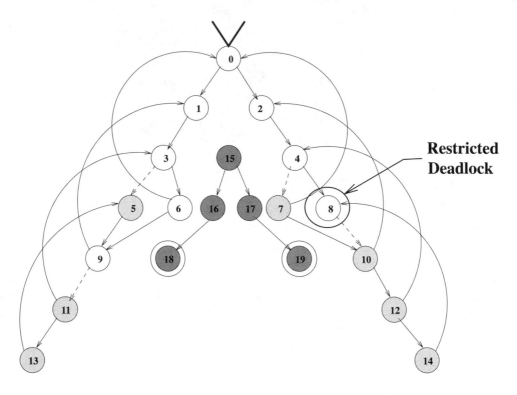

FIGURE 4.6 Example: a DAP inducing *restricted* deadlock.

Definition 4.6 Given an avoidance policy P, let $s^i \overset{P}{\leftarrow} s^j$ denote the fact that state s^i is reachable from state s^j through an event sequence which comprises policy-enabled events only. Let $S_r(P) = \{s^i : s^i \overset{P}{\leftarrow} s^0\}$ and $S_s(P) = \{s^i : s^0 \overset{P}{\leftarrow} s^i\}$. Then, policy P is correct iff $S_r(P) \subseteq S_s(P)$.

In words, a DAP is correct iff the policy-reachable subspace $S_r(P)$ is a strongly connected subgraph containing the idle and empty state s^0. Notice that the reachable and safe subspace of the uncontrolled system, S_{rs}, possesses this property. In fact, this is the maximal subspace possessing this property. This leads us to the concept of the optimal DAP. A correct avoidance policy P^* is optimal if the policy restriction on the S_{rs} subspace of the FMS disables only those actions that result in unsafe states.

Definition 4.7 The correct avoidance policy P^* is optimal iff

$$\forall' s^i \in S_{rs} \qquad \forall' e \in F(s^i) \qquad e \in P^*(s^i) \Leftrightarrow f(s^i, e) \in S_s$$

This characterization of the optimal policy has the following three implications:

1. For a given FMS configuration, the optimal avoidance policy P^* is unique.
2. $S_r(P^*) = S_{rs}$. Establishing the optimal control policy P^* is equivalent to removing from the reachability graph those transition arcs that belong to the cut[7] $[S_{rs}, \overline{S_{rs}}]$. For example, in the STD of Figure 4.5, the optimal control policy P^* consists of removing the arcs that emanate from lightly shaded solid nodes and crossing the twisted dashed line.
3. In References [38] and [39], it is shown that, in the general case, the problem of determining the safety of a RAS state is NP-complete. Because the inclusion of a transition to the optimal avoidance policy P^* depends on the safety of the successor state, it follows that obtaining policy P^* is a NP-hard problem [26].

---

[7]For a definition of this concept refer to Reference [69].

4.4 Single-Unit RAS Admitting Polynomially Computable Optimal DAP

The concluding remark of the last section regarding the complexity of the optimal deadlock avoidance policies in single-unit RAS made researchers of the manufacturing system deadlock — including our group — focus their efforts, for a long time, in obtaining good suboptimal policies, that is, policies that are computationally tractable and still relatively efficient with respect to the overall system performance. However, some recently obtained results establish that for a large subclass (in fact, the majority) of single-unit RAS, the optimal DAP is polynomially computable and, therefore, implementable in real-time. This section develops the relevant theory.

Specifically, the identification of the aforementioned SU-RAS subclass admitting polynomially computable optimal DAP resulted from the following two key observations.

1. While tlhe decision problem regarding the safety of a SU-RAS state is, in general, NP-complete, the deadlock detection problem — that is, whether a given SU-RAS state is deadlock or not — is polynomially computable.
2. There are SU-RAS models in which the set of unsafe states coincides with the set of the deadlock states. In other words, in this class of SU-RAS, there are no unsafe states which do not already contain a deadlock.

Observations 1 and 2 together imply that for the considered subclass of SU-RAS, optimal deadlock avoidance can be obtained by testing for deadlock through one-step look-ahead, that is, the satisfaction of a pending request is simulated and, if the resulting RAS state is deadlock-free, it is concluded that granting the request is a safe step. In fact, this result provides theoretical justification for a claim made by a number of researchers in the field (e.g., References [35], [37]), that in the majority of the cases, deadlock avoidance policies based on a finite horizon of look-ahead steps, even though not provably correct, are well behaved.

Deadlock Detection in Single-Unit RAS

To show the polynomiality of the deadlock detection problem in SU-RAS, it is sufficient to provide an algorithm resolving the problem in polynomial time. Such an algorithm is depicted in Figure 4.7, and was initially developed in Reference [30]. The algorithm logic is based on the concept of the Resource Dependency Graph (RDG). This is a graph with each vertex V_i, $i = 1,\ldots, m$, corresponding to a system resource type R_i, $i = 1,\ldots,m$, and with each edge E_{ij} implying that one of the units of R_i is currently allocated to a process j_k requiring a unit of R_j for its next processing stage. Hence, a partial deadlock in a SU-RAS is depicted by a closed subgraph of the RDG in which every node V_i has a number of emanating arcs equal to the capacity of the corresponding resource R_i; such a subgraph is typically called a knot in the deadlock literature. The algorithm stores the information contained in the RDG of a given state in a data structure, known as the Resource Dependency Vector (RDV). RDV has one component for every RAS resource type R_i, containing the number of R_i units allocated to processes in state s and a list of the resource types required by the processes currently allocated to R_i for their next step. Notice that this information can be easily obtained from the state vector and the process routes. Then, the algorithm considers the RDV and tries to identify a subset of resources entangled in a knot.

Initially, all system resources are possible candidates. Subsequently, the algorithm goes through a number of scannings of the RDV and, at every scanning, it eliminates a number of resources from further consideration. The resources eliminated are those which either are not filled to capacity or have units allocated to processes which are ready to leave the system or to move to an already eliminated resource. Proceeding in this way, the algorithm either eliminates all resources, in which case the considered state s is deadlock-free, or at a certain scanning no resource is eliminated, in which case state s is a deadlock.

Example To provide a more concrete example of the algorithm logic, consider the SU-RAS state depicted in Figure 4.8. During the first iteration of the algorithm, resources R_1 and R_4 are eliminated

```
algorithm DDA
begin
  / * Initialize */
  CANDIDATES := RDV;
  STUCK:= FALSE;

  / * processing step */
  while (not-empty(CANDIDATES) and not (STUCK)) do
  begin
    STUCK:= TRUE;
    for (i:= 1 to cardinality (CANDIDATES)) do
    begin
      if  (deleteable($R_i$))
      begin
        eliminate($R_i$);
          STUCK:= FALSE;
      end
    endfor
  endwhile

  if (not-empty (CANDIDATES))
  state s is a deadlock with resources in CANDIDATES being the deadlocked subset;
  else
    state s is deadlock-free;
end
```

FIGURE 4.7   A polynomial deadlock detection algorithm for single-unit RAS. (From © 1997 IEEE. With permission.)

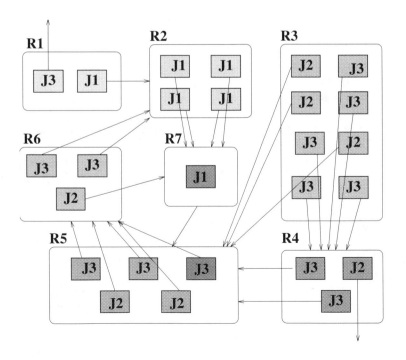

FIGURE 4.8   Example: using the Resource Dependency Graph to detect deadlock.

from the CANDIDATES list because each contains a job ready to leave the system and, therefore, they cannot be entangled in a deadlock. After the second iteration, resource R_3 is also eliminated because it contains jobs waiting upon resource R_4, which was shown to be able to obtain free capacity during the first run. However, during the third iteration, no other resources are eliminated and, therefore, it is concluded that the subset of resources R_2, R_5, R_6, and R_7 are permanently blocked owing to a partial deadlock.

The following proposition, proven in Reference [30], establishes the polynomial complexity of the DDA algorithm.

Proposition 4.3 *The complexity of the DDA algorithm is* $O(m_2 \cdot \overline{C})$, *where m is the number of the RAS resource types and* $\overline{C} = \max_{i \in \{1,...,m\}} C_i$ *(i.e., the maximum resource capacity).*

Single-Unit RAS Models with No Deadlock-Free Unsafe States

The result regarding the existence of SU-RAS models for which the class of unsafe states coincides with the class of deadlock states is stated in the following theorem:

Theorem 4.1 *Let a single-unit RAS be defined over a resource set R, and a set of job types JT, such that each resource* $R_i \in R$ *satisfies at least one of the following conditions.*

1. $C_i > 1$.
2. *If* JT_{jk}, JT_{lm} *are two distinct job stages supported by resource* R_i, *then either the preceding stages* $JT_{j,k-1}$ *and* $JT_{l,m-1}$ *or the succeeding stages* $JT_{j,k+1}$ *and* $JT_{l,m+1}$ *are supported by the same resource or* JT_{jk} *(JT_{lm}) is the initial/final stage in* JT_j *(JT_l).*
3. R_i *supports a single job stage* JT_{jk}.

Then, $S_{ru} \cap (S \setminus S_{rd}) = \emptyset$, *that is, every reachable unsafe state is a deadlock state.*

A formal proof for this theorem is provided in Reference [56] and it essentially combines similar results initially developed in References [30] and [54]. The significance of this result is demonstrated by means of the flexible manufacturing cell depicted in Figure 4.9. This cell presents the typical layout of many contemporary automated manufacturing cells, consisting of a number of workstations, served by a central robotic manipulator. Each workstation possesses an input and an output buffer, while the manipulator can carry only one part at a time. For such a system, it is clear that deadlock-free operation can be established by controlling the allocation of the buffering capacity of the system workstations and treating the manipulator as the enabler of the authorized job transfers. Furthermore, the SU-(sub-)RAS defined by the system

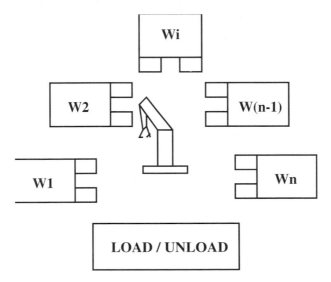

FIGURE 4.9 The Flexible Manufacturing Cell layout. (From © 1997 IEEE. With permission.)

workstations meets the specifications 1 and/or 2 prescribed in the conditions of Theorem 4.1 and, therefore, in such a system, the optimal deadlock avoidance can be attained through one-step lookahead on the allocation of the buffering capacity of the system workstations.

Taking this whole discussion to a more practical level, we would like to remind the reader of the fact that SU-RAS models apply to FMS environments in which the only cause of deadlocks is the limited buffering capacity of the FMS equipment (cf. Section 4.2; the RAS taxonomy). In the light of this remark, the practical implication of the combined results of Proposition 4.3 and Theorem 4.1 is that in these FMS environments, deadlock can be resolved optimally by ensuring that every FMS resource can accommodate at least two parts at a time. This can be easily achieved with the provision of an input, output, or auxiliary buffer. Hence, optimal deadlock-free buffer space allocation in contemporary FMS turns out to be a rather easy problem under the appropriate design of these environments. For the sake of completeness in the rest of this chapter, we discuss computationally efficient solutions to the problem of deadlock avoidance in SU-RAS for the cases in which the conditions of Theorem 4.1 are not satisfied. In addition to covering exhaustively the problem of deadlock avoidance over the entire class of SU-RAS, these results have provided useful insight for synthesizing deadlock avoidance policies for more complicated classes of the RAS taxonomy for which optimality results, similar to those of Theorem 4.1, are not available. The interested reader is referred to References [23] and [30].

4.5   Polynomial-Kernel Deadlock Avoidance Policies for Single-Unit RAS

To deal with the more computationally ill-behaved cases, we adopted an approach based on the concept of polynomial-kernel policies. Simply stated, this approach requires that, because the target set S_s is not polynomially recognizable, the system operation should be confined to a subset of these states which is polynomially computable. This state subset is perceived as an easily identifiable (polynomially computable) kernel among the set of reachable and safe states and gives the methodology its name. From an implementational viewpoint, this idea requires the identification of a property $H(s^i)$, $s^i \in S$, such that $H(s^i) \Rightarrow$ safe(s^i), $\forall s^i \in S_r$, and $H(\ )$ is polynomially testable on the system states. Then, by allowing only transitions to states satisfying H, through one-step lookahead, it can be ensured that the visited states will be safe.

An additional requirement is that the resulting DAP is correct, that is, the policy-reachable subspace must be strongly connected (cf. Definition 4.6). However, this characterization of policy correctness is based on a global view of the system operation and, given the typically large size of the system state space, it is not easily verifiable. A more operable criterion for testing the correctness of polynomial-kernel policies is provided by the following theorem.

Theorem 4.2 *A polynomial-kernel DAP is correct iff for every state admitted by the policy there exists a policy-admissible transition, which, however, does not correspond to the loading of a new job into the system.*

The validity of this theorem primarily results from the observation that at, any point in time, the system workload — in terms of processing steps — is finite, and every transition described in the theorem reduces this workload by one unit. Hence, eventually the total workload will be driven to zero, which implies that the system has returned to its home state s^0. A more formal statement and proof of this theorem, by means of the algebraic characterization of state safety provided by Eqs. (4.1) through (4.7) can be found in Reference [57]. Notice, that establishing the policy correctness by means of Theorem 4.2 resolves concurrently the validity of condition H as a polynomial-kernel identifier for state safety and the restricted deadlock-free operation of the controlled system.

In the rest of this section, we present three polynomial-kernel policies developed in our research program. For each of these policies, we present the motivation behind the policy defining condition, the policy defining logic, and an elucidating example. Formal proofs of the policy correctness and a complexity analysis can be found in the provided references. The efficiency of these policies with respect to the system operational flexibility is discussed in Section 4.6.

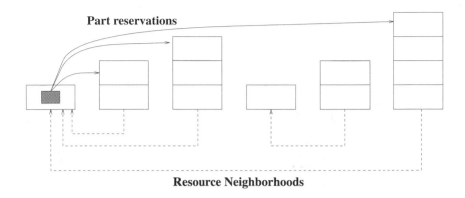

Figure 4.10 RUN motivation: the partial resource reservation scheme.

The Resource Upstream Neighborhood (RUN) Policy

It should be obvious that no deadlock would occur in a RAS, if every job were allocated all the resources required for its entire processing upon its loading into the system. In fact, this is a very general prevention scheme for deadlock resolution and, as such, it fails to take into consideration any existing information about the RAS structure, thus being overly conservative and underutilizing the system resources. The second remark that RUN builds upon, is that if at any point during its sojourn through the system, a job is allocated to a resource of very high — theoretically infinite — capacity, then, for the purposes of structural analysis, its route can be decomposed to a number of segments, each of which is defined by two successive visits to the infinitely capacitated resource(s).

Because in any practical context, resources have finite capacity, RUN exploits the existing job routing information to implement a nested, partial resource reservation system on the principle that, if there are some resources with higher capacity than others in the system, they can function as temporary buffers for the jobs that they support. A pictorial representation of RUN reservation scheme is provided in Figure 4.10. For the detailed formal statement of the policy, we introduce the concept of the resource upstream neighborhood.

Definition 4.8 [57] The upstream neighborhood of resource R_i consists of all route stages JT_{jk} which are supported by resource R_i, plus all the route stages belonging to the maximal route subsequences immediately preceding each of the aforementioned JT_{jk}, and involving stages JT_{jp} with $C_{R(JTjp)} \leq C_i$. A job instance j_j is in the neighborhood of resource R_i iff its current processing stage is in the neighborhood of R_i.

Then a formal definition of RUN is as follows:

Definition 4.9 RUN [57] A resource allocation state s is accepted by RUN DAP iff the number of jobs in the upstream neighborhood of each resource, R_i, does not exceed its buffering capacity, C_i.

Example We highlight the policy-defining logic and the resource neighborhood construction through an example. Consider the small SU-RAS depicted in Figure 4.11. This system consists of four resources R_1, R_2, R_3, and R_4, with a corresponding capacity vector $\mathbf{C} = \langle 2, 1, 1, 1 \rangle$. In its current configuration, the system supports the production of three distinct job types, with job routes: $JT_1: R_1 \rightarrow R_2 \rightarrow R_3, JT_2: R_3 \rightarrow R_4 \rightarrow R_3, JT_3: R_1 \rightarrow R_2$. By applying the logic of Definition 4.8 to this system, we obtain the neighborhood inclusions indicated by the following incidence matrix.

$$A_{RUN} = \begin{bmatrix} 1 & & & & & 1 & \\ & 1 & & & & & 1 \\ & 1 & 1 & 1 & 1 & 1 & \\ & & & 1 & 1 & & \end{bmatrix} \tag{4.8}$$

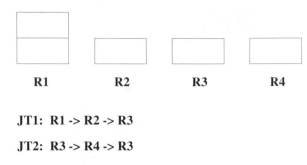

JT1: R1 -> R2 -> R3

JT2: R3 -> R4 -> R3

JT3: R1 -> R2

FIGURE 4.11   Example: a single-unit RAS for RUN and RO DAP demonstration.

Each row of the preceding array corresponds to a resource neighborhood, $N(R_1), \ldots, N(R_4)$. Each column corresponds to a route stage, starting with the stages of job type JT_1 and concatenating the stages of JT_2 and JT_3. Using the definition of the system state provided in Section 4.3 — that is, a vector $s(t)$ with components corresponding to the distinct job stages $JT_{jk}, j = 1,\ldots,3, k = 1,\ldots, l(j)$, and with $s_{jk}(t)$ being equal to the number of job instances executing stage JT_{jk} at time step t — it is easy to see that the policy constraints can be expressed by a system of linear inequalities on the system state.

$$A_{RUN}\mathbf{s} \leq \mathbf{C} \tag{4.9}$$

The policy correctness is established in Reference [57]. In fact, there it is also shown that if instead of the partial ordering imposed by resource capacities, any other (partial) ordering of the resource set is used in the neighborhood definition, the resulting policy is still correct. Therefore, RUN logic defines an entire family of policies for a given FMS configuration, with each member resulting from a distinct (partial) ordering of the system resources. The exploitation of this result for improving the policy efficiency is discussed in the Section 4.6. Finally, the policy is scalable (i.e., of polynomial complexity in the FMS size as defined by the number of resources and the distinct route stages of the underlying RAS), because

1. Construction of the neighborhood sets for a single job type is of complexity no higher than $O(L^2)$, where L is the length of the longest route supported by the system. Therefore, computing the complete resource upstream neighborhoods is of complexity not higher than $O(nL^2)$, where n is the number of job types supported by the system.
2. Evaluating the admissibility of a RAS state by the policy requires the verification of m linear inequalities in the D system state variables where m is the number of system resources and D is the total number of distinct job stages supported by the system.

The Resource Ordering (RO) Policy

To understand the logic behind the resource ordering (RO) policy, let us first concentrate on a subclass of SU-RAS with the special property that the RAS resources can be numbered so that all job routes correspond to strictly increasing or strictly decreasing resource sequences. We characterize this particular class of SU-RAS, as the class of "counterflow" systems [29]. It is easy to see that a sufficient condition for physical and restricted deadlock-free operation in counterflow systems is that no pair of resources (R_i, R_j), with $i < j$, are filled to capacity with the jobs in resource R_i corresponding to ascending resource sequences and the jobs in resource R_j corresponding to descending resource sequences. These remarks are proved in Reference [29] and are visualized in Figure 4.12.

Of course, the counterflow property is a very restrictive requirement, and a policy applicable only to this class of systems would not be of any practical use. It turns out, however, that the policy-motivating idea outlined previously can be extended to the more general class of SU-RAS. The solution

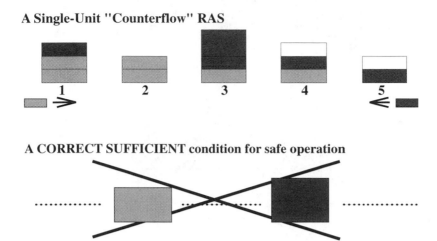

A Single-Unit "Counterflow" RAS

1   2   3   4   5

A CORRECT SUFFICIENT condition for safe operation

FIGURE 4.12   RO motivation: a correct sufficient condition for deadlock-free operation of single-unit "counter-flow" RAS.

is to double-count job instances for which the remaining route segment is nonmonotonic with respect to the resource numbering.

The complete policy definition is as follows:

Definition 4.10 RO [58]

1. Impose a total ordering on the set of system resources R, that is, a mapping

$$o : R \rightarrow \{1,\ldots, |R|\} \text{ s.t. } R_i < R_j \Leftrightarrow o(R_i) < o(R_j)$$

We say that R_i (R_j) is to the left (right) of R_j (R_i), iff $R_i < R_j$.
Furthermore, job stage JT_{jk} is characterized as right- (left-)directed if $R(JT_{j(x-1)}) < (>) R(JT_{jx})$, $\forall x > k$, where $R(JT_{jk})$ denotes the resource supporting stage JT_{jk}. A stage that is neither right- nor left-directed is an *undirected* stage.

   A job instance is characterized as right-, left-, or undirected on the basis of its running processing stage.

2. Let
   - $RC_i(t) = \{$right-directed + undirected job instances in R_i at time step $t\}$
   - $LC_i(t) = \{$left-directed + undirected job instances in R_i at time step $t\}$

3. A resource allocation state $s(t)$ is accepted by *RO* iff

$$\forall i, j : \; R_i < R_j \Rightarrow RC_i(t) + LC_j(t) \leq C_i + C_j - 1 \qquad (4.10)$$

Example   We elucidate the definition of RO DAP, by applying it on the small system of Figure 4.11. The ordering used in the policy implementation is the natural ordering of the system resources, that is, $o(R_i) = i$, $\forall i$. Furthermore, we observe that job instances executing the last stage of their route can never deadlock the system because their unloading from the system is always a feasible step. Hence, they can be ignored during the evaluation of the admissibility of a resource allocation state and, therefore, they are omitted during the definition of the content of $RC_i(t)$ and $LC_i(t)$.[8]

---

[8]Although not used in the previous example, a similar remark regarding the (in-)significance of last job stages in deadlock avoidance applies to the implementation of RUN DAP.

It is easy to see that job stages JT_{11}, JT_{12}, and JT_{31} are right-directed, while job stage JT_{21} is undirected, and job stage JT_{22} is left-directed. Hence, the contents of the RC_i and LC_i counts are defined by the following table:

Resource	RC_i	LC_i
R_1	JT_{11}, JT_{31}	
R_2	JT_{12}	
R_3	JT_{21}	JT_{21}
R_4		JT_{22}

Then, Part 3 of Definition 4.10 implies that this implementation of RO on the considered RAS imposes the following set of linear inequalities on the system state:

$$\begin{bmatrix} 1 & & & & 1 \\ 1 & & 1 & & 1 \\ 1 & & & 1 & 1 \\ & 1 & 1 & & \\ & 1 & & 1 & \\ & & 1 & 1 & \end{bmatrix} s \leq \begin{bmatrix} 2 \\ 2 \\ 2 \\ 1 \\ 1 \\ 1 \end{bmatrix} \tag{4.11}$$

In Eq. (4.11), each inequality corresponds to a pair (R_i, R_j) with $R_i < R_j$, and with all these pairs ordered lexicographically in increasing order [i.e., (R_1, R_2), (R_1, R_3), etc.].

The policy correctness is proved in Reference [58]. Given a certain FMS configuration, RO, like RUN, essentially defines a family of DAPs, generated by all the possible total orderings imposed on the FMS resources. The computation required for the initial setup of the policy (i.e., classification of the different route stages to left-, right-, and undirected) is of complexity $O(D)$, where D is the total number of distinct processing stages executed in the considered RAS, while the number of constraints on the RAS state to be checked online, is of the order $O(m^2)$, where m is the number of system resources. Therefore, the policy is scalable.

Ordered States and the FMS Banker's Algorithm

The classical Banker's algorithm [32] is based on the observation that a state is safe if its running processes can be ordered in such a manner that each process in the ordering can terminate using only its currently allocated resources, resources currently available in the system, and also, resources currently allocated to processes which are preceding it in the order. In the context of correct and scalable FMS DAPs, this idea leads to the concept of the ordered state, defined as follows:

Definition 4.11 [59]   Let $D = \Sigma_{j=1}^{n} l(j)$, the number of distinct route stages supported by a given system. RAS state $s(t)$ is ordered iff there exists an ordering of the set of distinct job stages, $o(\ ) : \{JT_{jk} : j = 1, \ldots, n, k = 1, \ldots, l(j)\} \rightarrow \{1, \ldots, D\}$, such that the resource requirements for processing to completion a job instance j_i in stage $JT^{(l)}$ can be satisfied by means of the free resources in state $s(t)$, plus the resources held by job instances j_q in stages JT^q, with $q \leq i$.

Let S_0 denote the set of ordered RAS states. In Reference [59], it is shown that S_0 is a strongly connected subspace of the RAS STD, containing the empty state and, therefore, the restriction of the system operation on this set defines a correct DAP.

Definition 4.11 also immediately provides an algorithm for testing whether a state is ordered or not.

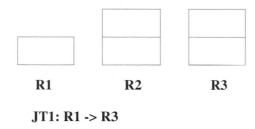

JT1: R1 -> R3

JT2: R3 -> R2 -> R1

FIGURE 4.13   Example: a single-unit RAS for Banker's algorithm demonstration.

Definition 4.12   FMS Banker's Algorithm [59]

1. set UJ := $\{JT_{jk}, j = 1,...,n, k = 1,...,l(j)\}$; $i = 0$; ORDERED := TRUE.
2. Repeat

   a. $i := i + 1$;
   b. Try to find a job stage $JT_{jk} \in$ UJ, the instances of which can terminate by using their currently allocated resources, plus the currently free resource units.
   c. If no such a job stage can be found, ORDERED := FALSE.
   d. else $JT_{jk} \equiv JT^{(i)}$; UJ := UJ $\setminus \{JT_{jk}\}$; release the resources held by the job instances of JT_{jk} to the pool of the free resource units.
   until (UJ = 0) $\vee$ (ORDERED = FALSE)

3. If (ORDERED = FALSE) return FALSE else return TRUE

Example   Consider the single-unit RAS of Figure 4.13. This system consists of three resources, R_1, R_2, R_3, with $C_1 = 1$ and $C_2 = C_3 = 2$. In its current configuration, the system supports two job types: $JT_1 = \langle R_1, R_3 \rangle$ and $JT_2 = \langle R_3, R_2, R_1 \rangle$. The reader should be able to verify that state $s = \langle 1\ 0\ 1\ 1\ 0 \rangle$ is ordered, with a valid ordering being $JT^{(1)} = JT_{11}, JT^{(2)} = JT_{22}$, and $JT^{(3)} = JT_{21}$. On the other hand, state $s' = \langle 1\ 0\ 2\ 1\ 0 \rangle$ is not ordered, even though it is safe: advancing one instance of job stage JT_{21} to its next stage allows the job instance in stage JT_{11} to run to completion, which further allows the remaining jobs to finish.

It is the inability of Banker's logic to discern the viability of the partial job advancements demonstrated in the previous example that renders the algorithm suboptimal. This is the price paid to keep the algorithm complexity polynomial in the system size. Indeed, similar to the "classical" Banker's algorithm, the correctness of the FMS Banker's results from the fact that the reusability of the system resources implies a monotonic increase of the pool of free resources whenever job instances terminate and makes back-tracking unnecessary during the search for a feasible job stage ordering. Hence, the complexity of the preceding algorithm is polynomial, specifically, $O(mD \log D)$.

4.6   Efficiency Considerations for Polynomial-Kernel DAPs

It has already been observed that polynomial-kernel policies attain their tractability at the cost of suboptimality. Hence, a valid point for this line of research is to try to reduce suboptimality as much as possible. Before, however, we start considering this problem, we must define a metric for measuring the efficiency of these policies. There are two general approaches to establish such a metric. The first approach compares the attainable system performance under the control of various DAPs with respect to typical performance measures like process throughput(s), waiting times, work-in-process, and resource utiliza-tions. Results along these lines for RUN and RO DAPs can be found in References [23], [60], and [61]. In this chapter, we shall focus on the second approach, which evaluates the efficiency of the different DAPs, based on the concept of operational flexibility. Specifically, the flexibility allowed by the evaluated policy is compared to the flexibility attained by the maximally permissive (optimal) DAP by assessing the coverability of the safe space S_s by the policy-admissible subspace, $S(P)$. More formally, consider the polynomial-kernel policy defined by property H and let $S(H) \equiv \{s^i \in S : H(s^i) \text{ is TRUE}\}$, that is, the policy admissible subspace. Then, a viable policy efficiency measure is provided by the ratio

$$I = \frac{|S(H)|}{|S_s|} \tag{4.12}$$

where $|S|$ denotes the cardinality number of set S.

Because of the typically large size of the $S(H)$ and S_s subspaces, their explicit enumeration will not be possible and, therefore, we must resolve simulation and statistical sampling techniques. Such a technique, known as the co-space simulation technique, is developed in Reference [62]. Briefly, this approach recognizes that the set of safe states of a given SU-RAS, Q corresponds to the reachability set of a "co-system," Q', which is defined from the original RAS Q by reversing its job routes. Hence, the operation of the co-system Q' is simulated until a sufficiently large sample of states is obtained. According to the previous remark, this sample set consists of safe states of the original system. In continuation, the condition H defining the evaluated DAP is applied on the extracted sample set and the portion of the sample states admitted by the policy is determined. This portion expresses the policy coverability of the extracted sample set, and constitutes a point estimate for index I. Application of this technique to the polynomial-kernel DAPs of Section 4.5, and experimental evaluation results can be found in References [58], [59], and [62]. In the rest of this section, we discuss some properties of polynomial-kernel DAPs which can be used to enhance the operational flexibility of these policies when implemented on any given FMS configuration.

Policy Disjunctions and Essential Difference

The first way to improve the efficiency of an FMS structural controller employing polynomial-kernel DAPs, with respect to the metric of Eq. (4.12) is based on the following proposition.

Proposition 4.4 *Given two conditions* $H_1( )$ *and* $H_2( )$ *defining correct polynomial-kernel DAPs, the policy defined by the disjunction* $H_1( ) \vee H_2( )$ *is another correct polynomial-kernel DAP.*

To see this, simply notice that acceptance of a state *s* by the policy disjunction implies that at least one of the two policy defining conditions, $H_1( )$, $H_2( )$, evaluates to TRUE at *s* and, therefore, state *s* is safe. Furthermore, if state $s \in S(H_i)$, $i \in \{1, 2\}$, then the correctness of the corresponding policy implies the existence of at least one feasible event *e*, which is enabled by that policy, and $\delta(e, s) \equiv s' \in S(H_i)$ (cf. Theorem 4.2). Then, $s' \in S(H_1 \vee H_2)$, and according to Theorem 4.2, the policy defined by $H_1( ) \vee H_2( )$ is correct.

It is also easy to see that the subspace admitted by the policy disjunction is the union of the subspaces admitted by the two constituent policies. If it happens that

$$(S(H_1) \not\subseteq S(H_2)) \wedge (S(H_2) \not\subseteq S(H_1)) \tag{4.13}$$

then $S(H_1) \cup S(H_2)$ is richer in states than any of its constituents. Therefore, the resulting policy is more efficient with respect to index I.

Two polynomial-kernel policies based on conditions H_1 and H_2 that satisfy Eq. (4.13) are characterized as essentially different. The essential difference of the polynomial-kernel policies presented in Section 4.5 is analyzed in Reference [59]. It turns out that RUN and the FMS Banker's algorithm are essentially different, while RO is subsumed by Banker's.

Optimal and Orthogonal Orderings for RUN and RO DAPs

A second opportunity for improving the efficiency of RUN and RO DAPs is provided by the fact that the defining logic of these two policies essentially leads to entire families of policies for a given FMS configuration. As we saw in Section 4.5, each member of these families is defined by a distinct ordering of the system resource set. Hence, a naturally arising question is which of these orderings leads to the most efficient policy implementation. In this way, an optimization problem is defined, which can be considered as a parameter-tuning (optimization) problem. This optimization problem has been characterized as the optimal ordering problem, and some further discussion on its formulation and solution can be found in References [58] and [62].

Furthermore, the aforementioned richness of RUN and RO implementations on a given RAS configuration raises the possibility of the existence of different orderings within the same policy family which lead to the admissibility of subspaces complementary to each other. This idea, combined with the closure of polynomial-kernel DAPs with respect to policy disjunction, has led to the definition of the orthogonal ordering problem. Details of the problem formulation and its solution can be found in Reference [23].

Combining Polynomial-Kernel DAPs with Partial Search

A last idea that can lead to efficiency improvement of polynomial-kernel DAPs is to expand the state space admitted by such a policy, by allowing transitions to states s which fail to satisfy the policy defining condition H( ) — that is, H(s) = FALSE — but for which their inclusion in the target set S_s can be established through controlled partial search, that is, n-step look-ahead schemes. Specifically, for an n-step look-ahead scheme, state s with H(s) = FALSE is admissible if there exists a sequence of feasible events w, such that $|w| < n$ and $\delta(w, s) = s' \in S$ (H). Because this new admissibility condition is of existential character, it is deemed that it can increase the policy efficiency with rather small computational cost for reasonable sizes of look-ahead horizons.

It is interesting to notice how the length, n, of the look-ahead horizon partitions the target set S_s accepted by the optimal DAP: $n = 0$ defines the kernel set S(H), while every time that the look-ahead horizon is increased by one step, say from n to $n + 1$, an additional ring of states is added to the set of states admitted when the horizon length is equal to n. Obviously, for finite state spaces this expansion continues only up to the point that the entire set S_s is covered for some maximal length N. This partitioning of the optimal subspace on the basis of the look-ahead horizon size, n, is depicted in Figure 4.14.

It might also be efficient to store some of the results of these partial searches in a lookup table. Specifically, some states s for which H(s) = FALSE, but which have been shown to belong to S_s through the search mechanism, can be stored to a table for future reference. In this way, the cost of a look-ahead search is spared for these states. However, given the large size of the underlying spaces, storing the entire set of states found to be admissible through n-step look-ahead search, in general, will not be practically possible. All the same, the idea can become practical by exploiting the following economies introduced by the system structure and operation.

1. From the FMS structural analysis of Section 4.3, it easily follows that

$$\forall s^1, s^2 \text{ s.t. } \forall i, s_i^1 \le s_i^2: \quad \text{safe}(s^2) \Rightarrow \text{safe}(s^1) \tag{4.14}$$

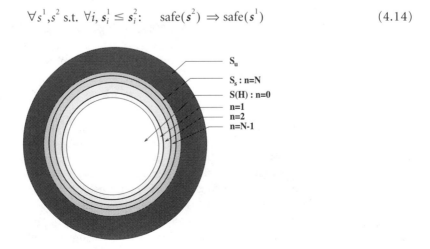

FIGURE 4.14 Expanding the admitted subspace of Polynomial-Kernel DAPs through n-step look-ahead search.

This result implies a covering relationship over the set of safe states admitted by a look-ahead policy, in the sense that an admissible state should be stored in the policy lookup table, only if its admissibility is not implied by an already stored state and Eq. (4.14).

2. Sometimes, the system operation presents considerable localization with respect to its underlying state space, that is, a rather small number of states is revisited a significant number of times. Then, an appropriately sized lookup table can function as a cache memory which holds information about the structure of the currently working subspace. Furthermore, a forgetting mechanism is required to update the set of stored states as the system "drifts" to new state regions. Such a mechanism can be provided, for instance, by time-stamping the stored states with the last time that they were visited during the system operation — a variation of a scheme known as aging [63].

An application of the combination of polynomial-kernel deadlock avoidance policies with partial search can be found in Reference [59].

4.7 Additional Issues and Future Directions in FMS Structural Control

The starting point of this chapter has been the observation that current trends in discrete-part manufacturing require extensive levels of automation of the underlying shop-floor activity and, in order to support this extensive automation, a more sophisticated real-time control methodology for these environments is needed. Specifically, the extensive automation, or even autonomy, required for the operation of these systems gives rise to a new set of problems concerning the establishment of their logically correct and robust behavior, which is collectively known as the structural control problem of contemporary FMS. In the rest of the chapter, we focus on one particular problem of this area, the resolution of the manufacturing system deadlock which has been a predominant FMS structural control issue. The details of the FMS deadlock problem and its solution depend on the operational characteristics of the system, as it was revealed by Section 4.2 and the RAS taxonomy. Restricting the subsequent discussion to the most typical case where deadlocks arise owing to ineffective allocation of the system buffering capacity (i.e., the first class of the underlying RAS taxonomy), we were able to show that, even though in the general case, the problem of optimal (maximally permissive) deadlock avoidance is NP-hard, when the system is configured so that every resource (i.e., workstation and/or MHS-component) can stage at least two parts at a time, we can obtain the optimal DAP with polynomial computational cost. Hence, the problem of FMS deadlock owing to finite buffering capacity is conclusively resolved for all practical purposes. Furthermore, to completely cover the class of single-unit RAS, we presented polynomial-kernel DAPs, a class of suboptimal yet efficient DAPs which are also computationally tractable.

In the rest of this section, we briefly discuss some other aspects of this work not covered in the material of this chapter. For one thing, it has already been mentioned that the polynomial-kernel DAPs of Section 4.5 have been extended to cover the single-type and conjunctive RAS of the aforementioned taxonomy. These results are straightforward generalizations of the policies presented in Section 4.5 and can be found in References [30] and [64]. At present, work is under development addressing the last class of the presented RAS taxonomy, that is, the integration of deadlock avoidance with flexible job routing. This problem is nontrivial, because, in addition to the increased complexity associated with the underlying state-safety decision problem. The introduction of routing flexibility results in very high space complexity owing to the exponentially large number of routing options associated with each job type. Some initial results investigating the trade-off between increased computational effort and the benefits of routing flexibility are presented in Reference [65]. A different approach on the issue is presented in Reference [66] acknowledging the increased complexity of the online job rerouting problem. The author suggests the exploitation of the system inherent flexibilities for effective accommodation of different operational contingencies like machine breakdowns and the arrival of expedient jobs. So, that work essentially deals with the effective reconfiguration of the system in the face of a major disruption, and "bridges" the existing results on deadlock avoidance with the second main area of FMS structural control, that is, the design of protocols for exception handling and graceful degradation. Finally, additional directions of extending the developed DAPs to new RAS classes include the study of hierarchically structured RAS, and RAS in which the length of the different job routes is not predetermined, but depends on the occurrence of various events. These results are currently under development.

FMS structural control and the developed policies have also significant repercussion on the various aspects of the system performance. In a sense, the constraints imposed by a structural control policy define the feasibility space for any performance-optimizing model. In other words, they significantly shape the system capacity.[9] As it is observed in Reference [67],

> [D]efining, measuring and respecting capacity are important at all levels of the [system] hierarchy. No system can produce outside its capacity, and it is futile at best, and damaging, at worst, to try. ... It is essential therefore to determine what capacity is, then to develop a discipline for staying within it.

Initial results for scheduling the capacity of the structurally controlled FMS are presented in References [23], [60], and [61]. An architecture for real-time integration of structural and performance-oriented control in contemporary FMS controllers is presented in Reference [23]. Essentially, the SC policy, at any point of the system operation, censors the decisions of the system dispatcher for action admissibility with respect to safety. In some more recent work [68], the effect of this censorship on the stability[10] of some well-known distributed scheduling policies has been analyzed. It turns out that even when maximally permissive deadlock avoidance control is applied, most of the distributed policies which have been formally shown to be stable under the assumption of infinite capacitated buffers, are not optimal (i.e., stable) any more. This result reiterates the significance of adequate modeling of the system structural aspects while analyzing its performance, and reintroduces the scheduling problem in this new environment. In a similar fashion, the entire FMS tactical planning problem (cf. Section 4.1) needs to be reconsidered.

A last contribution of this work, and a new horizon for additional research, is the systematic treatment of complexity in large-scale supervisory control problems. As it was stated in Section 4.1, current supervisory control theory has a very significant impact on the formulation and the rigorous analysis of the solvability (decidability) of problems rising in the area of logical control of discrete event systems [25]. However, most of the proposed algorithms, in their effort to remain generic enough, address these problems at a (formal language-theoretic) syntactical level which ignores the detailed structure of the controlled system. As a result, these solutions are of complexity polynomial to the size of the underlying system state space and, therefore, intractable for the cases where this size explodes badly (as in the case of the deadlock avoidance problem). The work presented herein demonstrates that much can be gained in terms of mastering this complexity if the undertaken analysis focuses also on the specifics of the system behavior (language semantics). It is our feeling that there are a lot of DES logical control problems currently begging computationally efficient (real-time) solution. Understanding the problem attributes determining the problem complexity and establishing rational trade-offs between computational tractability and operational efficiency are very important but also challenging problems for the field.

References

1. Drucker, P. F., The emerging theory of manufacturing, *Harv. Bus. Rev.*, May–June, 94, 1990.
2. National Research Council, *Information Technology and Manufacturing*, Tech. Rep., National Academy Press, Washington, D.C., 1993.
3. A survey on manufacturing technology, *The Economist*, March, 3, 1994.
4. Joshi, S. B., Mettala, E. G., Smith, J. S., and Wysk, R. A., Formal models for control of flexible manufacturing cells: physical and system models, *IEEE Trans. Robot. Autom.*, 11, 558, 1995.

---

[9]For a rigorous definition of this concept, see Reference [70].

[10]Practically, a DAP is stable if it can provide the maximum throughput attainable over the entire class of scheduling policies.

5. Suarez, F. F., Cusumano, M. A., and Fine, C. H., An empirical study of manufacturing flexibility in printed circuit board assembly, *Oper. Res.*, 44, 223, 1997.
6. Arango, G. and Prieto-Diaz, R., Domain analysis concepts and research directions, in *Domain Analysis and Software Systems Modeling*, Prieto-Diaz. R. and Arango, G., Ed., IEEE Computer Society Press, 1991, 3.
7. Bodner, D. A. and Reveliotis, S. A., Virtual factories: an object-oriented simulation-based framework for real-time fms control, in *ETFA'97*, IEEE, 1997, 208.
8. Stecke, K. E., Design, planning, scheduling and control problems of flexible manufacturing systems, *Ann. Oper. Res.*, 3, 1985.
9. Gershwin, S. B., Hierarchical flow control: a framework for scheduling and planning discrete events in manufacturing systems, *Proc. IEEE*, 77, 195, 1989.
10. Winston, W. L., *Introduction to Mathematical Programming: Applications and Algorithms*, 2nd ed., Duxbury Press, Belmont, CA, 1995.
11. Co, H. C., Biermann, J. S., and Chen, S. K., A methodical approach to the flexible manufacturing-system batching, loading and tool configuration problems, *Int. J. Prod. Res.*, 28, 2171, 1990.
12. Stecke, K. E. and Raman, N., Production planning decisions in flexible manufacturing systems with random material flows, *IEE Trans.*, 26, 2, 1994.
13. Srivastava, B. and Chen W. H., Heuristic solutions for loading in flexible manufacturing systems, *IEEE Trans. Robot. Autom.*, 12, 858, 1996.
14. Rodammer, F. A. and White, P., Jr., A recent survey of production scheduling, *IEEE Trans. Syst. Man. Cybern.*, 18, 841, 1988.
15. Connors, D., Feigin, G., and Yao, D., Scheduling semiconductor lines using a fluid network model, *IEEE Trans. Robot. Autom.*, 10, 88, 1994.
16. Dai, J. G., Yeh, D. H., and Zhou, C., The qnet method for re-entrant queueing networks with priority disciplines, *Oper. Res.*, 45, 610, 1997.
17. Kumar, P. R., Scheduling manufacturing systems of re-entrant lines, in Yao, D. D., Ed., *Stochastic Modeling and Analysis of Manufacturing Systems*, Springer-Verlag, New York, 1994, 325.
18. Kumar, P. R. and Seidman, T. I., Dynamic instabilities and stabilization methods in distributed real-time scheduling of manufacturing systems, *IEEE Trans. Autom. Contr.*, 35, 289, 1990.
19. Lu, S. H. and Kumar, P. R., Distributed scheduling based on due dates and buffer priorities, *IEEE Trans. Autom. Contr.*, 36, 1406, 1991.
20. Sharifnia, A., Stability and performance of distributed production control methods based on continuous-flow models, *IEEE Trans. Autom. Contr.*, 39, 725, 1994.
21. Kumar, P. R. and Meyn, S. P., Stability of queueing networks and scheduling policies, *IEEE Trans. Autom. Contr.*, 40, 251, 1995.
22. Kumar, P. R. and Meyn, S. P., Duality and linear programs for stability and performance analysis of queueing networks and scheduling policies, *IEEE Trans. Autom. Contr.*, 41, 4, 1996.
23. Reveliotis, S. A., Structural Analysis and Control of Flexible Manufacturing Systems with a Performance Perspective, Ph.D. thesis, University of Illinois, Urbana, 1996.
24. Lawley, M. A., Structural Analysis and Control of Flexible Manufacturing Systems, Ph.D. thesis, University of Illinois, Urbana, 1995.
25. Ramadge, P. J. G. and Wonham, W. M., The control of discrete event systems, *Proc. IEEE*, 77, 81, 1989.
26. Garey, M. R. and Johnson, D. S., *Computers and Intractability: A Guide to the Theory of NP-Completeness*, W. H. Freeman, New York, 1979.
27. Sethi, A. K. and Sethi, S. P., Flexibility in manufacturing: a survey, *Int. J. Flex. Manuf. Syst.*, 2, 289, 1989.
28. Peterson, J. L., *Operating System Concepts*, Addison-Wesley, Reading, MA, 1981.
29. Lawley, M., Reveliotis, S., and Ferreira, P., Design guidelines for deadlock handling strategies in flexible manufacturing systems, *Int. J. Flex. Manuf. Syst.*, 9, 5, 1997.

30. Reveliotis, S. A., Lawley, M. A., and Ferreira, P. M., Polynomial complexity deadlock avoidance policies for sequential resource allocation systems, *IEEE Trans. Autom. Contr.*, 42,1344, 1997.

31. Havender, J. W., Avoiding deadlock in multi-tasking systems, *IBM Syst. J.*, 2, 74, 1968.

32. Habermann, A. N., Prevention of system deadlocks, *Commun. ACM*, 12, 373, 1969.

33. Coffman, E. G., Elphick, M. J., and Shoshani, A., System deadlocks, *Comput. Surv.*, 3, 67, 1971.

34. Holt, R. D., Some deadlock properties of computer systems, *ACM Comput. Surv.*, 4, 179, 1972.

35. Viswanadham, N., Narahari, Y., and Johnson, T. L., Deadlock avoidance in flexible manufacturing systems using petri net models, *IEEE Trans. Robot. Autom.*, 6, 713, 1990.

36. Banaszak, Z. A. and Krogh, B. H., Deadlock avoidance in flexible manufacturing systems with concurrently competing process flows, *IEEE Trans. Robot. Autom.*, 6, 724, 1990.

37. Wysk, R. A., Yang, N. S., and Joshi, S., Resolution of deadlocks in flexible manufacturing systems: avoidance and recovery approaches, *J. Manuf. Syst.*, 13, 128, 1994.

38. Gold, E. M., Deadlock prediction: easy and difficult cases, *SIAM J. Comput.*, 7, 320, 1978.

39. Araki, T., Sugiyama, Y., and Kasami, T., Complexity of the deadlock avoidance problem, in *2nd IBM Symp. on Mathematical Foundations of Computer Science*, IBM, 1977, 229.

40. Dijkstra, E. W., Cooperating sequential processes, Tech. Rep., Technological University, Eindhoven, the Netherlands, 1965.

41. Murata, T., Petri nets: properties, analysis and applications, *Proc. IEEE*, 77, 541, 1989.

42. Zhou, M. and Dicesare, F., Parallel and sequential mutual exclusions for petri net modeling of manufacturing systems with shared resources, *IEEE Trans. Robot. Autom.*, 7, 515, 1991.

43. Giua, A. and DiCesare, F., Petri net structural analysis for supervisory control, *IEEE Trans. Robot. Autom.*, 10, 169, 1994.

44. Jen, M. D. and DiCesare, F., Synthesis using resource control nets for modeling shared-resource systems, *IEEE Trans. Robot. Autom.*, 11, 317, 1995.

45. Ferrarini, L., Narduzzi, M., and Tassan-Solet, M., A new approach to modular liveness analysis conceived for large logic controllers' design, *IEEE Trans. Robot. Autom.*, 10, 169, 1994.

46. Ezpeleta, J., Colom, J. M., and Martinez, J., A petri net based deadlock prevention policy for flexible manufacturing systems, *IEEE Trans. Robot. Autom.*, 11, 173, 1995.

47. Hopcroft, J. E. and Ullman, J. D., *Introduction to Automata Theory, Languages and Computation*, Addison-Wesley, Reading, MA, 1979.

48. Boel, R. K., Naoum, L. B., and Breusegem, V. V., On forbidden state problems for a class of controlled petri nets, *IEEE Trans. Autom. Contr.*, 40, 1717, 1995.

49. Sreenivas, R. S., On a weaker notion of controllability of a language k with respect to a language l. *IEEE Trans. Autom. Contr.*, 38, 1446, 1993.

50. Sreenivas, R. S., On the existence of finite state supervisors for arbitrary supervisory control problems, *IEEE Trans. Autom. Contr.*, 39, 856, 1994.

51. Cho, H., Kumaran, T. K., and Wysk, R. A., Graph-theoretic deadlock detection and resolution for flexible manufacturing systems, *IEEE Trans. Robot. Autom.*, 11, 413, 1995.

52. Wysk, R. A., Yang, N. S., and Joshi S., Detection of deadlocks in flexible manufacturing cells, *IEEE Trans. Robot. Autom.*, 7, 853, 1991.

53. Hsieh, F. S. and Chang, S. C., Dispatching-driven deadlock avoidance controller synthesis for flexible manufacturing systems, *IEEE Trans. Robot. Autom.*, 10, 196, 1994.

54. Fanti, M. P., Maione, B., Mascolo, S., and Turchiano, B., Event-based feedback control for deadlock avoidance in flexible production systems, *IEEE Trans. Robot. Autom.*, 13, 347, 1997.

55. Xing, K. Y., Hu, B. S., and Chen, H. X., Deadlock avoidance policy for petri net modeling of flexible manufacturing systems with shared resources, *IEEE Trans. Autom. Contr.*, 41, 289, 1996.

56. Kumar, P., Kothandaraman, K., and Ferreira, P., Scalable and maximally-permissive deadlock avoidance for fms, *Proc. of IEEE Intl. Conf. Robot. Autom.*, Leuven, Belgium, 1998.

57. Reveliotis, S. A. and Ferreira, P. M., Deadlock avoidance policies for automated manufacturing cells, *IEEE Trans. Robot. Autom.*, 12, 845, 1996.

58. Lawley, M. A., Reveliotis, S. A. and Ferreira, P. M., A correct and scalable deadlock avoidance policy for flexible manufacturing systems, *IEEE Trans. Robot. Autom.*, Vol. 14, 796, 1998.

59. Lawley, M., Reveliotis, S., and Ferreira, P., The application and evaluation of Banker's algorithm for deadlock-free buffer space allocation in flexible manufacturing systems, *Int. J. Flex. Manuf. Syst.*, Vol. 10, 73, 1998.

60. Reveliotis, S. A. and Ferreira, P. M., An analytical framework for evaluating and optimizing the performance of structurally controlled fms, in *1996 IEEE International Conference on Robotics and Automation*, IEEE Robotics and Automation Society, 1996, 864.

61. Reveliotis, S. A. and Ferreira, P. M., Performance evaluation of structurally controlled fms: exact and approximate approaches, in *Flexible Automation and Intelligent Manufacturing*, Manufacturing Research Center, Georgia Tech, 1996, 829.

62. Lawley, M., Reveliotis, S., and Ferreira, P., Fms structural control and the neighborhood policy. Part 2. Generalization, optimization and efficiency, *IIE Trans.*, 29, 889, 1996.

63. Deitel, H. M. *Operating Systems*, Addison-Wesley, Reading, MA, 1990.

64. Reveliotis, S. A. and Lawley, M. A., Efficient implementations of Banker's algorithm for deadlock avoidance in flexible manufacturing systems, *in ETFA '97*, IEEE, 1997, 214.

65. Lawley, M. A., Integrating routing flexibility and algebraic deadlock avoidance policies in automated manufacturing systems, *Int. J. Prod. Res.*, submitted for publication, 1997.

66. Reveliotis, S. A., Accommodating fms operational contingencies through routing flexibility. *IEEE Trans. Robot. Autom.*, Vol. 15, 3, 1999.

67. Gershwin, S. B., Hildebrant, R. R., Suri, R., and Mitter, S. K., A control perspective on recent trends in manufacturing systems, *IEEE Contr. Syst. Mag.*, 6, 3, 1986.

68. Reveliotis, S. A., The destabilizing effect of blocking due to finite buffering calacity in multi-class queueing networks, *IEEE Trans. Autom. Contr.*, Vol. 45, 585, 2000.

69. Ahuja, R. K., Magnanti, T. L., and Orlin, J. B., *Network Flows: Theory, Algorithms and Applications*, Prentice-Hall, Englewood Cliffs, NJ, 1993.

70. Gershwin, S. B., *Manufacturing Systems Engineering*, Prentice-Hall, Englewood Cliffs, NJ, 1994.

5

The Design of Human-Centered Manufacturing Systems

Dietrich Brandt
University of Technology (RWTH)

Inga Tschiersch
University of Technology (RWTH)

Klaus Henning
University of Technology (RWTH)

5.1   Introduction

Inga Tschiersch

Results of indiscriminate attempts to replace humans by machines have led to serious difficulties with machine-dependent systems, which are vulnerable to disturbance and frequently lack robustness and flexibility. There is a need to examine the scope of anthropocentric technologies which link the skill and ingenuity of humans with advanced and appropriate forms of technology in a true symbiosis of work and technology. Europe should concentrate on those areas that require order-bound manufacturing and, therefore, experience in flexible specialization, with an appropriate flexible production system and skilled workforce. This would shift Europe to a comparatively favorable position that could assume considerable competitive advantages (Cooley [6], p. 4).

This quotation characterizes the beginning of a widespread European discussion about the concept of anthropocentric — or human-centered — technology. In 1989, Cooley and others suggested this concept as a strategy to overcome some of the difficulties which have become evident in European countries over the past years, mainly the problems of mass production, which is not sufficiently flexible to respond to the dynamics and challenges of the world market.

In the following chapter, this concept of human-centered technology will be described in its sociotechnical context. It will be considered with regard to the different dimensions of workplace, groupwork, and networks and in terms of the frameworks of both society and the natural environment. These different aspects of human-centered systems will be illustrated by a series of case studies from industry, the service sector, and research representing several European countries.

The report covers a wide range of research fields. Therefore, each of the following case studies is reported by the authors who have themselves performed most of the research and development work described.

The emphasis of this report is on technology: the roles of control and information technology in enterprises today and the strategies of designing and implementing technology, taking into account the specific aspects which characterize human-centered systems.

5.2   Concept, Implementation and Evaluation of Human-Centered Systems

Dietrich Brandt and Paul Fuchs-Frohnhofen

This section gives a brief overview of the current use of the concept of human-centered technology in European industry. There are two main aspects of this concept to be considered: the sociotechnical system theory, providing the framework for today's human orientation, and the paradigm of human-centered technology design in its original sense.

This section is primarily based on the research of the Department of Computer Science in Mechanical Engineering at the University of Technology (RWTH) in Aachen, Germany, which has significantly contributed to implementing these concepts in German industry through numerous cooperative university–industry projects in technological and organizational innovation. Partners have included steel companies (e.g., Thyssen, Mannesmann), car manufacturers (e.g., VW, Mercedes), chemical and pharmaceutical enterprises (e.g., Hoechst), large hospitals, and many small and medium-sized enterprises throughout

Germany. This section integrates these experiences. It also incorporates a wide range of related experiences from several other European countries.

The Concept of Human-Centered Systems

Human-Centered Systems as Sociotechnical Systems

Industry today is increasingly shaped by technology, e.g., the automation of production, computer-supported information and cooperation networks, multimedia applications in production planning, and control. Such systems, however, are not merely technical systems. They are also sociotechnical systems comprising, besides technology, people and their personal communication and tasks, views, organizational structures, cooperation, etc. Therefore the basic concept of such sociotechnical systems is briefly described here.

The OSTO approach can be used as a method for analyzing, designing, and monitoring sociotechnical systems [14,17]. OSTO stands for open, sociotechnical-economic system, i.e., the open system comprises social, technical, and economic components. It builds on the sociotechnical system theory which was considerably influenced by members of the London Tavistock Institute of Human Relations including Trist [24] and Emery [7].

OSTO is particularly suitable for large, complex organizations. It focuses on the analysis of high-quality work processes which are strongly interrelated. OSTO understands systems as living systems (open cybernetic systems). Such systems include human beings together with their processes of work and life (Figure 5.1).

The system transforms input into output through a transformation process. The mission is the reason for the existence of the organization. It represents the unwritten contract between the system and its environment. However, all processes within the system only become core processes if they are oriented towards the mission. In the long run, living systems cannot survive without a reason for existence or core processes.

Moreover, it is necessary for survival of the system that the purpose (or meaning) of the system be future oriented. Only then can the members of the organization maintain both motivation for and identification with the system. Thus, the system can maintain its acceptance in a wider social context.

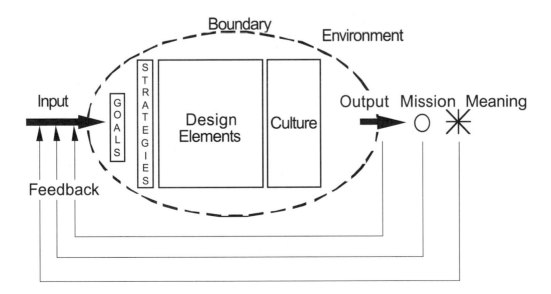

FIGURE 5.1 The OSTO System approach.

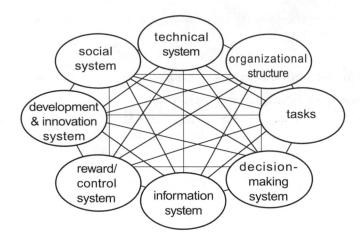

FIGURE 5.2   Design elements of an organization.

Further elements of cybernetic systems are feedback loops which are either negative or positive (weakening or enforcing). They give the system the qualities of stabilization and renewal. In this context, the shaping of feedback processes is an important managerial task in terms of survival of the system.

System output and behavior are regarded as consequences of the structure or design elements of the system, where these are in turn interpreted as system strategies and objectives put into practice. This system structure is shown in Figure 5.2. The eight design elements of a system are linked to each other by dynamic and complex interrelations.

These eight design elements characterize any sociotechnical system today, be it production or service enterprises, factories, banks, hospitals, airlines, or community administrations. They have to be taken into account when designing or analyzing such systems. Thus, they are the basic terms needed to specifically describe human-centered systems as they have been developed and implemented in Europe during recent years. All case studies reported here are based on this sociotechnical concept.

Technology, People, and Organization

Of the eight design elements, the technical, social, and organizational system elements—technology, people, and organization (Figure 5.3) — are most important when describing sociotechnical systems. These three terms are widely used to understand the behavior of such systems. They are, however, to be understood as being embedded into the overall network created by the complex links and interactions of all eight design elements. Without reducing this complexity, the three system elements of technology, people, and organization are at the core of this report because they have the most relevance for all professional activities within any manufacturing system, whether it concerns individual work, cooperation in groups or networks, etc. The following links in such manufacturing systems can be described in more detail:

- People refers to the dimension of the workplace: the individual employee and his or her working environment.
- Organization refers to the two dimensions of groupwork (teams) and networks.
- Technology refers to all three dimensions of the individual workplace, groupwork, and networks of people.

This view leads to defining different kinds of technology according to the three different dimensions — different functionalities and interfaces, etc. Some aspects are as follows.

Technology of the workplace concerns the human-machine system in its narrow sense. For example, technology would need to be designed so as to support the workers' skills and competencies. These aspects are mainly discussed in the case studies reported in Sections 5.3 and 5.4.

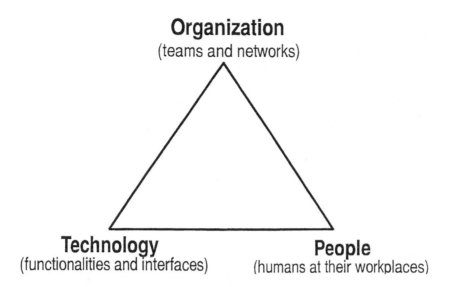

FIGURE 5.3 People, organization, and technology (From Brandt, D. and Cernetic, J., *J. Artif. Intelligence Soc.*, 12, 2, 1998. With permission.)

Technology of the group means, for example, the grouping of machines in a way which allows for groupwork. It may include suitable handling devices and robots, as well as software to support planning and scheduling, as a group responsibility. These aspects are mainly illustrated in the case studies in Sections 5.5 and 5.6.

Technology of networks mainly means information, communication, and cooperation. It allows for decentralized flexible structures to support and strengthen group autonomy within larger systems. Elements of this technology include the transportation of goods and people. The different aspects of networks are demonstrated in the case study in Section 5.7. It is important, however, to look at the workplaces of individual employees in groupwork as well as in networks. In this way, the different aspects of systems are taken into consideration. In addition to these aspects, the issues of society and the natural environment have to be taken into account; for example, production systems can justly be called human-centered if the effects of both their processes and products on society and on the environment can be considered acceptable. These aspects are examined in more detail in the last two case studies in Section 5.8 and 5.9.

The Three Dimensions of Human-Centered Systems:
Workplace, Groupwork, Networks

As referred to in the previous paragraph, human-centered systems can be considered in terms of three different dimensions: workplace, groupwork, and networks.

The First Dimension: The Workplace
The first of the three dimensions mentioned concerns people: the individual worker or employee and his/her work environment and workplace. The workplace can be characterized as the human-machine system in its narrow sense. There are certain conditions which a workplace has to fulfill if it is to be called human-centered. These conditions, or aspects, were suggested by, among others, Hacker [12], Rosenbrock [23], Rauner, Rasmussen, and Corbett [22], Brödner [3], and Hancke [13]. They have been the basis of several programs within Europe aimed at improving working conditions and humanizing work. They are particularly important with regard to women who are part of the labor force on equal terms.

The workplace has to be designed so as to make it possible for the users to actually use the technological system. This sounds simple, but many technological systems are initially designed without genuine concern for the prospective user. For example, the control instruments of certain complex processes (e.g., power stations, chemical plants) may be located in such a place as to make it impossible to watch, in parallel, the different dials which correspond to related process parameters. Thus, the process controller is unable to construct an appropriate mental model of the ongoing process. He or she is, in fact, unable to control the process, as was the case in the well-known Harrisburg incident [19]. Other examples are that task-related signals are below the threshold of perception and that the time available for making decisions (e.g., in an emergency) is shorter than the reaction time of human beings. For example, a twin-engine jet crashed during the approach for an emergency landing near Kegworth (U.K.) in 1989. One of the reasons for the accident was that the fault in the turbine system was only displayed for a few seconds. Within these seconds, the pilots had to decide which of the two engines was to be switched off. They switched off the wrong engine [13]. Taking these aspects into account may mean involving potential users in the system design process at a very early stage of development.

The health of the user has to be protected. This aspect entails the basic assumptions of ergonomics, including the physical safety of users (e.g., risks of injury or even death, illness through exposure to toxic chemicals and fumes), and also avoiding eye damage, extreme mental workload, or postural damage through working extensively with computers.

Mental and physical stress are to be avoided as they may be caused by overburdening, or by monotonous and repetitive short-cycle tasks. Therefore, the users should be allowed to plan the use of their time for themselves, including relaxation periods. The workplace should allow the workers to control stress, e.g., by giving them the opportunity to move about and to actively handle objects. Furthermore, it is necessary to offer direct interpersonal communication and social relations. Many workplaces in industry do not comply with this aspect of human-centered systems. The workers appear to be trapped in a working process in which technology and its rhythm seem to govern people.

The most complex aspect of human work takes into consideration the intrinsic need for people to develop themselves further and to experience challenges, motivation, success, and satisfaction from their work. It also requires that the worker makes use of his or her tacit knowledge, i.e., experiences, ingenuity, creativity, and skills which cannot be put into words or equations. Furthermore, human judgment needs to be requested at work, and it needs to be supported by the transparency of both systems and processes. The importance of this aspect becomes particularly obvious if people create a work of art (e.g., a sculpture or drawing) or perform as musicians or actors. Further aspects include being able to take responsibility for the working process, (semi)autonomous decision-making and planning procedures, and purposeful operator control of production, shop floor programming, and flexibility in controlling procedures. This should include system control under social interaction. Options should be offered for further qualification and promotion. Hence, technology should be designed to provide tools to support the workers' skills and ingenuity and to allow control of product quality. The worker expects to be involved in product or process innovation and to contribute to solving problems as they arise in the production process. These problems may be machine failures, system breakdown, productivity losses, hazards and dangers for life and health of people, etc.

It may be considered a fundamental element within our society that people want to fulfill meaningful and rewarding tasks which also take into account the individuality of the human operator. Our society relies on people who desire to contribute to improving conditions of work and life through their own work. However, these aspects should be compared with the expectations and competencies of those workers who are to fill those workplaces. Task demands and worker qualifications have to be compared with each other in order to avoid overburdening. Stress may increase due to new challenges and the growing intensity of high-quality work. This system requires continuous observations and negotiations concerning working conditions. But the broad variety of what people can do makes it easier to compensate for the negative consequences of challenging work.

These aspects have been developed in order to characterize the individual workplace. According to Hancke [13], they can be summarized as follows: purpose, skill, judgment, creativity, individuality, and social interaction. These are reflected in all the case studies reported here.

So far, these aspects basically refer to the individual human operator. Beyond these aspects, it is important to consider cooperation in groups as a feature of human-centered systems. Thus groupwork is the second dimension to be discussed.

The Second Dimension: Groupwork

The second dimension refers to work organization: a group of workers cooperating on the same task. Groupwork has become an important feature of all approaches to establish more human-centered systems. The concept is largely based on the traditional production model of the craftshop. Groupwork in industry today may follow this traditional model to a certain extent.

The shop floor can be structured by groupwork so as to provide a well-defined and comprehensive environment. The production focuses on one main type of product or on a small family of related products. The various skills required to produce the final product are found among the people working in the group at different levels of craftsmanship. They work as a team. Their tasks include the planning and allocation of work, quality, and performance control, even the acquisition of raw materials and the dispatching of finished products. For the individual worker within the team, the widening of his or her task scope may lead to job enrichment and job enlargement. This process is supported by job rotation among the group members. The group may have a team leader who may be a master craftsman or a foreman. Thus, the implementation of group technology leads to a change in the conventional organizational hierarchy. Some of the tasks which previously had been allocated to management personnel or to non-production departments may be taken over by the group.

The integration of new technologies and groupwork leads to new concepts of reorganization of a company toward a decentralized structure. Groupwork includes the concept of workshop production with (semi)autonomous groups. Beyond that, integration concepts may imply breaking down organizational barriers and divisions. Hence, through decentralization, the most important aim of improving flexibility in production and service can be achieved.

For instance, it is increasingly necessary to look at shop floor personnel with a new vision of work: skilled shop floor workers control both highly complex machine tools and cooperate across the production hall as a network of workgroups. Computer-based information support is needed for these operators. One software tool to be offered to the skilled worker may be as follows: the tool enables the worker to record in a specific database information such as all activities needed to execute a certain production task, e.g., to mill a certain complex workpiece in several steps using a sequence of special tools. These records are available to him and his workgroup for future use if identical or similar workpieces are to be produced [20]. The basic component of this tool is the template. It offers a tree-like structure on the screen which can be organized and filled in by the user. It describes all activities to be performed by the user when fulfilling a certain production task. The activity tree can be restructured using the template editor. In addition, the template manager can be used to manage the information needs of the shop floor workers by networking the different templates created by the individual group members in parallel. This tool can contribute to enhancing and supporting purpose, skill, and judgment as well as social interaction and cooperation within a workgroup. Similar tools are described in the case studies in Sections 5.4 and 5.6. They are widely offered today on the European market.

The Third Dimension: Networks

The third dimension to be discussed here is that of establishing organizational networks of groups within a company or across companies. As Gill [10] states with regard to the individual workplace in production and moving on to complex networks today:

> "The notion of human-centredness, defined in terms of "human-machine" symbiosis, is rooted in the production culture of the industrial society. It seeks collaboration between the human and the machine on the basis that the machine supports the human skills in performing human tasks and decision-making processes. The world of work and living are, however, changing rapidly. In this age of information networks, the symbiosis is not between the single machine and the single user, but is a matter of symbiotic relationships between the network of users and the network of machines. It is a matter of communication between groups and between the human and machine networks. It is not a matter of the interaction between the skilled worker and the machine, but of a world of collaborating users, who have a variety of functional and cognitive levels. The "tacit" knowledge no

longer resides in the individual artisan but resides in the community of users in the form of the social knowledge base or a network of social knowledge bases" (Gill [10], pp. 61–62).

Today, such complex networks can be observed around the world. Information technology links are their main means of technological support. The links include means of informal and personal communication as described in some detail, in the case studies in Sections 5.5 and 5.6. In addition, however, such networks need efficiency and safe transportation of goods and people. Hence, the concept of human-centered systems includes providing well-designed systems of communication and transportation.

Human-Centered Systems in Practice: Some Observations

So far, the concept of human-centered systems has been briefly described. In this section, some observations are reported on how this concept has been used, and which other developments have taken place which may counteract the implementation of this concept on a larger scale.

The recent developments in industry have shown that changes in workplaces and working life have, on the one hand, frequently led to considerable improvements experienced by the workers and employees: many unhealthy and dangerous jobs have disappeared in all industrialized countries, groupwork and decentralization have been introduced in many branches of industry, automation has removed many burdens from workers, etc. On the other hand, jobs have vanished worldwide in the wake of these changes at an alarming rate. The following problems arising from these changes, have been observed:

- The increased efficiency of groups may lead to making certain members of the group redundant.
- Management personnel may worry about being made redundant because of the increased responsibilities allocated to these groups at a lower level.
- Tensions among and between groups may contribute to decreasing efficiency.
- Uneven distribution of work may develop within groups and networks.

In addition, there is growing concern about the increase in decentralization and networks being a burden to society and to the natural environment. This is caused by the growing use of natural resources through transportation systems (fuel, traffic, buildings, etc.). Stress seems to increase if people work in decentralized structures linked by networks. The experiences of remote-control mechanisms, strict time-tables, and unyielding rules of delivery seem to contribute to reducing obvious gains of decentralization. These difficulties have to be taken into account if decentralization is expected to improve long-term productivity and efficiency.

Furthermore, changes of company ownership have made entire work sites disappear, although considerable improvements in the economic viability of these plants had been achieved through human-centered design of technology and work. The concept of shareholder value has become more important than long-term viability concepts.

These and other developments seem to counteract attempts to further improve production and service performance towards human-centeredness. Dissatisfaction, however, is already spreading. It concerns the general economic strategies which are frequently termed rationalization, or lean production, downsizing, etc. Today, rationalization may instead be understood as the continuous effort of humans to achieve optimum system performance through the most careful consumption of resources (labor, materials, energy, etc.) and without damage to the environment. It may not always include using more technology; it may rather concentrate on redesigning work organization to achieve its aims. The unique opportunity exists today to overcome the antagonism between rationalization on the one hand, and designing human-oriented work environments on the other hand [25].

As an example, these concepts have been followed further in Germany through the research programs Humanization of Work and Work and Technology, both funded by the German Federal Government. There are also several similar state-funded and Europe-funded research programs. These research programs characterize the governments' strategies to support and assist European industry in its difficult transition from manufacturing to automation. At present, these programs in Germany are carried out under the heading Production 2000 (1995–1999). As Martin [21] points out, "Production 2000 is a

German Federal framework program which supports the development of technology, organization, and qualification for production taking into consideration both social and environmental conditions."

A set of criteria was defined within this project which are to be met as a condition for support and possible funding: e.g., future-oriented, complex, interdisciplinary problem solutions; socially and environmentally sustainable developments; transfer concepts for broad application domains; combination with skill and knowledge enhancing strategies; strengthening small and medium-sized enterprises (SME); cooperation of industry and academia, as well as international cooperation, etc. Some of the first projects to be granted funding have dealt with methods of product development and production techniques information technology for production, general aspects of training, and further education. The overall number of such individual industry-based projects may presently be about 1500. This program symbolizes the importance of linking societal needs and expectations to industrial development and government decisions.

Preliminary results are as follows.

Some projects evaluate, based on company case studies, the framework conditions for successful production and performance on the market, considering comprehensive aspects of working time, personnel development, management principles, controlling, budgeting, and remuneration. They find that, to be successful, structural innovation must be based on sustained organizational learning leading to dynamic production and organizational structures.

Another area of research is globalization. The major motive of successful companies for moving production sites abroad is not only the search for cheap labor but also to secure and enlarge sales on different international markets. Even SMEs can become global through networking without losing either workplaces or local expertise. New forms of cooperation and networking of SMEs as regional and Europe-wide supplier chains are another crucial area of research.

Environmentally friendly production focuses, for instance, on the prolonged use of products, reparability and reuse, and effective material flow management as a circular economy model, with examples in several product areas.

These comments and observations may suffice to characterize recent developments in industry and services in Germany and Europe today. They make obvious the necessity to integrate technological change, quality of working life, and organizational development — including business process reengineering and other strategies — into the wider context of society and the natural environment. The sociotechnical systems concept may be the most promising and adequate strategy to achieve this challenging aim [4]. The case study in Section 5.5 describes its translation into industrial reality.

As mentioned before, the emphasis of this report is on technology development within the context of sociotechnical systems: the roles of technology today and the processes of designing and implementing technology in enterprises under the challenges of human-centeredness. Thus, the question arises regarding how to design such human-centered systems in practice. Some approaches are discussed in later sections. The criteria of human-centeredness described so far provide the basis and starting point for these design processes.

Designing and Evaluating Human-Centered Systems

The Dual Design Approach as a Concept of Human-Centered Design

Today, the design of a technical system is often understood as the design of a system to be automated as much as possible. The vision of the robot-run factory may be popular. It appears, nevertheless, impossible to implement this vision as many experiences have shown. The design of a technical system will always be the design of a human–machine system. As Rauner, Rasmussen, and Corbett [22] put it, the human–machine system needs, in reality, to be considered a human-centered system:

"The efficiency of a human-centered system is based on the complementarity of man and machine. Because of unforeseen disturbances that may enter the system, the operator must be able to control all tasks that contain choice-uncertainty via an interactive interface. However, an operator cannot control a system unless he comprehends its functioning. A system should support the operator's model

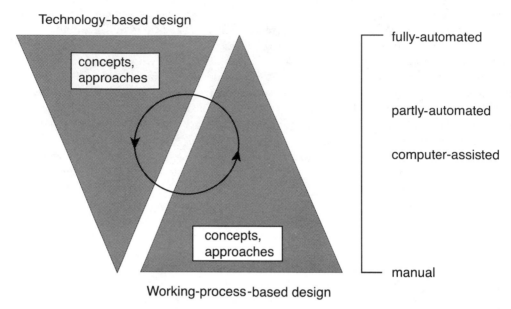

FIGURE 5.4   The dual design approach of designing human-machine systems. (From Brandt, D. and Cernetic, J., *J. Artif. Intelligence Soc.*, 12, 2, 1998. With permission.)

of its functioning so that the knowledge that is needed during infrequent task activity is obtained during general activity" (Rauner et al. [22], p. 55).

The aim is thus to design the system so as to find a balance of human action and machine action which makes the best use of human skill, judgment abilities, and experiences. Thus, the question to be dealt with here refers to the process of designing such a system. One approach is the dual design approach as illustrated in Figure 5.4 [18].

The dual design approach is a set of principles to ensure appropriate development of both technical and human aspects of human–machine systems. Usually, project engineers tend to head for fully automated concepts. This kind of approach, the technology-based design, is represented by the left-hand triangle of Figure 5.4. Here, the major part of design efforts, creativity, and research is used to obtain a fully automated system. However, at a certain stage of the development, it becomes obvious that some elements of the system cannot be fully automated. This may be due to economic reasons or to limited technical possibilities. It means that humans are becoming part of the concept at a late stage. However, it is very difficult, if not impossible, to obtain jobs which take advantage of the strengths and competencies of people.

Therefore, it is necessary to introduce a second approach, the working process-based design, in order to consider the human work situation as well. Contrary to the technology-based design, a working process-based design raises the issue of how to solve a problem with a lower level of automation or computer use. This will result in a concept where tasks are mainly performed by people. It means that the main part of design efforts, creativity, and ideas will be put into this approach (see Figure 5.4, right-hand triangle). It will help the systems designer to gain a better understanding of the system to be designed.

Both the technology-based design and the working process-based design should be used in parallel to obtain an optimum. This is the basic idea of the dual design approach. It is complementary in the technology-based design and in the working process-based design. The advantages and disadvantages of both concepts have to be compared and analyzed. A continuous exchange between both design processes is indicated by the two arrows in Figure 5.4. It leads to concepts of human–machine systems which correspond to the demands of both the technical processes and the process of human work. Concepts created by this approach make the best use of both the technical and the human resources of a company. The approach may thus be considered as one way to achieve economic advantages combined with designing meaningful workplaces.

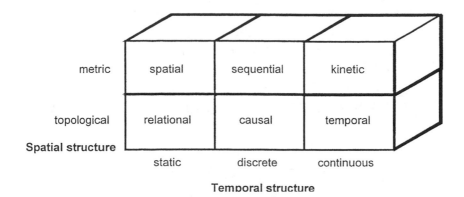

FIGURE 5.5   Taxonomy of physical mental models.

This approach supports the design of human-centered systems as described in the six case studies in Sections 5.3 through 5.8 of this chapter. One important question that arises in these case studies is as follows: how can the concept of the working process-based design be used to develop different technological systems? To answer this question, the concept of mental models has proved useful; it is described in the following section.

The Concept of Mental Models

Hartmann and Eberleh [16] proposed a taxonomy of mental models in order to describe the views of skilled shop floor workers of their working tasks in operating and controlling machine tools today.

Six basic types of physical–mental models can be distinguished by their methods of representing spatial and temporal structures, as shown in Figure 5.5. Physical–mental models deal with the real world. The columns of the matrix combine the different types of physical–mental models according to how they deal with time: there may be no temporal variation at all (static models), a temporal sequence of static structures (discrete models), or a completely continuous process, comparable to a movie, which may or may not be in real time (continuous models).

The lower line of the matrix contains mental models which represent the world in a nonmetric fashion (i.e., without regard to actual size and location); their spatial structure is topological. Three types of models can be distinguished. Relational models represent properties of objects and relations between objects. A good example is the organizational structure of a company which shows only the functional relations between departments but not the physical location of these departments on the company site. In causal models, these objects and relations can be represented at several discrete points in time. Temporal models allow for the same effect in continuous time.

Models in the upper line of the matrix contain metric information, e.g., the spatial model representing the mental image of a location visited on holiday. Similar to causal and temporal models, sequential and kinetic models add the possibility of representing discrete and continuous time, respectively [15].

One example of applying this concept may briefly be described as follows: on the shop floor, research has shown that the most common mental model of skilled workers is to view working processes as processes — sequential (discrete) or kinetic (continuous) working steps of machine motions, hand movements, workpiece handling, etc. The CNC control most common today tends to offer mainly causal or alpha-numerical interfaces which show discrepancies with regard to the workers' mental models. These discrepancies would need to be bridged in new interface design [9].

Additional strategies are needed to put this concept into practice. For this purpose, one fundamental strategy has been suggested by Carrol et al. [5]. It is described in the following section.

The Task-Artifact Cycle to Design Human-Centered Technology

Carrol et al. [5] describe a circular paradigm of human–machine interface design (Figure 5.6). It starts with an analysis of existing technology by looking at technological artifacts in use (step 1). Every artifact (technical system, tool, work of art, etc.) corresponds to the hypotheses of those who designed

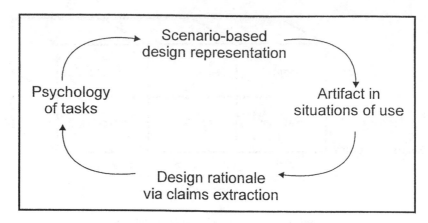

FIGURE 5.6   The task-artifact cycle. (From Carroll, J. M. et al., in *Designing Interaction,* Carroll, J. M., Ed., Cambridge University Press, Cambridge, 1991. With permission.)

it about the future users. Thus, it is an operationalization of (implicit) hypotheses about its users. The human–machine interface of a technical system, for example, is designed according to the designer's mental model of how the system will be used in practice. This mental model includes assumptions about the users, their tasks and their ways of working and thinking.

In step 2 of the cycle, these assumptions can be inferred from an analysis of this technology in situations of use, and the design rationale, the aims, purpose, and claims, can be reconstructed as a claims extraction. The design rationale describes how the system specifically fulfills the requirements of its application in practice. Using concepts from psychology, especially German work psychology [11,12], a vision of desired future work in this application area is generated in step 3; the psychological concept of future tasks is generated. From this vision, a scenario is derived as a representation of the future system in situations of use (step 4). As an example, the development of a new CNC interface will be briefly described here.

In large companies today, the production process is frequently controlled by the programming department rather than by the skilled workers on the shop floor. This development has caused criticism and rejection among skilled workers in many factories. The workers demand to be given control of the production process while making use of the advantages of CNC technology.

Research has recently been performed on CNC human–machine interfaces in use today. It has supported this experience of the workers. The design rationale of most CNC systems derived in this research shows the following results: The alphanumerical data input frequently contradict the mental models of the skilled operators. Their mental models are rather sequential and process-oriented corresponding to the structures of their daily work at the machines, as mentioned in the previous section. This research has led to the development of a vision of how CNC operators in the future should be enabled to work: by using, as in the past, analog input devices, e.g., conventional control systems, such as manually controlled hand wheels. Their movements correspond to the production process, and thus, to the mental models of the operator, more closely than today's digital/graphical input devices. The new analog input devices read data electronically and transfer them to the graphical representation of the workpiece and to the simulation of the production process. In this way, the newly developed CNC system allows one-of-a-kind production as well as mass production based on record-playback options (Figure 5.7). The system has been successfully introduced into the European market [19].

Similar experiences and developments are reported in the case study in Section 5.3. The application of the task-artifact cycle on networked systems is described in Section 5.7.

Once a new technological system (or a prototype of it) has been implemented, studies in working environments will allow for evaluation of the design scenario. Such strategies have been extensively discussed, e.g., in the framework of the Swiss-based project KOMPASS. It suggests a comprehensive set of criteria to be used by practitioners in order to evaluate automated production systems in terms of

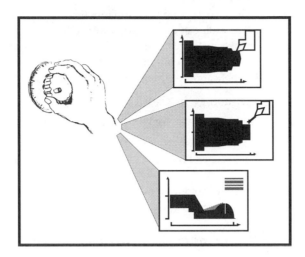

FIGURE 5.7 Modes of operation of a new CNC interface corresponding to different mental models of the human operator.

r. = responsibility	m. = monotony
m.d. = mental demands	t.p. = time pressure
p.d. = physical demands	e.s. = ergonomic stress
f. = flexibility	p.o. = physical obstruction
s.d. = social demands	

human-centered design. The KOMPASS tool aims at participatory design within a multidisciplinary design team [27]. The KOMPASS approach closely corresponds to the complementary strategy of the dual design approach (Figure 5.4) and the task-artifact cycle (Figure 5.6).

As the example of KOMPASS shows, the application of these strategies in practice relies on having sets of criteria available in order to assess the degree of human-centeredness of sociotechnical systems. Thus, the concept of the sociotechnical system to be designed has to be broken down into sets of questions. These questions can be asked by the design engineer during the design process or subsequently, during implementation and first use of the system designed (or its prototype). Such a set of questions is described in the following section based explicitly on the dual design approach.

Four Aspects of Applying the Dual Design Approach

The dual design approach — embedded into the task-artifact cycle — can be structured by the following four aspects:

1. The degree of automation achieved through implementing this technology
2. The degree of networking implemented with this new technology
3. The degree of dynamics of changes, e.g., in work and society, accompanying the implementation of this technology
4. The degree of formalization of human communication and cooperation to be accepted by humans using this technology

These four aspects of applying the dual design approach are the basic concepts of designing technology today, and they also form the basic concept of evaluating this technology in use. For this purpose, criteria have been developed which make it possible to apply the evaluation approach to certain technologies in use [9, 26]. These criteria can be weighted, for example, in the following way:

- This criteria is sufficiently dealt with in the design of this technology.
- In terms of this criteria, the new technology is basically acceptable.
- In terms of this criteria, the new technology entails unresolved problems.

Another approach may be to weight the criteria from 1 to 5. From these values, a mean value can be derived which may be additionally weighted according to the importance or relevance of the aspects considered.

The criteria suggested here have been derived from the different theoretical concepts of work and organizational psychology as well as from observations in practice, as described in this report. The case study in Section 5.8 illustrates the application of these criteria. They are as follows:

The Criteria for Assessing Technology in Use

Criteria concerning the degree of automation achieved through implementing the new technology system:

1. Comprehensive workplaces: Are the tasks of the users of this technology integrated and comprehensive and do they include straight-forward as well as challenging task components but avoid new burdens and stress?
2. Easy-to-use systems: Does the technology allow for easy use by different users?
3. Flexibility of use: Can the users use the system in different ways according to their individual competencies and skills?
4. Reality problem solving: Does the system allow the users to solve problems in reality, e.g., in the case of a system breakdown?
5. Work rhythm definition: Can the users at least partially define their own rhythm of work?
6. Working conditions: Does the new system contribute to improving conditions at work, e.g., health conditions?

Criteria concerning the degree of networking implemented with the new technological system:

1. System reliability: Is the system designed to be reliable and safe?
2. User groups access: Does the system offer fair rules for different users concerning access and interventions?
3. System stability: Is the system stable even under conditions of user error?
4. Decentralized structures: Is the system structure sufficiently decentralized to allow independent operation of subsystems if needed?
5. System transparency: Is the system as a whole transparent to different user groups?
6. Freedom for human decisions: Does the system allow the users to decide their ways of working with the system?

Criteria concerning the degree of dynamics of changes accompanying the implementation of the new technological system:

1. Ergonomic design: Does hardware and software design comply with basic rules of ergonomics, etc.?
2. System compatibility: Are the system components to be implemented compatible with the mental models of the prospective users?
3. System consistency: Does the system respond in a consistent and predictable manner if used in different situations and tasks?
4. Learning support: Does the system allow and support learning processes and further development of the users?
5. First use transparency: Is the process of learning how to use the system transparent to new users?
6. Societal impact: Does the new system avoid far-reaching changes of employment, qualifications, etc.?

Criteria concerning the degree of formalization of human communication and cooperation to be accepted by the users of the new technological system:

Human-human communication (closely related to the degree of automation):

1. Human communication: Does the system allow for and support continued human–human communication as well as human–technology communication?
2. Cooperative problem solving: Does the system expect the users to become involved in real, cooperative problem solving as well as cooperation through the technological network?

Human–machine hierarchy (closely related to the degree of networking):

3. Data security: Are safety and security of use and transfer of personal data controlled by the individual users?
4. Networked cooperation: Are the users allowed to organize their cooperative work processes on networks independently from technological constraints?

Human–machine interactions (closely related to the degree of dynamics):

5. System feedback: Does the system offer the users feedback and information on its technological constraints and possible faults?
6. System development: Can the users develop the system further as they develop their own competencies and skills in using it?

The following aspects are not directly included in these four sets of criteria:

- Economic viability of the new technological system
- Environmental impact in terms of both production and use of the new system
- Fall-back options in case of system failure

They are to be applied additionally according to existing assessment approaches.

The dual design approach and the task-artifact cycle require the integration of people into the design process at a very early stage. These people are the best judges of the degree of human-centeredness of the sociotechnical system considered. One strategy of assessing the human role in sociotechnical systems is described in the following section.

VEMAS: Assessing the Human Role in Action

VEMAS [2] is a concept for assessing the degree of human-centeredness of a workplace in terms of the human role within the human–machine interaction. The dimensions of assessment are clustered in two groups:

1. Demands (which may enrich or overburden the human operator):

   - Responsibility
   - Mental demands
   - Physical demands
   - Flexibility
   - Social demands

2. Stresses (which are also needed in working life but, if they occur excessively, may overburden the human operator):

   - Monotony and boredom
   - Time pressure
   - Ergonomic stress
   - Physical obstruction and obstacles

These nine aspects have been derived from ergonomics, psychology, and sociotechnical research (e.g., KOMPASS [27]). To apply these aspects in workplace assessment, each function and activity of the human operator is considered separately.

Subsequently, the different functions comprised by the workplace are rated following all nine aspects: from 1 (underchallenged/no stress) to 5 (overburdened/extreme stress). These ratings are weighted by multiplying each of them with the time of duration needed by the human operator to perform the activity or workplace function. The means of the weighted ratings are used as indicators in two different diagrams which characterize the human-centeredness of workplaces in action (or even before they have been implemented and finalized).

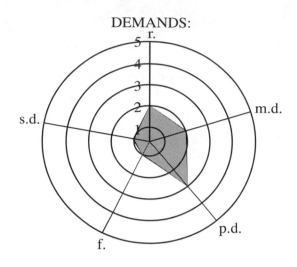
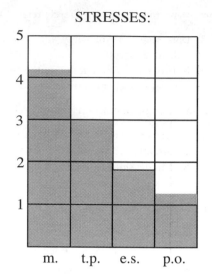

r.   = responsibility
m.d. = mental demands
p.d. = physical demands
f.   = flexibility
s.d. = social demands

m.   = monotony
t.p. = time pressure
e.s. = ergonomic stress
p.o. = physical obstruction

FIGURE 5.8   Evaluation of a production line. (From Brandt, D. and Cernetic, J., *J. Artif. Intelligence Soc.*, 12, 2, 1998. With permission.)

As an example, the following two evaluation results are presented: Figure 5.8 shows the evaluation of the more or less conventional production line of a car manufacturing plant. The shop floor tasks do not appear well balanced and the stresses can be fairly high. Team responsibilities are low.

Figure 5.9 shows the case of a highly automated production shop floor. The stresses can be rated very low. The demands, however, are very high. Responsibility is at a maximum and can be seen as a burden; all the responsibility for setting-up and running the system is concentrated on one person. Physical and social demands can be rated quite low [28].

After such a comprehensive evaluation of the sociotechnical systems to be developed, simulations of the system and its prototype are performed on site, integrating the users. This strategy step is illustrated in the next section.

Simulation as an Evaluation Strategy

Usually, evaluation of newly developed technology takes place during prototype implementation and testing. In this report, a different strategy is suggested which has been derived from the concept of laboratory experimentation and testing, which is common in natural sciences and engineering. The emphasis of this evaluation strategy is on integrating the users into the design and evaluation processes.

It is important to integrate people into designing and simulating complex networked systems at a very early stage of development. Only then can technology be expected to be sufficiently user-oriented to be finally accepted by the users. Complex human–machine systems need to be tested in the working process in real time. This can be achieved in the laboratory situation: the users perform their usual tasks while the technical components of the new system are integrated step-by-step into the working process. Thus, the simulation comes very close to reality. It is an aim of this strategy to test new technology in terms of how much they are designed around people in order to comply with the concept of human-centeredness.

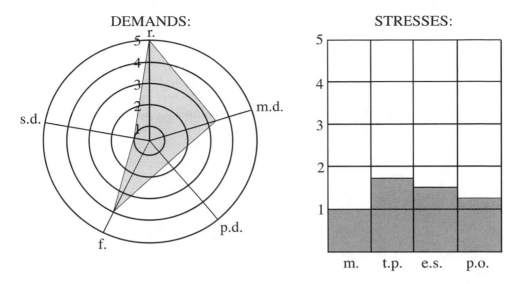

FIGURE 5.9 Evaluation of a one-man automated production system. (From Brandt, D. and Cernetic, J., *J. Artif. Intelligence Soc.*, 12, 2, 1998. With permission.)

FIGURE 5.10 Human-integrated simulation: testing new software at the container terminal. (From Brandt, D. and Cernetic, J., *J. Artif. Intelligence Soc.*, 12, 2, 1998. With permission.)

As an example, the following human-integrated simulation is described; it was used to evaluate the design of a new computer-based information system on a train-truck container terminal. The laboratory setting consisted of wooden mock-ups of cranes, trains, trucks, containers, etc., as well as the real prototype of the computer system. The mock-up terminal had a genuine computer link between the gate where the trucks arrive and the gantry crane which loads the containers from the trains onto the trucks. At the simulated gate, the truck driver of the simulation was told where to get his container. The crane operator saw, on the real screen, which container was needed next. He could also see the waiting times of all trucks within the terminal. Hence, he tested the technological system in real time for its qualities to support his decisions about which truck to deal with next (Figure 5.10). In the laboratory evaluation runs, these different roles were played by the future users in real time using genuine terminal data. Thus, they evaluated the technology they may work with in the future. Different strategies of structuring and controlling the container movements, and the sequencing crane were tested and evaluated [1].

This example of using human-integrated simulation may illustrate how the first prototype evaluation of any newly developed technology will take place in a laboratory setting. The laboratory has to be close to the real working setting of the cooperating company in its networked context. After the laboratory evaluation, different phases follow the evaluation of the technological system in the reality of the company. All evaluation should be strongly based on the criteria and the framework of the dual design approach and the sociotechnical system concept as described here.

So far the basic concepts of human-centered systems, their design, and evaluation have been described. In the following sections, several case studies are described which illustrate these concepts. They are taken from recent research and industrial developments in Europe. As mentioned before, the authors of these case studies have largely been in charge of the projects described.

References

1. Beister, J. and Henning, K., *Human Integrated Simulation and Petri-Net Simulation to Design Transportation Systems*, Augustinus Publishers, Aachen, Germany, 1995.
2. Bohnhoff, A., Video Identification of Freight Containers, Ph.D. thesis, RWTH, Aachen, 1991.
3. Brödner, P., Tentative definition of AT, *AT&S Newsletter*, 1, 1990.
4. Butera, F. and Schael T., The renaissance of sociotechnical system design, in *Self-Organisation: A Challenge to CSCW*, Mambrey, P., Paetau, M., Prinz, W., and Wulf, V., Eds., Springer-Verlag, London, 1997.
5. Carroll, J. M., Kellogg, W. A., and Rosson, M. B., The task-artifact cycle, in *Designing Interaction*, Carroll, J. M., Ed., Cambridge University Press, Cambridge, 1991.
6. Cooley, M. *European Competitiveness in the 21st Century*, FAST, Brussels, 1989.
7. Emery, F. E., Characteristics of sociotechnical systems, Tavistock Institute, doc. 527, London, 1959.
8. Fuchs-Frohnhofen, P., *Designing User-Oriented CNC Production Systems*, Augustinus Publishers, Aachen, Germany, 1994.
9. Fuchs-Frohnhofen, P., Hartmann, E. A., Brandt, D., and Weydandt, D., Designing human-machine interfaces to match the user's mental models, *Control Eng. Pract.*, 41, 13, 1996.
10. Gill, K. S., The foundations of human-centered systems, in *Human-Machine-Symbiosis*, Gill, K. S., Ed., Springer-Verlag, London, 1996, 1.
11. Greif, S., The role of German work psychology in the design of artifacts, in *Designing Interaction*, Carroll, J. M., Ed., Cambridge University Press, Cambridge, 1991.
12. Hacker, W., *Arbeitspsychologie*, Huber, Bern, 1986 (in German).
13. Hancke, T., *The Design of Human-Machine Systems: The Example of Flight Decks*, Augustinus Publishers, Aachen, Germany, 1995.
14. Hanna, D. P., *Designing Organizations for High Performance*, Addison-Wesley Publishers, Reading, MA, 1988.
15. Hartmann, E. A., *The Design of Cognitively Compatible Human-Machine Interfaces*, Augustinus Publishers, Aachen, Germany, 1995.

16. Hartmann, E. A. and Eberleh, E., Inkompatibilitäten zwischen mentalen und rechnerinternen Modellen im rechnergestützten Konstruktionsprozeß, in *Software Ergonomie,* Ackerman, D. and Ulich, E., Eds., Teubner, Stuttgart, 1991 (in German).

17. Henning, K. and Marks, S., Application of cybernetic principles in organizational development, 2nd International Symposium on Systems Research, Informatics, and Cybernetics, Baden-Baden, 1989.

18. Henning, K. and Ochterbeck, B., Dualer Entwurf von Mensch-Maschine-Systemen, in *Der Mensch im Unternehmen,* Meyer-Dohm, P. et al., Eds., Bern, Stuttgart, 1988 (in German).

19. Kraiss, K. F. *Anthropotechnik in der Fahrzeug- und Prozessführung,* RWTH Aachen, Germany, 1988 (in German).

20. Louha, M., A do-it-yourself expert systems for shop floor use, *Proceedings of the 5th IFAC Symposium Automated Systems Based on Human Skill,* Vol. II, 58, Sept. 1995, Berlin.

21. Martin, T., Production 2000 — trying to jointly improve technology, organization, and qualification, 6th IFAC Symposium on Automated Systems Based on Human Skill, Sept., 1997, Kranjskagora, Slovenia.

22. Rauner, F., Rasmussen, L. B., and Corbett, M., The social shaping of technology and work — a conceptual framework for research and development projects in the CIM-area. ESPRIT-project 1217 (1199), Bremen, 1987.

23. Rosenbrock, H. H., *Machines with a Purpose,* Oxford University Press, Oxford, U.K., 1990.

24. Trist, E., Human relations in industry, paper presented at the Vienna Seminar: Congress for Cultural Freedom, 1958.

25. VDI, *Rationalisierung Heute,* VDI, Duesseldorf, 1988.

26. Veldkamp, G., *Future-Oriented Design of Information Technology Systems,* Augustinus Publishers, Aachen, Germany, 1996.

27. Weik, S., Grote, G., and Zoelch, M., KOMPASS-complementary analysis and design of production tasks in sociotechnical systems, in *Advances in Agile Manufacturing,* Kidd, P. T. and Karworski, W., Eds., IOS Press, Amsterdam, 1994, 250.

28. Wortberg, A., A survey on human-orientation of selected industrial production systems in the U.K., Research report, HDZ/IMA, RWTH, Aachen, Germany, 1996.

5.3   Shop Floor Control: NC Technology for Machining of Complex Shapes

Henning Schulze-Lauen and Manfred Weck

The first case study reported here describes a research project aimed at giving new production responsibility to the skilled worker through new control technology for milling: empowerment through technological innovation. This project has been shaped by the cooperation of two university departments from Germany (RWTH Aachen) and Switzerland (ETH Zurich), and industrial enterprises from Switzerland (Grundig-ATEK-Systems) and France (Aerospatiale). The project was funded in part by the European Commission [1].

A major contributor to this project has been the Laboratory for Machine Tools and Production Engineering (WZL) at the University of Technology (RWTH) in Aachen, Germany. It is represented here through the two authors. This unique institute has a worldwide reputation and is highly recognized for its research. It works mainly in close cooperation with industry to develop new technology and new organizational concepts for industrial production. All research takes place in close cooperation with the most important industrial enterprises in Germany and Europe.

The Scope of the MATRAS Project

Modern Numerical Control (NC) technology has made the manufacture of surfaces of utmost complexity and fanciful design possible. Techniques like five-axis milling would not even be thinkable without high-performance numeric controls. A human operator is simply unable to coordinate the simultaneous movements of three, five, or more machine axes. Examples of relevant workpieces are molds and dies for complex pieces, e.g., car body parts, perfume bottles, and aircraft or car models for aerodynamic testing or turbine blades.

While broadening the designer's scope, however, this development has come at a price for the manufacturing department. NC programming for sculptured surfaces is a complex and time-consuming task, and the result is an NC program which demotes the skilled machine operator to a mere onlooker while his machine is (one hopes) following the desired milling paths.

There are two major drawbacks to current NC technology when milling sculptured surfaces. They involve the processing of geometry information in the NC and process flexibility at the shop floor. These two problems are briefly described as follows.

The first problem deals with today's NC programming data interface (DIN 66025/ISO 6983), which provides no means for directly exchanging and processing information relating to complex shapes. Thus, the desired machine movements are linearized and transmitted to the machine control as a cloud of small linear axis movements. With the advent of highly dynamic axis drives, these linear movements cause machine vibrations and inferior surface quality, especially during high-speed cutting operations.

To resolve this problem, a new NC structure has to be found which is able to process continuous curves as milling paths, e.g., in the form of splines. This requires, on the one hand, a new data exchange interface with the NC programming system which must provide such data. On the other hand, a new NC kernel functionality is needed to make consistent use of the provided curves from the interpreter down to the interpolator.

The second problem mentioned above is closely coupled to the question of the NC programming data interface. In today's NC process chain, all complex calculations are done in the NC programming system. The machine operator receives a program which contains only axis movements in the form of a long list of numbers and letters. All information about the workpiece shape, the structure of the NC program, or the decisions made by the programmer are lost and cannot be retrieved from the low-level information reaching the shop floor. Therefore, the machine operator can only guess what will happen at his machine when he starts the NC program. He is unable to flexibly react to faults or perform last-minute changes. He cannot even exchange one tool for another if, by chance, a certain tool is not available as needed; this tool exchange would require a rerun of the NC programming post-processor.

To improve this situation, any attempt to redesign the NC programming data interface has to take into consideration the information requirements at the shop floor. A new interface will have to transmit data of higher quality which allows the machine operator to perform certain modifications at shop floor level.

The main results will be described in this report: the design of new kernel functionalities and the design of a new NC programming data interface [1, 2].

NC Kernel Improvement

The basic requirements for any improvements in the quality of milling complex sculptured surfaces have been pointed out above. They concern the improvement of the data to be processed by numeric controls. This refers both to the format and the level of data fed into the controller. In terms of format, it has been widely discussed that using linear segmentation vastly increases the amount of data to be transmitted and does not allow proper control of machine movements. Functions like look-ahead or jerk control can be implemented much more efficiently if longer segments, described as mathematical curves, are provided as input data.

During the MATRAS project, several different modules of an NC controller were implemented. They allowed for milling path data in the form of different spline curves (Bézier, or B-Splines) to be processed. The data are not simply passed through a front-end processor creating the ordinary DIN/ISO machine code, but instead all modules concerned, down to the interpolation of the position command, are based on spline algorithms.

In terms of the level of data used for operating the NC, it had to be considered that the resulting system was intended to be more user-friendly, namely, giving more flexibility to the machine operator. Today, one of the major obstacles in the way of flexibility is the NC programming post-processors. Such post-processors are needed to transform machine-independent milling path information, like APT, into machine-specific axis movements. While the former describes the cutter contact path in three-dimensional space on the workpiece surface, the latter specify the linear or rotational movements of three to five machine axes which determine the cutter location. For this machine-specific conversion, a transformation has to be implemented in the post-processor.

The problem is that this transformation has to take into account all details of the manufacturing process, like the exact diameter and length of the tool tip and the special shape of the cutting edge of any tool. In three- and five-axis milling, these data cannot be altered at the machine. The exchange of similar but not quite identical tools would, therefore, require a new post-processor run, as mentioned above. These operations can take several hours, thus inhibiting flexible reactions. This is even more true if, due to a programming error or other circumstances, the machine operator is required to change the process sequence at the machine. If the operator wants to skip a certain area on the surface to be milled, he has to guess in which line out of several megabytes of NC program code he has to restart the program, which is practically impossible.

Therefore, one main requirement for more flexibility at the machine is the consideration of the transformation of the cutter contact path into axis movements. This transformation needs to move from the post-processor into the NC controller. The MATRAS project has thus implemented a universal five-axis transformation for the two NC controllers involved in the project (Figure 5.11). This development effectively makes the post-processor obsolete and allows for machine-independent code to be fed into the NC.

However, it should be mentioned that the term machine-independent has to be assessed carefully. Collision avoidance, especially in five-axis machining, is a major problem which has to be solved in the NC programming system. Therefore, NC programs have to be tested and checked based on a specific machine environment. But still, a machine-independent NC format has a lot of advantages, including saving time for unnecessary post-processor runs. The possibility of exchanging a smaller, collision-safe tool for the originally planned one already solves a large share of the problems caused by the low flexibility of today's process chain.

In addition to removing the post-processor from the process chain, the new technology also provides for a much higher quality of actual machine movement. The processing of splines and the integrated

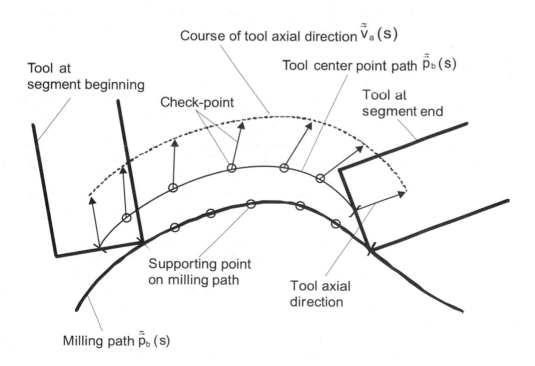

FIGURE 5.11 Cutter path and tool geometry.

five-axis transformation allows for the implementation of powerful look-ahead and jerk control modules. Based on the calculated acceleration data for the individual axes, the path velocity is adapted to avoid jerks and to remain within the acceleration capacity of all the axes. In doing so, the spline-based look-ahead avoids unnecessary acceleration and deceleration. The results are more constant cutting conditions and better surface quality while achieving the highest possible feed rates.

A New NC Programming Data Interface

Program Optimization on the Shop Floor

Flexibility not only means flexibility in process technology; it also means flexibility in relation to unforeseen organizational or design changes. Here, the current NC programming interface has probably its most disastrous effects. It is bad enough for the operator when a tool breaks while milling a $100,000 die, but it is even worse if, after replacing the tool, the operator has to restart the entire 5 megabyte NC program from the beginning and run in fast-forward mode until it comes close to the place where the machine stopped.

Similar problems occur when a certain tool is not available as required. The operator would like to go ahead and mill another region of the workpiece while waiting for the right tool to become available, with current technology, however, this is not possible. The operator depends on the sequential processing of the given NC program. This is especially disappointing because the NC programmer, when creating the program, probably distinguished between several logical blocks of machining which he may have grouped at random. Thus, the sequencing could easily be changed without affecting the resulting shape if the program were designed to allow it.

Furthermore, the NC programmer will not always have chosen the optimum machining data. It might be desirable for the operator to skip every second milling path because he is just doing a test workpiece in soft material, or, from his experience, he knows he would prefer to change the tool selection for a specific area during the finishing cut.

Considerations like these have led to the design of a new NC programming data interface which is well-structured, hierarchical, and object-oriented. It provides the information needed at the shop floor without requiring the full performance of a shop floor programming system at each NC controller.

The Concept of WorkingSteps

The central structural element of the new interface is the so-called "WorkingStep." A WorkingStep is defined as a specific region of a workpiece for which one tool and one set of technological parameters is valid. The WorkingSteps, therefore, form the logical building blocks of a complete NC program which is described in the form of a work plan (Figure 5.12).

The WorkingSteps serve both the NC programmer and the machine operator. For the NC programmer, they provide a good opportunity to structure his thoughts when designing the NC program. It is thus possible for the machine operator to understand the logic behind the NC program and to actively influence and optimize the machining operation. The simplest change of this kind could be to exchange the sequence of two WorkingSteps or to skip one WorkingStep if the overall working process requires this change.

Within the WorkingSteps, axis movements are transmitted to the machine. Furthermore, all relevant technological information is provided to the shop floor to enable quick verification of the data. The WorkingSteps even allow for information about the region's geometry or the shape of the raw piece to be included. Whether the shop floor can take more or less responsibility for the machining process, including the replanning of the milling paths if desired, may depend upon the capabilities of the NC controller and the extended permissions.

The new data format also includes all information needed for the design of a comfortable human–machine interface. Conventionally, the progress of the NC program would be shown by a one-line display of the executed NC block. MATRAS makes possible a graphical visualization of the currently milled

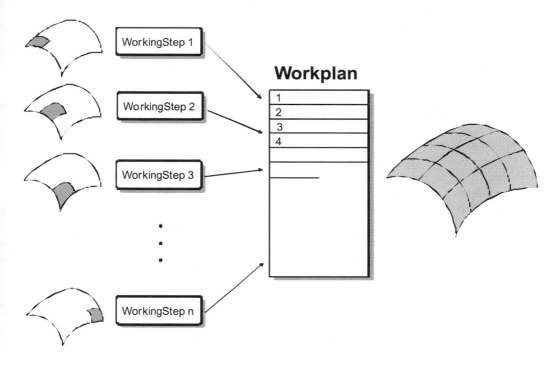

FIGURE 5.12   The concepts of WorkingSteps and the Workplan.

region and the progress of the operation as well as the next steps. This makes the machining process much more controllable for the experienced machine operator and reduces the probability of errors caused by negligent NC programming or simply by wrong data.

It should be noted that the intention of the project was not to completely shift the NC programming process to the shop floor, as has been suggested. While shop floor programming may be adequate for many simple operations, the generation of NC programs for complex shapes tends to require a great deal of time and concentration. An optimum solution has, therefore, been found in a data interface which allows performance of each task of the overall program where it is best performed (Figure 5.13).

This approach includes the possibility of feeding back information from the shop floor to the NC program. The goal of the MATRAS data interface is to build a consistent information base from CAD via NC programming down to the shop floor, and vice versa. The necessary extensions have already been considered in the design of the interface.

Conclusions

The MATRAS project is an example of how new technical solutions can improve both technological issues and human–machine interaction simultaneously. In this project, the technology for high-performance NC controllers has been greatly developed and improved. The new NC programming data interface has been designed to better incorporate the workers' experiences into the manufacturing process. As a result, the workplace on the shop floor becomes more attractive, and the overall process becomes faster, more reliable, more flexible, and, thus, more cost-effective.

The improved NC core has been implemented both in a commercial and an academic NC machine tool system, and its advantages have been proved impressively. Its NC programming data interface has become part of the activities in ISO TC 184 SC 1 WG 7, which is working on a new consistent data format for physical devices in manufacturing automation.

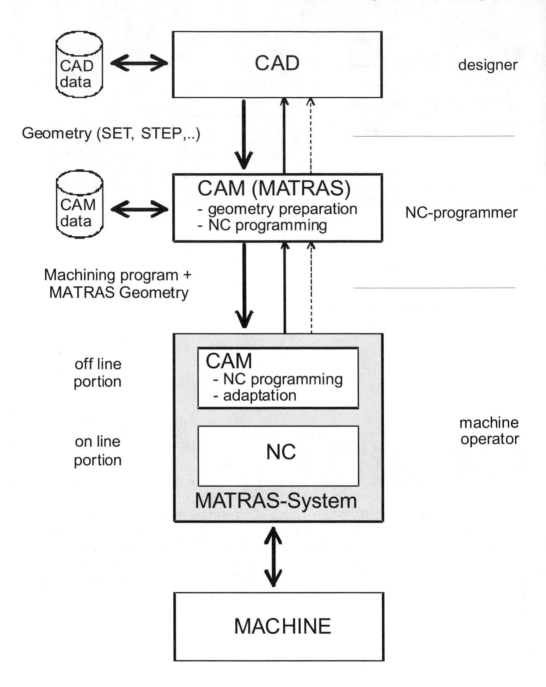

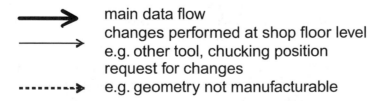

FIGURE 5.13 MATRAS information flow and feedback loops.

References

1. ESPRIT, Project 6245 MATRAS, Final report, Bruxelles: European Commission, To be published.
2. Fauser, M., Steuerungstechnische Maßnahmen für die Hochgeschwindigkeitsbearbeitung, Ph.D. thesis, RWTH, Aachen, Germany, 1997.

5.4 Shop Floor Information Support: Developing User-Oriented Shop Floor Software

Axel Westerwick, Mourad Louha, and Christian Wenk

The research project reported here characterizes one of the main university–industry projects funded by the German Federal Government to rebuild German industry in view of the world-market challenge. The emphasis of this research and development is on giving new information technology tools to the skilled shop floor working teams. The aim is, thus, to improve efficiency of both performance within the teams and cooperation between the teams. The project is called InnovatiF.

This project represents the cooperation of two different research departments of the University of Technology (RWTH) in Aachen: the well-known Laboratory for Machine Tools and Production Engineering (WZL) and the Department of Computer Science in Mechanical Engineering (HDZ/IMA).

The Concept of Shop Floor Software Development

The implementation of new data processing and information technologies on the shop floor leads to difficulties for the workers in using the computers. Systems would need to be easy to use in order to avoid loss of efficiency through computer control and automation. Thus, an important purpose of software development is to find a balance between the interests of the users and the expected software development time.

In this report, the software tool to be developed is aimed at closely corresponding to the needs and mental models of the users on the shop floor. In order to incorporate this correlation into the prototype, a design strategy has been employed which was originally suggested by Ainger [1]: the helical approach. It specifically takes into account criteria and concepts of human-centered systems design.

The design of software usually follows the waterfall approach, which starts with the definition of specifications for the new system and leads to a software prototype which is tested by the user. Frequent criticism of this strategy tends to point out that user expectations and needs are not sufficiently taken into account; the user is mainly exposed to the new system in use only when the software is about to be finalized.

The definition of a software project is often based only on a vague idea about the actual requirements for future use. However, failure to get the requirements right is one of the biggest risks in project development. Thus, an effective way must be found to analyze the requirements which necessitate the step-by-step development and presentation of the software concept. Thus, the approach can make the user aware of what the system to be developed will be able to do and how it will be used.

Therefore, the helical approach has been suggested. It aims at integrating the user into an interactive process of both defining systems specifications and designing systems models. Figure 5.14 shows the spiral of this approach oscillating between system specifications (left-hand side) and system models (right-hand side).

The approach starts with brief requirements statements and moves through several modeling stages which show increasing complexity and closeness to the final tool. The different models are (see Figure 5.15):

- Scenario: a description of a sample case, where the use of the new system is written down on paper as a story, thus allowing for the first specifications to be derived.
- Paper screens: screen drafts of how the scenario could be implemented on the computer.
- Noninteractive cardboard model: based on a software drawing package, producing drafts of the actual screen designs.

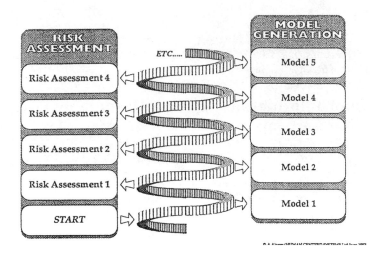

FIGURE 5.14   The Helical Project Life Cycle. (Courtesy of A. Ainger, Human Centred Systems Ltd., 1992).

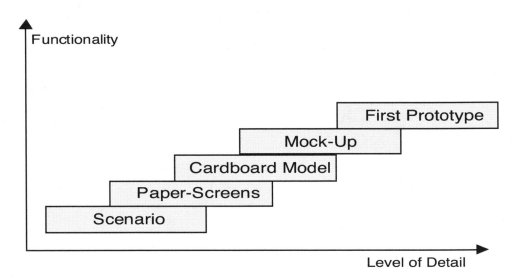

FIGURE 5.15   Steps in software modeling.

- Semi-interactive mock-up: based on a software drawing package with animation; computer screens are linked to each other to represent a model of the whole system. The system functions are simulated.
- First prototype: the first real product.

The tool development described in this report has followed the helical approach up to the stages of mockup and first prototype. This prototype is explained and illustrated in the following paragraph.

The Software System to Support Work Planning

The Concept of the System

Industry today increasingly tries to meet customer demands. Groupwork is one possibility for designing efficient information and decision handling to obtain flexible work processes. Working in groups means that the skilled workers, as members of shop floor groups or production islands, receive production orders which they are responsible for managing. In this organization, the skilled workers have to perform new tasks, e.g., communication, cooperation, and work planning. In order to be able to achieve their tasks, the skilled workers have to be requalified. Furthermore, new technologies have to support the new group tasks.

One objective of the research project InnovatiF, as reported here, has been to develop a software support system for work planning on the shop floor, similar to the system described in the case study in Section 5.5. The system enables the skilled workers to manage their production orders and to distribute the work load of a production cell both among their machine tools or among other workplaces. The main concept is that the system is designed for skilled workers based on their ways of thinking rather than for specially qualified programmers or information engineers. Therefore, the participation of the final users in the development process of the new tool is essential. The following points have to be considered in the development process:

- A special design which supports groupwork
- The ways of thinking and the working approach of the users
- Object-oriented design and programming
- Modular design
- Networking possibilities

The members of a workgroup have different qualifications and experiences; for instance, some members of the group may be trained in working with computer systems, others may not. Therefore, the design of the system has to be clearly structured so that the users can understand and control every step of their operations independent of previous computer experience. The consideration of the users' thought and work processes is part of the design process: it helps to realize the user-friendly functionality of the new tool as suggested by the dual design approach (see Section 5.2). Object-oriented design and programming (using the C++ programming language) supports the learning process, particularly for untrained users. The result of the modular design is a variable system for several production environments and different workplaces. Customers can install individual modules of the system as they need them for actual production. At a later stage, they can complete the system with other modules without having to install an entirely new system.

In all production processes, breakdowns and failures take place nearly every day. In groupwork, the skilled workers have to manage these problems. Thus, the new work planning system has to support the workers during emergencies. Therefore, the new system can communicate with any networked workplace supported by a shop floor information system called InfotiF. It is described in some detail in later paragraphs. When a worker has a problem with one of the machines, he reports to the system that the working process will be completed, for example, 30 minutes later. This message will be transferred to the main system and produce a warning. The members of the group working on this order then redistribute the work load among their machine tools. This new information will be transferred to all the workplaces involved. Figure 5.16 shows the concept and the functionality of the work planning system APS supported by the information system InfotiF.

Description of the System Modules

The software is divided into four modules. The planning module is used to plan the different tasks which have to be achieved by the workgroup. The machine list module allows for management of all the machines by the group. The network module implements the network functions. The setup module is used to set up the software options. All modules can be accessed by the user through different sheets. The following paragraphs will detail the different modules. Note, however, that the software is in the Alpha-state, i.e., it is still in the process of being designed.

The Planning Module

Figure 5.17 shows a screenshot of the actual module. On the left-hand side of the window, the user finds a list of all the machines available. With the timetable across the screen, the user gets a fast overview of the purpose of each machine. Several colors enable the user to distinguish between the different

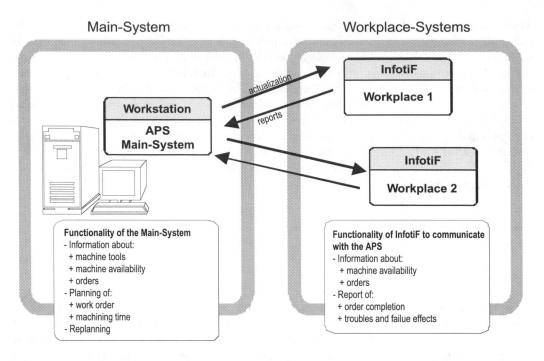

FIGURE 5.16   Concept and functionality of the Work Planning System (APS).

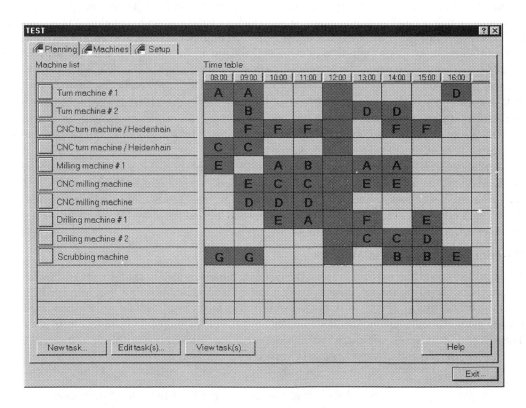

FIGURE 5.17   The planning module.

states, known as used, free, or unavailable. The time unit, 1 hour in this case, can be changed to zoom in or to extend the overview. Each planned task is represented in the timetable by a unique identifier. Using the two buttons at the bottom of the window, the user switches between two operating modes as follows:

1. Overview of the planned tasks: In this mode, the user can view all tasks planned for the time period selected. He would be able to call up some functions such as printing, zooming, sending messages, etc. However, all edit functions are disabled.
2. Edit planned tasks: In this mode, the user can create new or modify existing tasks. First, he may select a task previously created or he may input a new task to be planned. Then he can select the machines to be used and set them up in the timetable. The software supervises his inputs and suggests corrections and improvements if needed.

The Machine List Module
This module allows for the list of the available machines to be managed in a very simple way (Figure 5.18). Each time the user selects a machine from the list on the left-hand side of the window, he gets the related machine properties on the right-hand side. The user can group machines with similar properties, e.g., he may group all milling machines together as shown in Figure 5.18.

The Network Module
This module manages the network functions of the software. On the client systems at the users' workplaces, it disables the edit functions and enables the messaging functions of the planning module. Thus, the user is able to send messages to the main system, e.g., a request for more time to complete his task. On the main system, the network module collects the messages from the clients, informs the other members of the group, and may request a joint decision before proceeding.

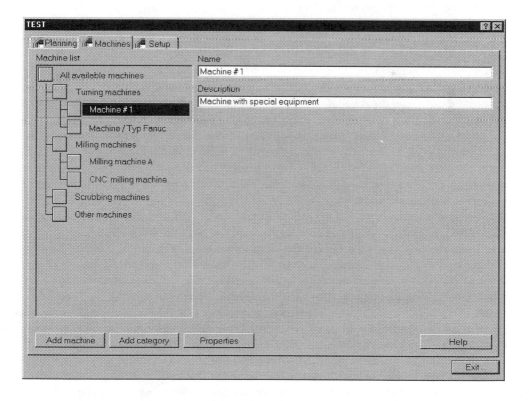

FIGURE 5.18 The machine list module.

The Setup Module

This module is normally accessible to a supervisor. It enables the software options to be set up. These are mainly the associated status — enabled or disabled — for each time base item, the defined colors for the different machine states, the default time base, and the path for data storage. The built-in clock and diary can be individually configured, e.g., work time settings and machine repair times.

The Software Implementation

The software has been implemented in C++, using object-oriented programming. The following class hierarchies have been used:

Basic Classes

Most of the data structures are organized in trees or double lists. These basic data structures are implemented in template classes. They are subsequently used by linking their parameters to more specific classes.

Process and Data Classes

These classes map, as closely as possible, the user's mental models and their associated processes. They are continuously improved by the users in the different workshops based on their newly acquired knowledge.

GUI Class Hierarchy

The GUI used is Microsoft® Windows '95 or Windows 3.11. The user interface of the software has been created by extending the existing Borland® object Windows class library. An important objective here has been to make the GUI classes as independent as possible from the data classes.

An additional tool to support groupwork on the shop floor has been developed by the authors in parallel to the tools described so far and it is described in the following paragraph.

The Software Tool to Support Group Communication

As already discussed in the previous sections, the utilization of manufacturing and production data for machine tool control is, today, only supported to a limited degree. Insufficient use is therefore made of the personal experiences of the skilled workers. The creation and dissemination of feedback or experiences are still insufficiently supported in terms of manufacturing processes or manufacturing documents in pre- and post-production areas [2]. Thus, a shop floor information front-end tool has been developed as part of the project described here. It is called InfotiF (Figure 5.19).

InfotiF provides the possibility of calling up, completing, and transmitting manufacturing-related information to any networked shop floor workplace. As the basis for this functionality, the appropriate areas, preproduction, design, NC programming, shop floor, assembly, etc. must be integrated into a companywide information network. On the basis of this network, the shop floor information front-end user has access to any manufacturing-related information that has been made available for this purpose. The control-integrated user interface is created using Internet and WWW technologies. By this means, it is independent of a specific control manufacturer. The software is based on a client-server approach using the object-oriented programming language Java (Java is a registered trademark of Sun Microsystems, Inc., United States and other countries). It is accessible via a PC-terminal or PC-based NC control at the shop floor workplace. Thus, the worker has access to all information relevant to his production tasks such as the NC program, work plan, tooling sheet, parts drawings, complete drawings, documentation concerning clamping, etc. He can comment or edit information or initiate feedback between different shop floor groups and to pre- or post-production areas.

Comments in text or code can be stored and transmitted so that the experiences of other workers and groups can be used again later. This information can be stored either related to certain projects (task dependent) or as general information (task independent). The user can make his information available

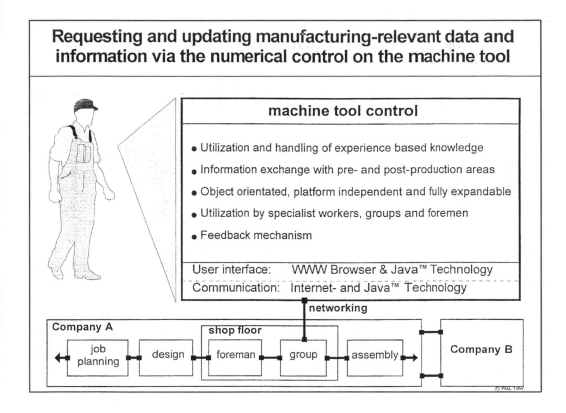

FIGURE 5.19 Outline of the InfotiF tool.

to other users in a controlled manner. The shop floor information front-end can also be used by workgroups to communicate with supervisors in the shop floor area. Remarks, comments, suggestions for improvements, feedback, or questions concerning the production documents can be forwarded to the appropriate experts by the shop floor workgroups or directly by the skilled workers at their workplaces.

Conclusions

One of the main challenges of today's industrial production is coping with increasing needs for flexibility and contingency handling. This fundamental problem can only be tackled if responsibility is moved to where production actually takes place: the shop floor. The skilled workers of today's shop floors expect to be challenged by such responsibilities which are related to, for example, running the machine tools, organizing work schedules, machine utilization, and managing the production flow. These workers need computer support for these tasks. InnovatiF aims at developing a set of software tools for this purpose. The tools consist of:

- The APS main system, offering information about machine tools and orders and planning support for work orders and machining times
- The InfotiF communication network within and between working groups, on manufacturing-related data, orders completion, or emergency events

Furthermore, it enables workers to both record and call-up their own working experiences and activity structures. The software has been developed as a human-centered development project involving the users right from the initial project stages.

References

1. Ainger, A., The helical project life cycle, ACiT documentation, Human Centred Systems Ltd, Hemel Hempstead, U.K., 1991.
2. Daude, R., Schulze-Lauen, H., Weck, M., and Wenk, C., Machine tool design for skilled workers: information to empower the machine operator. Proceedings of the 5th IFAC Symposium Automated Systems Based on Human Skill, Sept. 1995, Berlin.

5.5 Shop Floor Cooperation Networks: A Shop Floor Production Planning System for Groupwork

Thomas Schael

Production planning is becoming an important factor on the factory shop floor, as has been discussed in the previous sections. Here, the role of the computer is that of a tool in the decision-making process. It provides background information and enables the operator to estimate and possibly simulate various alternative options before making a decision. This concept is valid for individual operators as well as shop floor teams. At present, computer systems are being developed in Europe which are intended to support the cooperation of shop floor teams, workgroups, and departments across companies, or between companies linked through customer–supplier chains. The software systems needed for these tasks are described by the term computer supported cooperative work (CSCW). These concepts are discussed in this section.

This report is based on several recent research projects of CSCW performed by a European consortium integrating researchers from London, Aachen, and Rome/Milan. Similar systems are being developed and implemented in industry and administration in many different countries around the world. Through his own research and publications, the author presents this development.

The Need for Computer Supported Cooperative Work

In the current business environment, manufacturing enterprises have to cope with shorter product life cycles, roaring product diversification, minimal inventories and buffer stocks, extremely short lead times, shrinking batch sizes, concurrent processing of multiple or different products and orders, etc. [3]. A work organization operating in this environment cannot rely on advanced planning for task allocation and task articulation. The compilation of tasks into jobs must allow for a high degree of flexibility and, in order to be able to adapt to unforeseen contingencies, task description must basically rely on local control. A particular challenge is represented by emergency handling as a normal, daily life activity.

In response to the turbulent manufacturing reality, feed-forward oriented coordination mechanisms have been introduced for local control. Manufacturing companies are changing toward an order-driven mode of coordination such as Kanban or Just-In-Time systems. These production-related models can be augmented with mechanisms of interaction for coordination, collaboration, and co-decision in manufacturing, and the design of customer–supplier chains for the material exchanges and the linking of production islands or semiautonomous working groups. Along with the need for better understanding of such cooperation in manufacturing, the need for computer support of cooperation becomes evident.

The strategic means suggested to meet the present challenges include advanced manufacturing technologies and information systems, summarized under the concept of computer integrated manufacturing (CIM). But most of the present concepts of CIM and production planning and control (PPC) have been designed according to a Tayloristic, centralized organizational model. This results in systems which have a deterministic view of the working processes and information flows in production enterprises. The working processes, as supported by these systems, are rigidly controlled and fed back to the central control units of the company.

The example of the European production industry, in contrast to this development, has, until today, not been fully based on a Tayloristic large-scale production principle, but on differentiated production plans, personal control, and intrinsic motivation. Furthermore, it is based on having workers available with experience, skill and expertise, informal cooperation, and trust between different divisions involved in production. The missing reflection of the sociotechnical setting at present results, however, in the implementation of centralized computer control systems for production at an increasing scale.

This report discusses the development of a model based on workflow management for shop floor production planning, which uses interpersonal communication in order to integrate the normal production procedures with the continuous management of chaos and breakdowns. It has been derived from research which originated in the ESPRIT Project 1217 (1199) on human-centered CIM. It follows explicitly the sociotechnical systems concept as outlined in Section 5.2.

Human-Centered CIM

Esprit Project 1217 (1199)

The human-centered approach to the development of technology was researched in the ESPRIT project 1217 (1199) which characterizes the visible beginnings of the European debate on human-centeredness. It is briefly referred to in the first, introductory section of this chapter. The aim of this ESPRIT project was to demonstrate that CIM, in view of the philosophy on human-centeredness, is significantly different from a traditional Tayloristic division of labor. The human-centered approach is focused on people both working within and managing manufacturing cells. The development of multiskilled personnel is a key concept: much of the production planning and scheduling activities that take place in offices are intended for transfer to the shop floor. A working prototype of such a human-centered production cell was developed by a company based in London. It may serve as an example of how to translate into practical production the challenges of both improving human-centeredness and increasing flexibility. A brief description of this project is presented here.

In this project, the cells on the production shop floor produce high-frequency connectors: the company involved in the project specializes in the manufacture of precision connectors for communication signals at frequencies up to 46 GHz. The connectors can be described as consisting of an outer body and a central pin which is surrounded by a PTFE insulator. The parts are of some millimeters in size. The whole variety of the company's production is approximately 44,000 items. Batch quantities are mostly in the range of 200 to 1000 parts. The production process for the final connector consists of several main steps. The metal and plastic parts are turned on a lathe (30 minutes for a batch of 250 parts). Thereafter, the main production steps are secondary operations, heat treatment, gold plating, inspection, and assembly.

Design of a Cellular Factory

Schael [3] suggested an overall factory model for the cell-based production of the connectors following the underlying philosophies of group technology, period batch control, and optimized production technology. The model of the factory is briefly described here as an example of how these philosophies have been put into practice. The factory consists of two cell types with different skill levels and a constant number of people required. The cell with the higher skill level (machining cell) produces finished parts which are later assembled in the second cell (assembly cell) to make up the final products. Three machining cells and one assembly cell work together within a product family.

- Machining cell: The order of a connector consists of the pin, insulator, and body. All operations and processes to finish these parts are done within the machining cell. The machining cell consists of four lathes and one multipurpose machine for second operations. The required number of operators is six.
- Assembly cell: In the assembly cell, the machined components are assembled. The cell dispatches the finished connectors. Four operators are needed for six assembly and dispatch workplaces.

- Plating cell: The connectors are plated in a centralized process which represents a bottleneck in the production flow. Thus, this flow is organized by smoothed period batch control, adjusting the starting times of production of each batch smoothly along the time scale through the working week.

Information and Control

Based on this concept of factory planning, information flow and control have to change in this factory in comparison to traditional production planning practices. The planning scenario is as follows: The material manager receives all orders from customers. He identifies the product families and sorts them by groups. The resulting job list is transferred to the cells via the computer network. Subsequently, an operations list is scheduled with assistance by the computer system and/or manually. Basic rules for an optimized sequence of orders are available in the cell software.

Once the load is in the range of 100%, the cell members microschedule the operations list. They check and adjust all the used data, paying special attention to the setup times. These times are important for small- to medium-sized batch production because they reduce the utilization times of the machinery. Once the operator thinks that he has found the best sequence, he simulates it on the computer.

The material manager at the factory level does not control the decisions in the cell. His task is the balancing of orders between the available cells; if a cell rejects a planned job, he would have to find a different cell to do the job.

Shop Floor Tasks

According to this cell-based production concept, there are many new tasks which have to be performed by the cell team on the shop floor. The team working in the cell may have to perform the following tasks on the shop floor:

- Planning the work for the week
- Preparing the work and setting the machines
- Producing the different parts
- Quality and performance control
- Management of tools, raw materials, etc.
- Maintenance and minor repairs of machine tools

Thus, the strategy aims at giving shop floor personnel an increasing degree of responsibility — not only for technical aspects of production, but also for organizational and planning tasks. This strategy is at present being implemented on a large scale all over Europe, characterized by its enterprisewide introduction.

Workflows for Shop Floor PPC

The Concept of Workflows

This example has shown that a procedural and top-down control perspective does not fit in with the concept of today's flexible cell-based manufacturing organizations. Therefore, a new concept is needed for looking at production planning and control (PPC) across the plant: e.g., linking the different production cells as customer–supplier chains. This concept is called workflow. A workflow, similar to activities in coordination and control, is a unit of work that happens repeatedly in an organization of work. However, there is a difference from traditional control concepts; in workflow concepts, processes get completed that are related to, or result in, customer satisfaction. In fact, every workflow has a customer who may be an external customer of the company or another workgroup or individual within the organization (an internal customer). The resulting enterprise model sees the entire organization as a chain of customers and suppliers, starting from the external customer (market), through the front-office units, to the internal production cells and to other back offices and service departments [4].

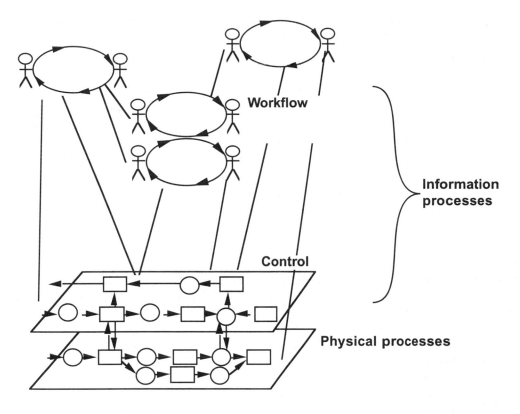

FIGURE 5.20 Three enterprise views.

Three Enterprise Views

In order to understand the different dimensions of the coordination mechanisms in such an enterprise model, one has to distinguish three different views on processes within organizations [2]. Each of these views highlights a specific interest in understanding the nature and characteristics of how an organization functions; the different perspectives describe the same phenomena by reducing the complexity to a specific subset of interest. The three different views concentrate on the material, informational, and business perspectives (see Figure 5.20, from bottom to top [5]).

First View: Material Processes
Material processes relate to human activities which are rooted in the physical world. An observer notes the movements and changes of physical things. In this kind of process, materials are transported, stored, measured, transformed, and assembled, for example, on the production shop floor.

Second View: Information Processes
Information processes control the material processes. They also overcome the failure of material processes to capture what is important about everyday activities in the 20th century. Theorists and information technology providers have developed sophisticated ways of analyzing and facilitating the flow of information. Some main approaches include data flow analysis, entity-relationship models, database storage and retrieval, transaction processing, network communication, etc. These techniques are the basis for applications offered by computer industries today, including CIM environments. What is lost in this perspective of information processes is the recognition that "information in itself is uninteresting. Information is only useful because someone can do something with it [2]." Information is used to make business, or useful information produces action. This is very much related to the philosophical foundations of the

language-action perspective [6] which has been deeply discussed in the CSCW community [4]. This perspective is specifically dealt with in the next view.

Third View: Business Processes

Business processes improve the way workers interact and coordinate their activities to produce a better service for their customers. At the same time, business processes are implemented in information processes, just as information processes are implemented in material processes. However, the focus is being placed on the business perspective and customer satisfaction, rather than on material treatment, logistics, forms, or database transactions. In this way, a higher level of organization is being revealed. Business processes are modeled as workflows, putting emphasis on the communicative relation of people in organizations [1].

The Language–Action Perspective of Workflows

The modeling of communication in the business processes view follows the language-action perspective, as suggested by Schael [4]. The basic unit of such processes is a four-step action workflow protocol, shown in Figure 5.21. In this figure, black arrows represent the requester's speech acts, and gray arrows represent the performer's speech acts. In the first phase of the loop (request phase), the customer asks for a service or product. In the second phase (commitment phase), the supplier promises to fulfill a specific request. The supplier's agreement with the customer's request is possibly modified during the commitment phase through negotiation. Thus, the second phase is not always straightforward. The negotiation in the commitment phase includes the possibility of a counteroffer. In this case, there are two paths for concluding the commitment phase: the customer accepts the counteroffer, or the customer makes a second counteroffer which the supplier accepts. In all other cases, the commitment phase leads to states where no further moves are possible.

In the third phase (performance phase), the supplier fulfills his or her work which leads to the delivery of the requested service or product. The final phase (evaluation phase) closes the loop and involves the customer's acknowledgment or formal declaration of satisfaction (or nonsatisfaction) after he or she receives the service or product. At its simplest, this declaration is a thank you or payment for the service/product.

In Figure 5.21, closure of the loop means that the link between one customer and one supplier for a specific service is completed and has achieved customer satisfaction. Thus, the loop represents the four transitions, request, commitment, performance, evaluation.

It is important to observe that in such negotiations, values and goals of the enterprise, through all levels of activities, are negotiated in parallel to straight business transactions. These soft goals in particular are the basis of all interactions of human operators and human agents; they cannot be avoided.

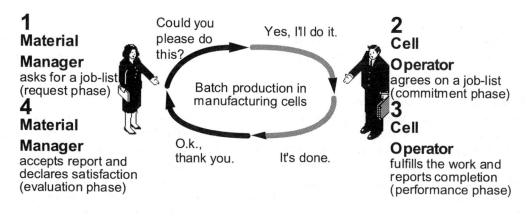

FIGURE 5.21 Basic workflow model describing customer–supplier relations in the language-action perspective. (From Medina-Mora, R. et al., in *Proc. Conf. Computer-Supported Cooperative Work*, ACM, 281, 1992. With permission.)

By acknowledging the existence of soft issues in business transactions, people become empowered on all levels of the company hierarchy, including the shop floor.

Conclusions

In today's manufacturing, the relationship between the material manager and the shop floor operator is difficult to support in conventional CIM environments because CIM tools tend to concentrate on the information perspective. They are missing the cooperative and communicative dimensions of the business perspective. However, this more informal discussion about workloads and commitments is in the scope of research on understanding cooperation and the design of support tools for cooperative work in the concept of computer-supported cooperative work (CSCW). CSCW offers a new understanding of cooperative work settings and a complementary set of applications in addition to CIM environments. The need for supporting cooperative work in advanced manufacturing is obvious. The example for improving shop floor production planning and control through workflow management technology is one example of much wider perspectives of the future use of CSCW systems in terms of human-centeredness.

References

1. Butera, F. and Schael, T., The renaissance of socio-technical system design, in *Self-Organization: A Challenge to CSCW*, Mambrey, P., Paetau, M., Prinz, W., and Wulf, V., Eds., Springer-Verlag, London, 1997.
2. Medina-Mora, R., Winograd, T., Flores, R., and Flores, C. F., The action workflow approach to workflow management technology. *Proceedings of the Conference on Computer-Supported Cooperative Work*, ACM, 281, 1992.
3. Schael, T., Human-oriented CIM-concepts for flexible manufacturing, *Research Report 100*, MWF Publishers, Duesseldorf, Germany, 1991.
4. Schael, T., Workflow management systems for process organizations, *Lecture Notes in Computer Science*, Vol. 1096, Springer-Verlag, Berlin, 1996.
5. Wollenweber, D., Systemic Concept of Transportation Chains, Ph.D. thesis, RWTH, Aachen, Germany, 1996.
6. Winograd, T. and Flores, Understanding Computers and Cognition — a new foundation for design. Ables Publisher Corp., Norwood, New Jersey, 1986.

5.6   Process Control: Human-Process Communication and Its Application to the Process Industry

Michael Heim and Norbert Ingendahl

Chemical engineering and process control represent important areas in today's production. This field has seen the development of the need for operators in the control room to be given more process-related information than has been the case until now. The operators need to know about process details in relation to the different process stages as well as in relation to time, e.g., across different operator shifts. A new information system for the control room has been suggested here. It allows the operators to call upon information and guidance in certain difficult situations and also to input their own observations, recent experiences, and actions into the system. An operator can call up all information by himself, or it may be done by other operators working in subsequent shifts. In this sense, control personnel today are always incorporated into teams of different people: their colleagues who are responsible for other areas of the process plant, their production supervisors, maintenance staff, etc.

The research reported here characterizes the work of the Department of Process Control Engineering at the University of Technology (RWTH) in Aachen. It is well known for its excellent cooperation with leading manufacturers and users of process control technology including Bayer, Siemens, and Foxboro.

The Concept of Human-Process Communication

Human-process communication has recently become an important research area due to the rapid developments in process automation. The concept of this human-process communication has to be understood as embedded into the human–human communication among the different teams running the plant. This view is particularly important if shift hours are taken into consideration: different teams are separated not only in space but also in time, working different shift hours. Efficient plant performance, however, can only be guaranteed if each of these different teams knows what the other teams have done, experienced, or observed during their shift. Thus the concept suggested has to be considered as a team-oriented support system for human-process communication.

In today's process industries (chemical and pharmaceutical production, sewage treatment, metallurgy, and mining industries, etc.), the actual process is hidden and concealed from the operator behind layers of impenetrable plant technology. Nevertheless, the operator should still be familiar with the process as it is ultimately his job to:

- Control the operation according to set rules
- Detect critical developments early in the process
- Develop potential improvements for both the process itself and process control

Therefore, the operator requires a picture of reality, generated by a suitable process-control system which is capable of delivering:

- The necessary information about the industrial process itself
- The possibility of controlling the process

In other words, the human operator must be able to communicate with the process. Faerber, Polke, and Steusloff [2] present a comprehensive approach to such a future human-process communication which is based on three fundamental elements. They are shown in Figure 5.22, which illustrates the flow of information from the process to the operator via sensors and back into the process via actuators. Sensors today are capable of delivering an almost unlimited amount of information about the process, which cannot be fully perceived by the operator. To this end, information reduction techniques have to be used to handle this flow of information. The reduced information has to be brought within the range of

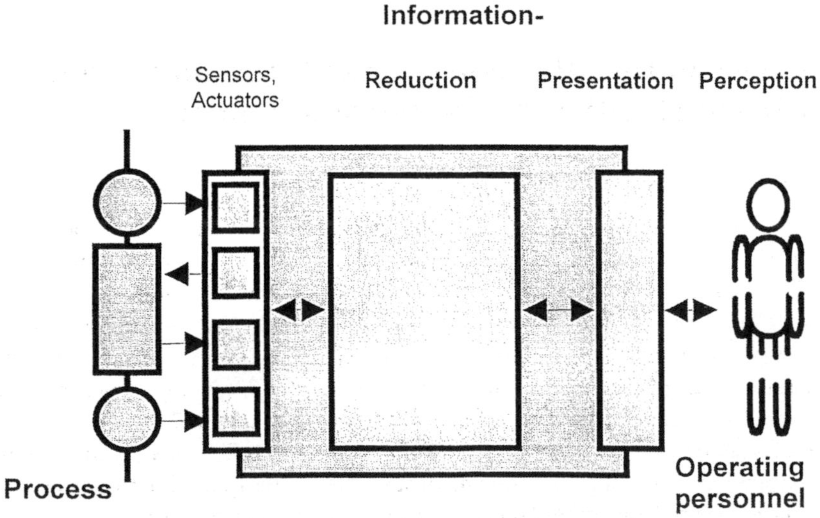

FIGURE 5.22 Information flow from process to operating personnel.

Information structuring   Requirements

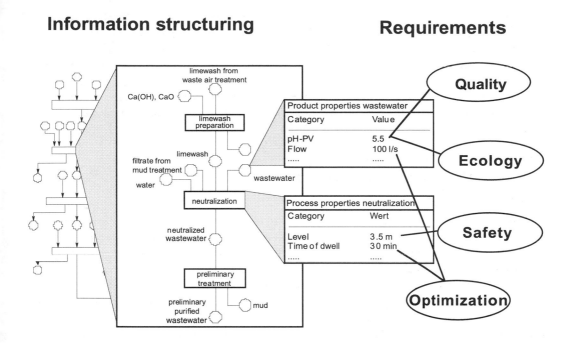

FIGURE 5.23 Phase model of production for wastewater treatment.

perception by the human senses. This is the job of presentation. The presentation sets the conditions for the transfer of information to the human consciousness. To this end, the possibilities and capabilities of human perception have to be taken into account.

Hence, a prerequisite of all human-process communication is the structuring of the process itself in forms of information reduction, presentation, and perception by the human operator. It includes the systematic analysis of production demands (quality, safety, etc.) and involves the structuring and definition of information relating to product and process properties and its dependency on time and space [1] as shown in Figure 5.23. A good example is wastewater treatment, in which the products and process elements represent the causal network of process/production in a very general way. Thus, this report demonstrates human-process communication as applied to wastewater treatment. The wastewater treatment process is described, and important operational situations are explained. Subsequently, different aspects of human-process communication are introduced. The solutions shown in this report have been implemented in several process control systems in industry [1, 3].

Characteristics of Operational Situations

Wastewater is treated by sequentially performing different physical, chemical, or biological processes. The operating personnel may influence the individual processes mainly by varying the mass-flow rates. As an example, biological decomposition of wastewater is controlled by the mass-flow of recirculation water, returned mud, and oxygen. Moreover, nutrients, chemicals, and special feeds may be added. The complexity of operating wastewater plants results from variations in boundary conditions of the process itself. For example, the effects of altering the inflow (quantity and concentration of substances) on purification processes are often known only in qualitative terms. The operating personnel find themselves confronted with the task of deciding on the probability of certain process developments. At the same time, they have to prevent any deterioration of bacteria population by intervening with appropriate control measures. Competing goals (minimal deposits vs. economic and energetic restraints) and interdependencies between product and process properties frequently call for compromises.

In view of this complexity, today's design of monitoring and control panels shows obvious weaknesses. For example, typical process diagrams tend to follow the conventional structure of control room implementation of single-loop controllers. Thus, all relations between individual signals must be extracted by the operator. The enormous power of today's graphic presentation systems is used merely for presenting three-dimensional apparatus pictures instead of three-dimensional relations between relevant properties! Thus, the operational staff has a hard job in reacting to critical situations, such as load surges, mud flotation, or lack of nutrients. Important indicators and characteristic factors are often not presented at all.

Presenting Human-Process Communication

Creating the Process Overview

To ensure smooth operation, the personnel first need to obtain a general overview of the whole process. In Figure 5.24, such an overview is shown for the biological unit of a wastewater treatment plant. The process structure is represented by essential input products, intermediate products, recycled products, and final products, as well as by the respective process elements. The structure of the process, as shown in the overview, does not necessarily correspond to the apparatus structure. In biological clarification processes, for example, denitrification and nitrification are conducted in different basin sections. Therefore in an apparatus-oriented representation of this process, the important analysis of the individual processes cannot easily be carried out because essential intermediate products (e.g., nitrate) are displayed at unsuitable places separated from each other.

In the general overview suggested here, elements are characterized by their attributes. They considerably improve the perception of the overview representation. These attributes can be either direct measurement data, such as BSB (required oxygen for biological decomposition), CSB (required oxygen for chemical decomposition), and TOC (total organic carbon), or indirect (i.e., calculated) characteristic values such as period of dwell or mud age. It should be noted that it is necessary to achieve a presentation which can be easily perceived. Thus, the essential attributes must be presented for each element of the process structure.

During the operation, it is important to compare the actual process data with the set-point data. In this way, a comparison between actual and desired values can be carried out and the current process can be

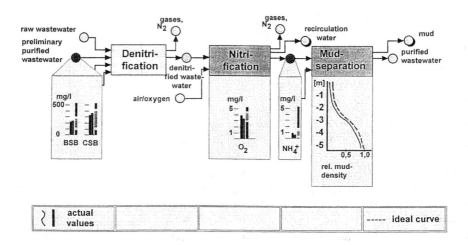

FIGURE 5.24 Process-oriented overview.

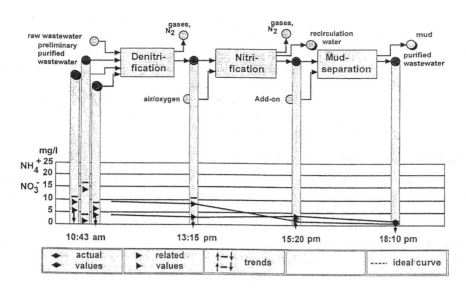

FIGURE 5.25 Process structure and nitrogen balance.

evaluated with regard to the product specifications (e.g., the fulfillment of legal requirements). By choosing certain forms of graphical presentation, the personnel's capability for recognizing complex process conditions is enhanced by typical patterns.

Creating the Time-Space Dependencies of the Process

In order to understand the process, it is important to present the characteristic dependencies of the process on time and space. This is shown in Figure 5.25 for the nitrogen balance. The ammonium and nitrate concentrations of vital products are presented according to their relative quantity in the inflows along the process chain. Thus, the process (here the ammonium decomposition and the resulting formation of nitrate) is made clear. The slope of the correlation lines is a distinct indication of the reaction rate. Moreover, the change of the slope provides the operating personnel with a clear indicator of the dynamics of the process. The presentation described is well suited to understanding the process structure. It allows for early detection of critical developments in the process, such as problems with sensitive nitrificants.

Integration of Operational Experience

Recording Operator Experience

Aside from routine tasks, however, process control requires personnel experienced in making decisions and setting priorities with the objective of making the best contributions to the production objectives. Human-process communication places the responsibility for this objective with the human operator. Such systems of human-process communication combine human knowledge and experience on the one hand, with automated support on the other hand, while the operator remains in control of the system. This view corresponds to the issues discussed in Section 5.2 of this chapter.

Important aspects of operator control of the system are:

- What-if scenarios based on simulations
- Integration of operational experience by the operators themselves
- Online configurable presentations, which are adjustable to the user's information needs

The operator plays an important role in operational situations when there is not enough information available on either the process or production properties to allow for automated control support. In such situations, the operator has to intervene with the process. Quite often, these decisions must be made while the operator is being subjected to considerable time pressures. It is important that his assumptions and experiences are suitably documented for later evaluation. The experiences of the operators may subsequently be used to more accurately determine the product/process relationships allowing more detailed knowledge of the process. In addition, the operator can add his own observations and experiences during his shift into this event-browser in a similar way to writing a diary of events.

Configuration of Presentation by the Operator

It is important to design the interface in a way that allows easy access, input, and recall of such information by the operator. The system described here has been designed so that the user only has to answer some application-oriented questions in order to relate the relevant information to appropriate forms of presentation. These questions deal with the characteristics of the information such as scale, type, and time and space dependency. The structuring of each presentational form of this information is defined with the aid of the usual inference mechanisms and knowledge structuring methods.

In the first step, the user selects the information from the process structure which he wants to be displayed. Subsequently, the properties chosen will be suitably fed in or defined by the user. Finally, the selection process is started, and a suitable presentational form is suggested. The selection process is suitable for engineering purposes as well as for the adaptation of human-process communication during operation. The information to be displayed to the operator is easily accomplished through a suitably designed interface display.

Conclusions

The system described has been designed to allow easy information handling and presentation in the process control room. The control room personnel themselves are expected to use this system continuously in order to control and optimize the processes. Furthermore, the system allows the operators to record their own observations and experiences during shift hours for subsequent recall by themselves or by their colleagues in another shift or at another site. Through this system, the control room personnel are able to create their own expert system, incorporating all data and relations of both production process and products needed for optimum process control.

The different process control tasks can be hierarchically structured according to their degree of abstraction. According to certain situations and to the system's state, the operators may switch between these different levels. Due to the different responsibilities, the technical staff have to be able to intervene with the process at each level and at any time. Consequently, process-oriented functional units [4] need an interface which enables control to be switched between manual and automatic.

References

1. Arnold, M., Heim, M., Ingendahl, N., and Polke, M., Human-process communication and its application in chemical industry, power plant control, and mining, in *Advances in Agile Manufacturing*, Kidd, P. T. and Karworski, W., Eds., 469, IOS Press, Amsterdam, 1994.
2. Faerber, G., Polke, M., and Steusloff, H., Human-process communication, *Chem. Eng. Tech.*, 57 (4), 307, 1985.
3. Heim, M., Ingendahl, N., and Polke, M., Human-process communication, in *Process Control Engineering. Ullmann's Encyclopedia of Industrial Chemistry*, Vol. B6, Polke, M., Ed., VCH, Weinheim, 338, 1994.
4. Polke, M., Ed., *Process Control Engineering. Ullmann's Encyclopedia of Industrial Chemistry*, Vol. B6, VCH, Weinheim, 1994.

5.7 Enterprise Networks: The Reengineering of Complex Software Systems

Horst Kesselmeier, Inga Tschiersch, and Sebastian Kutscha

Current technology design can no longer be understood as design for a green site. Design and implementation of new technology always depends on existing technology and the way it is used by people. In this respect, software engineering has changed the characteristics of normal technology design, taking into account existing computer systems. Experiences show that the conditions and needs of such software reengineering projects are highly complex and differ in their special characteristics ranging from aspects of quality of existing system documentation to organizational structures of the computer departments concerned. The task-artifact cycle presented here gives a suitable approach, emphasizing both analysis and design in software reengineering.

The authors of this report represent partners in the excellent cooperation of the University of Technology in Aachen with industry — in this case, one of the leading German software development companies, SD&M, in Munich. This cooperation includes software reengineering projects with some of Germany's largest companies.

Problems of Software Reengineering: The Example of a Tourism Network

The implementation of new control technologies usually leads to far-reaching changes. In aircraft and air traffic control, substantial effort is invested into retraining all users in order to guarantee the smooth and safe transition from old to new work structures. In the production industry or service enterprises, training is considered much less important, as is the strategy to adapt technology as much as possible to the existing work structures. Neglecting the smoothness of this transition frequently leads to dangerous developments in enterprises because software changes may paralyze the company for days (or even longer) before the new system works properly. Companies, however, are dependent on their computer systems in a very delicate way. Hence, such a system must be reengineered in a way which does not threaten the profits, or even the existence, of the company.

One specific software reengineering strategy has been applied and described in a recent large-scale project which aimed to smoothly replace the complete software system of a large tourism enterprise without one day of system failure [5]. In this project, the emphasis has been on user involvement and the integration of the entire software development process.

Tourism companies carry out their daily business on a trade-wide booking and information network. The structure of the software and the hardware of these systems has often grown together with the organizational structure of the enterprise over the years. Computer systems and organizational structures today form highly interlinked and complex networks. New software components are being developed mostly under the high pressure of innovation while the business is actually running. There is no time or pressing need to sufficiently document the various steps of new developments. Thus, instead of well-thought-out solutions, as was intended when the architecture of the system was originally established, temporary arrangements serve as the basis for the development of further system components. Today one can find efficient systems, yet the inner structure is no longer comprehensible. Further development of the software is shaped by unpredictable and unintended effects. Market demands, however, lead to new functional requirements of computer systems. International networks also require structural changes to the system architecture.

It is almost impossible to resolve this problem by replacing the software systems since they are so deeply linked with the organizational structure of the company. Common methods of software engineering are not sufficient since the conditions for software reengineering are very different from forward-engineering.

In the process of reengineering complex information systems, previous system solutions have to be taken into consideration. A reengineer needs an abstract view of how the work is currently being done and which conditions have led to the present software solution. The subsequent design and implementation of new structures is strongly affected by technical and organizational conditions within the company concerned. The computer system used defines, to a great extent, the technical tools to be used. Organizational structures of the computer departments have their employees, specific documentation of software systems, specific programming guidelines, and specific procedure models for the software development; they often form a culture of data processing in the company that is rich in tradition.

Software reengineers get little support from methodology. In this section, such an approach for software reengineering is suggested. It is based on observations and practical experiences gained in the software reengineering project of the touristic enterprise mentioned before.

The Reengineering of a Tourism Booking and Information Software System

NUR Touristic is one of the biggest tourism companies in Europe, with 600 employees in its headquarters in Germany. About 100 employees develop and maintain the computerized information and booking system, integrating around 12,000 travel agencies and some 400 internal users of NUR Touristic.

The development of the system's structure began in the late 1960s with assembler programming and simple databases. During the 1980s, new components programmed in COBOL were linked to the old assembler structures. The old and new world were linked by bridges which transferred data in both directions during the night; new and old generations of databases formed heterogeneous data structures. At the end of the 1980s, the need to restructure the system's architecture arose with new tourist demands caused by the opening of the European market. First attempts showed the dangerous dependence of the company on its complex software structures. It was obviously impossible to restructure the booking and information system without endangering the running of the business processes.

At the beginning of the 1990s, a specific software reengineering project began in cooperation with the software consulting company SD&M in order to gain new insight into the system architecture and the process of further development. The project ran for 3 years and had a scope of more than 100 man-years. It consisted of several subprojects with up to 50 internal and external project engineers [5]. The following main results were achieved:

- Replacement of the old assembler system
- Restructuring and redefinition of the company's data structures
- Creation of a restructured system architecture as a basis for further software developments

Some strategies appear to have been essential for the success of the project [3]:

1. The system structures were merely reengineered with just a few functional extensions (1:1* strategy): separation of structural and functional changes.
2. External SD&M and internal (NUR) experts formed a series of heterogeneous teams incorporating in-depth working knowledge related to the existing system and applied new strategies in managing the software reengineering process.
3. The project was divided into small steps with highly efficient mechanisms of controlling the success (e.g., continuous review of subproject concepts in the heterogeneous teams).
4. Decisions were discussed with the concerned tourism and technical departments of NUR Touristic in order to obtain continuous user acceptance.

The project mainly aimed at restructuring technology, but it also influenced the organizational structures of the NUR Touristic computer department, for instance, by including participation of all the departments involved in the design process. A new organizational framework within the computer department at NUR was developed over the course of the project, and was intended to last beyond is finish. The technical structures of the software systems turned out to be the basis and the framework of the living structures within the NUR organization: those people who were developing, maintaining, and using the system. The achievement of the project is, thus, not only the improvement of the software, but also the creation of a suitable environment for the continuous software development process leading to long-term solutions both for technological and organizational development.

At present, a second project with a similar budget is running. It aims at functional extensions, e.g., the integration of European subcompanies into the booking dialogue and the integration of new customer requirements.

The Methodological Approach to the Software Reengineering Project

Forward and Reverse Engineering

According to the experiences mentioned above, a viable method of software reengineering should provide:

- A global orientation for software reengineers or project managers
- A set of tools within a flexible methodical framework to support their daily work

For this new kind of software design, the following strategy is suggested which has been developed in the context of the software reengineering project mentioned above.

The reengineering process is to be structured in three distinctive steps [3]:

1. The step of reverse engineering: Analyzing and documenting the existing system, the tasks and needs of system users in situations of use, the system context, e.g., the work environment, modularizing the system, and defining new, transparent interfaces and links between these new modules. This step is also the most important strategy of user participation and user training to guarantee smooth working of the new system to be developed.

2. The step of forward engineering: Developing concepts of what the system has to do and drafting this system. In a complex reengineering project, it does not mean the addition of any new functionalities but merely the incremental replacement of the old system with the new system as a basis for future developments (the user may actually not realize that, during this step, the system is changing at all). This process step can be called 1:1* transition. This means that the old system's functionality is kept (1:1) within the new system. Only small extensions (*) may be added to the system which can be realized without risk to the project and the company.

3. The step of system extension as the continuation of forward engineering: Only now can the new system be extended to integrate and offer the user new functionalities which were the initial aim of the process.

These distinctive steps can be related to the following model of reengineering: the task-artifact cycle.

The Task-Artifact Cycle

The methodological approach of software reengineering suggested here is related to the approach of technology design suggested by Carroll et al. [1], which is the task-artifact cycle. It is described in Section 5.2 of this chapter. In the tourism project described here, it has been translated into the needs of software reengineers. The approach forms a framework for suitable tools which are the subject of further research within the project reported here [4].

This cycle gives a methodological orientation when designing software technology within existing system structures. It stresses the understanding of how the existing technology is used in practice: what users think while using it and which conditions or decisions have led to its actual design. The task-artifact cycle is comprised of the following two phases structured into four separate steps:

- Analysis (step 1) and recovery (step 2) of the existing system (reverse engineering)
- User task definition (step 3) and scenario-based design (step 4) of the new system (forward engineering)

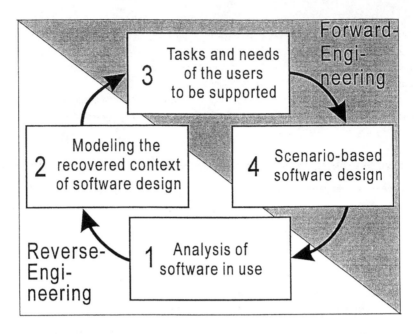

FIGURE 5.26   The suggested approach of software reengineering.

The analysis phase (reverse engineering) fills the gap which characterizes the apparent weakness of the conventional waterfall model of software development; the waterfall model only comprises the design phase, as discussed in Section 5.4 of this chapter.

The Approach of Software Reengineering in Practice

Figure 5.26 shows the approach of software reengineering which is inspired by the task-artifact cycle. In the reengineering project at NUR Touristic, the phases of the task-artifact cycle can be found both on the abstract level of project management and within different subprojects. In practice, the sequence of analysis and design phases has been followed through repeatedly as an iterative process.

This analysis and design cycle is oriented around the users tasks and needs. Figure 5.26 shows how the cycle is aimed at the end-users of system functionality and their requirements (e.g., the travel agency giving tourist information). Besides considering these end-users, software reengineering also has to consider as users the teams of computer departments who are developing and maintaining the software system and architecture. Their tasks and needs in terms of software development and maintenance differ from the requirements of the end-users of system functionality. Thus, the cycle has to be used twice in two different ways in order to correspond to the obvious differences of these two user groups.

Conclusions

Today, technology design can no longer be understood as design for green site systems. Design and implementation of new technology are always dependent on existing technology and the way this technology is being used by people. For design processes in modern industry, it has become normal to develop and implement new technological systems or subsystems for production within technical and organizational structures of existing production processes. In this respect, software projects seemed, for some time, to be different because software systems were developed in areas without previous computer use. Over the last years, this situation has changed rapidly. Software engineering has developed the characteristics of normal technology design, taking into account existing computer systems; brown site design is becoming normal for software development.

The methodological approach has been suggested to start software reengineering through a reverse engineering strategy of analyzing and documenting the existing system. This strategy is to be followed by the forward-engineering strategy. Both strategies are integrated into the task-artifact cycle which includes the four steps of:

- Analysis and recovery of the existing system (reverse engineering: steps 1 and 2)
- User task definition and scenario-based design of the new system (forward engineering: steps 3 and 4)

The question remains to be investigated whether these new concepts of software reengineering offer a solution to bridge the gaps and discrepancies between what software engineers today generally produce and what practitioners in industry and service enterprises really need and want.

References

1. Carroll, J. M., Kellogg, W. A., and Rosson, M. B., The task-artifact cycle, in *Designing Interaction. Psychology at the Human-Computer Interface*, Carroll, J. M., Ed., Cambridge University Press, Cambridge, 74, 1991.
2. Chikofsky, E. and Cross, J., Reverse-engineering and design recovery: a taxonomy, *IEEE Software*, Vol. 7, 1, 13, 1990.
3. Eberhardt, K. and Kutscha, S., Obsolete systems inhibit innovation (in German), *Computer-Woche Extra*, 1, 17.2, 14, 1995.
4. Kesselmeier, H., Methodological Approaches to Software Reengineering, Ph.D. thesis, RWTH, Aachen, Germany, 1997.
5. Surrer, H. and Taubner, D., Object oriented specification through a CASE tool for structured analysis/structured design (in German), Info. Spektrum, 19(4), 196, Springer, Berlin, 1996.

5.8   Assessing the Human Orientation of New Control Technology: The Example of Slovenia

Janko Cernetic, Marjan Rihar, and Stanko Strmcnik

Significant differences can still be observed between European countries in terms of industrialization and automation. Those countries which until now may have been considered less industrialized are now dealing with the issue of automation in production as one of their main tasks. Control of chemical processes requires particular attention because of the following problems: low efficiency of resource utilization, high risk of failures, environmental damage and other hazards, high work load for plant operators, etc. The highest degree of automation and control available today may represent the most intriguing technical solution to be implemented at a certain chemical production plant. A different design of the control system, however, may represent a superior solution if it takes into account the specific aspects of organization and plant personnel, e.g., their skills, experiences, and needs. The example of the design and implementation of process control in Slovenia is described here. Slovenia is one of the countries in the former Eastern Block bordering, among other neighbors, Austria and Italy. The country is fast developing toward integration into the European Union.

The authors represent the leading research institute in Slovenia, the J. Stefan Institute, named after the well-known Slovenian–Austrian physicist Josef Stefan (1835–1893). The J. Stefan Institute cooperates very closely with all the main industrial companies in Slovenia as well as with the corresponding departments of the Slovenian universities. This report gives some insight into current developments in countries moving toward increased industrialization.

The Concept of Success Factors

It is well accepted within the engineering community that the introduction of advanced control technology into industrial practice is a very demanding process due to its deep impact on people and the work organization. In accordance with this, system designers as well as users are often faced with severe difficulties and problems [4]. Many of these problems have been encountered by the authors

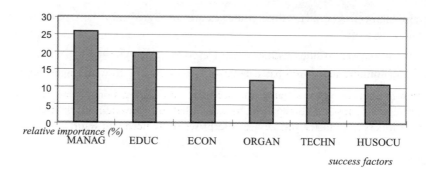

FIGURE 5.27 Relative importance of six success factors of introducing automation. (From Brandt, D. and Cernetic, J., *J. Artif. Intelligence Soc.,* 12, 2, 1998. With permission.)

during their work in the area of control and information systems development and application in Slovenia. Usually, the reason problems arise is related to the nature of the approach used. Very often, the approach is partial, only taking into account the technical aspects of control system implementation. Thus, the approach neglects human and organizational aspects of change in enterprises. In two previous papers, the authors [2, 6] report on the opinion of a group of leading industrialists in Slovenia in response to the following question: Which are, in your opinion and according to your experience, the most important factors affecting the success of introducing computer-based automation, e.g., in the case of your company?

The respondents were subsequently asked to assign an assumed relative weighting (percentage) to each factor mentioned with all weightings adding up to 100%. The questionnaire listed the following success factors:

- Clear vision of enterprise development and support of the management (MANAG)
- Proper knowledge and education (EDUC)
- Economic justification and monitoring of effects (ECON)
- Proper organization (ORGAN)
- Proper technology (TECH)
- Consideration of human, social, and cultural aspects (HUSOCU)

Results of the survey are given in Figure 5.27.

The Importance of Human Orientation as a Success Factor

The results of this survey clearly show that the respondents assigned considerably higher importance to management issues than to purely technology issues. They assigned relatively low importance, however, to human orientation and organizational issues relating to control technology. This opinion may reflect the present state of general development in Slovenia. Obviously, the problems related to other success factors (particularly those of management and education) drastically outweigh the problems of human orientation and organization. It may be envisaged that the latter problems will be addressed once the present problems are better mastered.

In order to prepare for this future, a new research direction was initiated in 1994 by the authors, integrated with their research and development work into computerized process control systems at the J. Stefan Institute. One important goal of this research program is to prepare guidelines for the introduction of such computer-integrated production systems (CIPS) in Slovenian companies. It is planned that these guidelines will, in particular, deal with the organizational and human-oriented aspects of controlling and managing production processes. A new (integrated and interdisciplinary) methodological approach is currently being developed by the authors to deal with these aspects.

How to Integrate New Technology, Human Aspects, and Organizational Change

The search for a new methodology was based, on the one hand, on the existing approach of the J. Stefan Institute for the development of control systems complemented by, on the other hand, a method for dealing with organizational, human, and business issues. This search has led to the USOMID method [5] which specifically supports the integration of the soft part of system design into the change process. Thus, computerized control systems have been developed by the J. Stefan Institute which take into account these different aspects of technology, organizational change, and human factors. In order to assess the success of implementing such systems, criteria were necessary to specifically evaluate the human and organizational aspects of change. For this purpose, the approach which was described in Section 5.2 of this chapter was chosen — the dual-design approach — and the criteria for assessing technology in use has been derived from this dual-design approach. These criteria can be used to assess the degree of human orientation or human-centeredness. The assessment system is based on the assumption that the human-centeredness of any control system can be analyzed by applying the following four aspects (dimensions) of new technology:

- The degree of automation achieved through implementing this technology
- The degree of networking implemented with this new technology
- The degree of dynamics of changes, e.g., in work and society, accompanying the implementation of this technology
- The degree of formalization of human communication and cooperation to be accepted by humans using this technology

To better understand the nature of the assessment procedure, each of these aspects can be regarded as an observation point. Each of these aspects (viewpoints) is assessed through answering six one-sentence questions. The complete set of (4×6) answers (or criteria) can be used with an appropriate scaling method; a value from 1 to 5 can be assigned to each answer.

An Example of Assessing the Degree of Human Orientation

In 1995, a preliminary test was performed by the J. Stefan Institute to evaluate how well these human-orientation criteria can be used as (systematic) guidelines during the design of complex and large-scale control systems. The questionnaire was applied to a pulp cooking control system in Slovenia which had recently been introduced as a newly developed system [1, 3].

The pulp cooking control system was developed entirely using domestic know-how at the J. Stefan Institute for the pulp and paper mill at Krsko, Slovenia, which produces about 130,000 tons of pulp per year (data for the period 1983–85). The objective was to upgrade the existing, primarily manually operated controls, for seven 175 m^3 batch digesters using a computer-based control system. The purpose of modernization was to increase the production rate, improve pulp quality, and reduce the steam consumption of existing facilities as efficiently as possible. Simultaneously, the new control system was to improve process safety in order to reduce the risks of environmental emergencies.

The design of the control system included some advanced functions, such as supervisory batch control, model-reference control, online estimation of model parameters, real-time cooking process simulation, coordination of multiple digesters on two parallel production lines, and the smoothing of steam consumption by a set of microcomputer-based controllers. In the context of this example, it is important to note that a relatively high degree of human–computer interaction was built into the design of the control algorithms. Thus, one aim was to keep the skilled human operators in the control loop.

Results of the assessment of the above control system are shown in Figure 5.28. The human-centeredness may be understood as being roughly proportional to the height of the columns. Note also that the four bar-chart headings in Figure 5.28 represent assessment aspects (or observation points), not the variable to be evaluated.

The results of the pulp cooking control system assessment can be described as follows for each of the four criteria groups. In terms of the degree of automation group, the flexibility of system use (criterion A3) is low because the variability of human–machine dialogues of the system is relatively narrow. On the

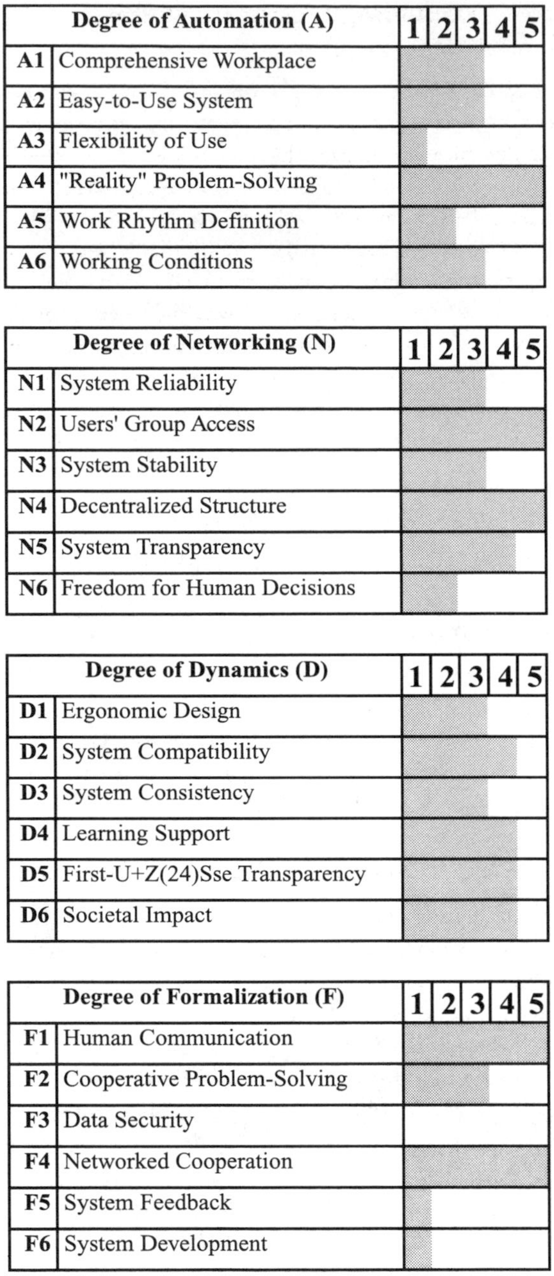

Degree of Automation (A)		1	2	3	4	5
A1	Comprehensive Workplace	▓	▓	▓		
A2	Easy-to-Use System	▓	▓	▓		
A3	Flexibility of Use	▓				
A4	"Reality" Problem-Solving	▓	▓	▓	▓	
A5	Work Rhythm Definition	▓	▓			
A6	Working Conditions	▓	▓			

Degree of Networking (N)		1	2	3	4	5
N1	System Reliability	▓	▓			
N2	Users' Group Access	▓	▓	▓		
N3	System Stability	▓	▓			
N4	Decentralized Structure	▓	▓	▓		
N5	System Transparency	▓	▓	▓		
N6	Freedom for Human Decisions	▓	▓			

Degree of Dynamics (D)		1	2	3	4	5
D1	Ergonomic Design	▓	▓			
D2	System Compatibility	▓	▓	▓		
D3	System Consistency	▓	▓	▓		
D4	Learning Support	▓	▓	▓		
D5	First-U+Z(24)Sse Transparency	▓	▓			
D6	Societal Impact	▓	▓	▓		

Degree of Formalization (F)		1	2	3	4	5
F1	Human Communication	▓	▓	▓		
F2	Cooperative Problem-Solving	▓	▓			
F3	Data Security					
F4	Networked Cooperation	▓	▓	▓		
F5	System Feedback	▓				
F6	System Development	▓				

FIGURE 5.28　Assessing the degree of human orientation of a pulp-cooking control system. (From Brandt, D. and Cernetic, J., *J. Artif. Intelligence Soc.*, 12, 2, 1998. With permission.)

other hand, the control algorithms support, to a great extent, the involvement of human competence in problem solving (A4).

In terms of the degree of networking, the control system appears, on average, to be moderately to highly anthropocentric or human-oriented, except for the last criterion in that group (N6), which refers to the user's freedom of decisions on how to work with the system.

The system has the highest positive assessment rate according to the degree of change dynamics. The reason for this rating may be that the design of control algorithms was made after careful consideration of existing manual (human) control procedures. Thus, the system integrates and supports human work patterns rather than replacing them through automation.

Within the last group of criteria (degree of formalization of human communication and cooperation), the third criterion (F3) referring to data security was not applicable to this system; the others were variably assessed, which can probably be explained by the fact that very different issues are addressed in this group. For this new control system, human–human communication (F1) and cooperation (F4) continue to be possible and encouraged, as well as real problem-solving (F2). This is obviously due to the specific choice of the degree of automation as shown in the response to question A4 (problem-solving in reality). From the users' point of view, however, there is not sufficient system feedback (F5) and human skills development (F6) built into the new control system. This corresponds to observations of most automation systems today and would need to be improved in the future.

Conclusions

The implementation of control technology in Slovenia today may serve as an example of how less industrialized countries are advancing in order to be equal with the whole of Europe. Slovenia has been very successful in this process. The country is highly regarded as an interesting industrial partner and is coping well on the world market. This is partly due to the careful consideration in companies of what kind of technology may be best implemented.

This report describes research at the J. Stefan Institute, Slovenia, which aims at improving the process of implementing new control technology in terms of human-centeredness. This example shows that some important aspects of human-centeredness have been taken into consideration by the development of one particular control system. Other aspects of human-centeredness must still be improved, as can be said about many other control systems worldwide. Further aspects which need particular consideration are the question of environmental protection and the cost-effectiveness of control systems in comparison to their degree of human-centeredness.

References

1. Cernetic, J. and Strmcnik, S., An approach to computer control of pulp cooking, 2nd Workshop on Process Automation, November 1987, 51.
2. Cernetic, J. and Strmcnik, S., Automation success factors as seen from a developing country. Proceedings of the 1st IFAC Workshop on Cultural Aspects of Automation, October, 1991, 34.
3. Hvala, N., Cernetic, J., and Strmcnik, S., Study of pulp cooking control using the Systems Approach. *Int. J. Syst. Sci.*, 24(4), 707, 1993.
4. Martin, T., Kivinen, J., Rijnsdorp, J. E., Rodd, M. G., and Rouse, W. B., Appropriate automation — integrating technical, human, organizational, economic, and cultural factors, *Automatica*, 27, 901, 1991.
5. Mulej, M., Cernetic, J., and Drozg, F., Dialectical systems theory helps the control technology consider the organizational aspects in practice. Proceedings of the 12th International Conference on Systems Science, September 1995, 175.
6. Strmcnik, S., Cernetic, J., Mulej, M., and Brandt, D., Organization as an important success factor in applying control technology. Proceedings of the 5th IFAC Symposium on automated Systems Based on Human Skill, September 1995, 195.

5.9   Clean Technology in Industry Today

Rita van der Vorst

Environmental protection activities in industry have rapidly increased in number over the last years. Additionally, surveys in environmental activities have identified a change in the type and approaches used in environmental problem solving. A new paradigm, clean technology, has been described which is gradually replacing the cleanup technology paradigm and the older dilute-and-disperse paradigm.

The new paradigm brings with it not only a new way of looking at environmental protection, but also a range of rules guiding the application of technology and the design of technological systems. This section presents a few case studies highlighting and evaluating clean technology activities.

The author is based at the Center for Environmental Technology at the Imperial College in London. This vantage point allows her to look at recent developments in British and European industry with a specific future-oriented perspective.

The Paradigms of the Environment

Environmental pressures on industry come from many different directions. Major sources of pressure to change are the regulative bodies and pressure groups, but pressure is also applied directly through market pressures, consumer choice, supply chain pressures, and stakeholder pressures. As previously outlined [7], ethical considerations of professional engineers acting in the industrial apparatus could also play a major role in the growing emphasis on environmental protection.

In addition to a mere increase in the number of environmental projects, a change in the kind of activities has been visible. A general move from cleaning up to prevention activities has been identified. Research has shown that this shift in method has not been reached through evolution. Furthermore, the shift resembles a revolutionary change such as that identified by Kuhn [4] for the development of science. Characteristically, such a change would move through stages which could be described as follows: the old paradigm is not sufficient to explain new findings; the abnormalities are accepted as exceptions to the rule; more and more of these exceptions accumulate and a crisis results; at this stage somebody (normally a younger person or a person who is less strongly rooted in the established paradigm [normal science]) will suggest a new explanation or a new pattern. Over time, the newly suggested pattern will form itself as a new paradigm with its own sets of rules.

Applied to the engineering activities described in this section, Kuhn's definition of a paradigm has been extended to mean a cognitive, perceptual, and behavioral framework. Thus, a change of paradigm implies a change of values guiding the translation of cognition into perception and of perception into behavior [8]. In this particular case, the new paradigm has not completely replaced the older paradigms. It will be argued that individuals in groups or organizations might follow different paradigms, and thus the resulting activity might not be a clear representation of any particular paradigm. Hence, in the work presented through the case studies in this paper, the identification of clean technology characteristics of an individual project are taken to illustrate that someone somewhere in the organization has made the shift. Mass persuasion in the form of education (together with generational changes) might eventually create a situation where only one single paradigm exists.

In the following, the main characteristics of the new clean technology paradigm are presented. This section will conclude with a few brief case studies to illustrate industrial environmental activity within the new clean technology paradigm.

The Clean Technology Paradigm

Clean technology could be described as the philosophy guiding the application of technology with the ultimate aim of achieving sustainable development. "The aim [of sustainable development] is to benefit from the advantages provided by ... resources to the point where the rate of 'take' equals the rate of renewal, restoration or replenishment [5]."

Clean technology has been defined as, "... a means of providing a human benefit which overall, uses less resources and causes less environmental damage than alternative means with which it is economically competitive [1]."

Although this definition is generally accepted in the engineering community and valid for most preventative engineering activities, the definition of clean technology should include those technologies which are less economical but have a potentially smaller environmental impact. However, it is obvious that only economically competitive technologies have a chance of implementation in the current economic climate.

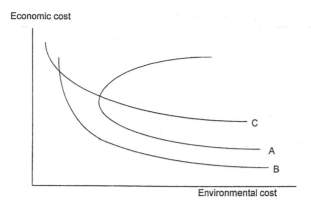

FIGURE 5.29 Clean technology vs. environmental improvement through clean-up technology. (Adapted from Clift, 1995.)

Nevertheless, such clean technology, or preventative engineering, provides the technical and philosophical guidance for the design and implementation of technology in a sustainable world, where preventative stands for "striking the fundamental causes of environmental problems [3]." Thus, the philosophy of clean technology, entailing systems considerations, life cycle considerations, and a focus on services, directs the design and planning of production systems (to include the strategic planning, management, and the environmental and social assessment). Clean technology represents the result of a paradigm shift, as introduced above, away from the cleanup technology paradigm or end-of-pipe engineering approach.

This change and redesign of technological systems for the purpose of environmental improvement is presented in Figure 5.29. Three types of investment into technology are displayed. Technology A is a given technological system (including organizational and managerial aspects) aimed at the provision of a product for the satisfaction of a human need. Investment can improve the environmental performance to a certain point. Then the environmental burden carried by the add-on technology itself will outweigh any environmental benefits. However, a change in technology, for example, to Technology B, could achieve similar environmental performance with lower running costs. Technology B is a clean technology system as defined above, if environmental improvement has been achieved through the redesign of the production system with a life cycle approach and exhibits a shift in focus away from the product to the provision of a service. Once the environmental performance becomes even more important, investment into Technology C will achieve even higher levels of performance. Depending on the foresight of the company or group of companies and the pressures for improvement, the more expensive option might prove to be the only competitive alternative in the long term.

To achieve improvements as reflected above, the preventative engineering paradigm is distinct from earlier paradigms in:

- Its approach to environmental problem solving (proactive, preventative, and holistic)
- The kind of activities (design, life cycle approach, and strategic and managerial development connected to organizational learning)
- The participation in its activities (across the organization, all employees or even all stakeholders involved)
- The focus of its activities (service focused instead of product focused)
- The breadth of its activities (global, cross-disciplinary, engineering and management, addressing social and environmental issues)
- Its outcome (organizational learning, participation, improvement of global environmental, and social performance with a reduction in life cycle costs, and continuing improvement).

Examination of the three paradigms, dilute and disperse, cleanup technology, and clean technology, and the quality of environmental action has provided three sets of characteristics, which will be used in the following paragraphs to describe the industrial case studies (see Table 5.1).

TABLE 5.1   The Three Paradigms of Industrial Environmental Action

	Paradigm		
Criterion	Dilute and Disperse	Clean-Up Technology (End-of-Pipe Engineering)	Clean Technology (Preventative Engineering)
Leitbild	Linear thinking, linear processes	Linear thinking, linear processes	Holistic, ecological thinking, cyclical activities
Motivation for action	No motivation for environmental protection	Extrinsic motivation for defensive and reactive actions	Extrinsic and intrinsic motivation for environmental protection, opportunity seeking
Kind of activity	None	Purely technical	Managerial, strategic, technical
Participants in activity	No activity	Specialists	Participation across the organization
Organizational development	No development	Some specialist learning, development of environmental specialists	High level of organizational learning throughout the enterprise and beyond (e.g., stakeholders)
Production	Continuous growth	Continuous growth	Minimized production with focus on quality
Production process design	Unchanged processes	Unchanged process, technology-centered solutions leading mainly to add-on technology	Changed processs, cyclical closed-loop, or cascaded production process
Product design	No concern for environmental impacts	No design changes, recycling of products	Design for the environment, design for minimization of life cycle impacts
Material consumption	Unrestrained	Unrestrained (possibly increased due to additional processes)	Minimized over the (product) life cycle
Pollution	Unrestrained, dilute and disperse	Restrained through additional process, concentrate and contain	Minimized over the (product) life cycle, if possible, rendered harmless
Global environmental impact	Not accounted for	Aim to reduce, is it successful?	Minimized through proactive activities
Industrial output	Product to be owned by user	Product to be owned by user	Provision of services, product on lease, servicing and recycling by producer
Financial cost for environmental protection	No investment	Additional cost for add-on technology	Investment into process change, saving of running and treatment costs
Profit	Largely based on non-accounting of environmental resources (free commodities)	Partly achieved through avoiding legal penalties, also through saving of secondary treatment costs	Largely achieved through innovation, savings for minimized material consumption, and pollution control and treatment

Two industrial case studies are presented. They were selected from a range of cases studied to identify and prove the existence of the three engineering paradigms introduced above. These cases have been chosen to illustrate the three different approaches to environmental protection outlined through the individual paradigms. Furthermore, the identification of clean technology characteristics in the cases selected has been taken as a proof that paradigms are changing in industrial practice.

Two Industrial Case Studies

Shell

A project was carried out with Shell on the abandonment of an oil platform [6]. The task was to evaluate possible abandonment options and suggest the best option to the project team. Criteria for the selection of the best option included economic considerations, safety issues, public acceptance, and environmental impact. These criteria received similar attention in the final assessment. However, difficulties for a balanced and scientific assessment were provided through the criterion of public acceptance, because of the company's experiences with the Brent Spar incident.

While the operation on the 20-year-old platform could be categorized as a dilute and disperse activity, the idea behind the abandonment could be called preventative (clean technology). The project reflects the careful assessment of the remaining life cycle stages for the particular oil platform. Thus, deconstruction, transport, and final disposal, as well as their social and environmental impacts have been considered. Possible solutions for abandonment, however, would, because of their nature, fall into the older paradigms. The discussed options of mothballing and leave *in situ*, as well as the disposal options at sea, could be classified as dilute and disperse activities, while the transportation and disposal onshore would represent the cleanup technology paradigm.

Thus, the company seems to be acting concurrently in all three paradigms; many of their activities might still be planned and executed as dilute and disperse activities, while their planning approaches and considerations about future North Sea platforms fall into the clean technology paradigm.

Lucas Industries

This research project [2] was initiated by academia, but has been fully supported by and stirred great interest in the industrial company, then called Lucas Industries plc. The agreed task was to identify possible ways of including the environmental issue into the company's own product introduction management (PIM) process. A design approach was chosen to have the best possible effect on the environmental impact created by the company's products.

Two possible approaches to the implementation of environmental design have been suggested: (1) a modification of the existing technical and commercial introduction processes could entail the redefinition of design goals and the inclusion of the environmental concerns into the checklists; and (2) the solution preferred by the academic partners meant the establishment of a new organizational unit within the PIM structure which would house the facilitating function for all designers and design teams. This unit would coordinate support, educate and train designers involved in environmental design, as well as collect and store relevant environmental and design information.

While the second solution meant to provide a new and integrated approach to design and environmental protection, i.e., a paradigm change, the first solution did not rely on such a cultural change. The adapted design checklists would point out possible opportunities for the designer to decrease environmental impacts. However, in the first case, environmental improvements would require time-intensive design-loops around the respective checkpoints. While the project findings had not been implemented at the time, Lucas has since been involved continuously in research into environmental improvement. Currently, they employ a researcher whose main task is to design an integrated system combining both environmental design and the Lucas product introduction management process. Many of the Lucas plans for the environment fall into the clean technology category. However, the difficulties of implementing a paradigm change across an organization have become visible. Lucas has to overcome natural resistance to change. Progress across the sites has been slow but is definitely visible.

Conclusions

Three engineering paradigms have been identified to guide environmental protection activities. While both the dilute-and-disperse paradigm and the cleanup technology paradigm reflect linear and product-focused thinking, the clean technology paradigm guides its followers into holistic considerations and life cycle thinking. The focus of production is the provision of services instead of the marketing of a product. Thus, clean technology could be described as a philosophy about the application and implementation of technology with the aim of achieving sustainable development.

In a further step of the research project reported here, industrial case studies on environmental improvement projects have been used to prove the existence of the three paradigms in practice. The two cases presented in this section demonstrate environmental solutions resulting from the different engineering paradigms.

It is also evident that the paradigm shift, while revolutionary concerning one individual, only gradually permeates through industrial organizations. Thus, industrial projects rarely show characteristics of only one of the paradigms. More extensive research into management of change and organizational development with a special emphasis on environmental pressures and paradigm shifts could potentially provide guidance on the (continuing) education and training of professional engineers (and other professions) pivotal to industrial development.

References

1. Clift, R., Inaugural lecture, University of Surrey, Guildford, U.K., 1995.
2. Giessler, T., Environment-friendly product/process design and development — a feasibility study on the integration of environmental issues into product introduction strategies, Research report, HDZ/IMA, RWTH, Aachen, 1994.
3. Jackson, T., *Material Concerns Pollution, Profit and Quality of Life,* Routledge, London, 1996.
4. Kuhn, T. S., *The Structure of Scientific Revolutions,* 2nd ed., University of Chicago Press, Chicago, 1970.
5. O'Riordan, T., The politics of sustainability, in *Sustainable Environmental Economics and Management,* Turner, R. K., Ed., Belhaven, London, 1993.
6. Rippon-Swaine, R., The abandonment of the Shell/Esso oil production facility — Auk Alpha, Final year project, Brunel University, Uxbridge, U.K., 1996.
7. Van der Vorst, R., Engineering, Ethics and Professionalism, Keynote lecture presented at the Annual Conference SEFI, Humanities & Arts in a Balanced Engineering Education, Cracow, Poland, 1997.
8. Van der Vorst, R., Clean Technology and Its Impact on Engineering Education, Ph.D. thesis, Brunel University, Uxbridge, U.K., 1997.

6

Model-Based Flexible PCBA Rework Cell Design

Necdet Geren
University of Çukurova

0-8493-0997-2/01/$0.00+$.50
© 2001 by CRC Press LLC

6.1   Introduction

The assembling of printed circuit boards (PCBs) is a complicated and defect-prone process because it often involves a multitude of component parts and a series of interdependent and sometimes repetitive assembly processes. There are three main types of defects: board and component defects (often quoted at 50 to 5000 ppm), placement defects (100 to 1500 ppm), and soldering defects (50 to 1000 ppm). For a board consisting of 100 components and 1000 solder joints, each has an average defect rate of 50 and 100 ppm, respectively, and the expected yield of the completed board would be 90% [22].

In most electronic companies, printed circuit board assembly (PCBA) rework is an unwelcome but routine activity. Rework is necessary not only because of the financial cost involved in scrapping faulty PCBAs, but also because of the strict contractual requirements of many assembly houses to deliver a fixed number of assembled boards based on a limited supply of bare boards and components.

PCBA rework is traditionally carried out by a group of highly-skilled operators who, with the help of magnifying glasses, identify and locate faults based on a test diagnosis. They then remove the unwanted protective coatings if any exist and solder joint connections and/or faulty components with various mechanical, chemical, or thermal techniques. The process is concluded by the replacement and soldering of appropriate wiring and components, and, perhaps, by another visual inspection and test diagnosis. Manual rework on PCBs demands strenuous visual effort and a combination of different skills. It is labor-intensive, time-consuming, and increasingly difficult to accomplish manually, even with the assistance of special equipment, because of miniaturistic components and the complexity of modern boards.

Although the number of rework systems available in the market has grown considerably over the last few years, the rework process itself has not been fully automated. The reasons for the limited automation in rework include:

- The necessity for flexibility to accommodate a wide variety of components and board designs and rather complex process requirements for their rework
- The relatively high cost of vision and associated equipment required for automated rework
- Rework is still viewed by most companies as an admission of failure in the design and manufacturing processes, even though defect rates can be as high as 35%.

An autonomous and fully automated printed circuit board assembly (PCBA) robotic rework cell has been developed, implemented, and manufactured based on system modeling techniques at the Aeronautical and Mechanical Engineering Department of Salford University in the U.K. This cell was designed to be an integrated part of a PCBA cell and to perform assembly, rework, and in-process inspection of both through-hole (TH) and surface mount (SM) electronic components with a cell controller coordinating in-process inspections and robot/tool operations. Thus, the solution could be structured and easily incorporated into existing robotic assembly cells economically.

A computer integrated rework cell was specifically designed to study the generic problems associated with the removal, replacement, and fastening of pre- and post-soldered SM and TH components. That study aimed to develop appropriate rework methods, software techniques, and hardware tooling to enable an autonomous, sensor-guided robot cell to automatically perform the rework of single-sided PCBAs in a batch size of one. The cell was designed to have maximum flexibility and be capable of:

- Identifying and locating faulty components based on automatic test equipment (ATE) and CAD/CAM data
- Planning appropriate rework operations according to inspection results
- Formulating the robot program

- Performing adequate in-process inspections
- Carrying out automated rework

To summarize, the ultimate goal was to automate entire manual rework process with maximum flexibility to cope with the rework of any single-sided PCBA on the work floor where the PCBA is already being manufactured.

The project was divided into three interdependent parts, with one research engineer concentrating on each part: (1) cell controller; (2) in-process inspection; and (3) development of methods and tooling.

As seen from the above structure of development, the cell consists of three interwoven parts that require software and process development and successive integration of these to achieve the proposed objectives. The development of the cell and the techniques used are discussed in the following sections.

6.2 Overview of Printed Circuit Board Assembly Technology

The purpose of this section is to give a general overview of printed circuit board assembly (PCBA) technology, because without knowing the current state of PCBA technology, describing and overcoming problems would be difficult. In this section, therefore, the most common types of PCBs, electronic components, and electronic assembly-related processes and techniques are described to establish an easy link to the following sections and to clarify the specific technical terms.

The automatic assembly of electronic circuits as a technology started in the late 1950s. Early versions were used primarily to insert components with axial leads into printed circuit boards (PCBs) and were ceased because of the unwanted drop-out during the assembly and soldering process, which is called through-hole (TH) technology. After the 1960s, surface mount (SM) technology evolved from the hybrid technology of TH components. In SM technology, components are placed on the printed circuit board surface and soldered. Even if TH and SM components perform the same logical function in the printed circuit board assembly (PCBA), their physical appearances are as different from one other as their manufacturing technologies.

Printed Circuit Board

Printed circuit boards may be classified into two basic categories based on the way they are manufactured: graphical and discrete-wire interconnection boards. A graphical interconnection board is another term for the standard printed circuit board, in which the image of the master circuit patterns is formed photographically on a photosensitive material such as a glass plate or film. The image is then transferred to the PCB by screening or photoprinting the artwork generated from the master. Discrete-wire interconnection does not involve an imaging process for the formation of signal connections. Rather, conductors are formed directly on the wiring board with insulated copper wire. Wire-wrap, Unilayer-II, and Multiwire are some of the best-known discrete-wire interconnection technologies [14]. The majority of PCBs are graphically produced rigid boards and may be classified into three classes as single-sided, double-sided, or multilayer boards. Single-sided boards have circuits on only one side of the board. Double-sided boards have circuits formed on both side of the boards. Multilayer boards (MLBs) have three or more circuit layers by definition. Some boards have as many as 60 layers.

Many types of copper-clad materials are available, and the features of base materials are strongly related to the PCBA process. The copper-clad laminates (base materials) most widely used in the manufacture of PCBs, however, are XXXP, XXXPC, FR-2, FR3, CEM-1, CEM-3, FR-4, FR-5, and GI [14,43]. Each of these materials has different operating temperatures (glass transition temperature) and different structures.

Surface Mount Components

SM components can be classified into two parts: passive components and active components. Monolithic ceramic capacitors, tantalum capacitors, and thick-film resistors form the core group of passive devices. Transistors, diodes, and integrated circuits form the core group of active devices. The most popular SM

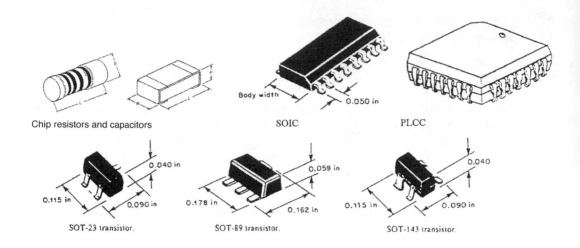

FIGURE 6.1   Examples of SM components.

packages for transistors are the small-outline packages (SOT-23, SOT-89, and SOT-143). The packages used for transistors are often used for surface mount diodes.

Surface mounting offers considerably more types of active packages than are available in TH technology. There are two main categories of chip carriers: ceramic and plastic. The plastic chip carriers are primarily used in commercial applications. The ceramic packages provide hermeticity and are used primarily in military applications. The most common type of chip carriers are small-outline integrated circuits (SOICs), plastic leaded chip carriers (PLCCs), quadpacks, leadless ceramic (LCCC), and ceramic leaded (CLCC) chip carriers. Figure 6.1 gives some examples of SM components.

Surface Mount Component Assembly Processes

The three principal SM process technologies in use (Figure 6.2) are surface mount devices (SMDs) single-sided, wave soldering, SMDs single-sided, reflow soldering, and SMDs double-sided [11].

Double-sided technology is the combination of the first and second techniques. In this approach, components are located on both sides of the PCB by using wave and reflow soldering methods. The whole manufacturing procedure is shown in Figure 6.2c.

Through-Hole Components

Through-hole components fall into categories of axial, radial (see Figure 6.3), and multiple lead components such as dual in-line (DIL), pin grid arrays (PGA), and sockets (Figure 6.4). There are other types of components such as connectors, sockets, transformers, and various passive devices as well as high-power semiconductors with their particular type of connection [14,42,43].

Through-Hole Assembly Processes

The procedure for one-sided through-hole technology is shown in Figure 6.5. As seen from the figure, the first step of the procedure begins with component feeding and orientation. When assembling new components into a board prior to soldering, the leads have to be cropped to length or, alternatively, cropped to a specific length, depending on the type of soldering machine or technique to be used. Furthermore, it is usually necessary to retain the component in the board prior to soldering, because the board will need to be transferred from the assembly machine to a soldering machine. To increase the security of those components, some form of retention is desirable. Various techniques for holding the component on the board have been

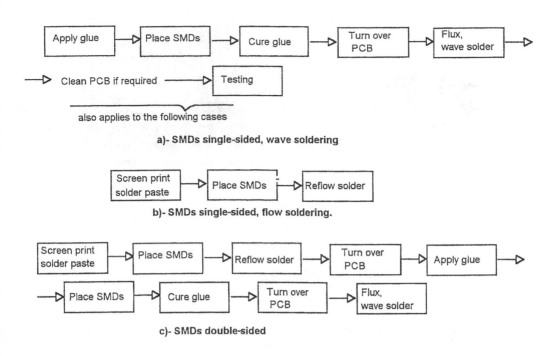

FIGURE 6.2 Possible assembly procedures for SMDs.

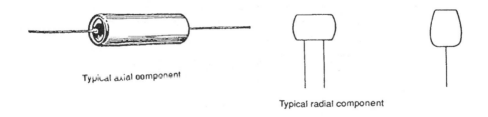

FIGURE 6.3 Through-hole axial and radial (passive) components.

tried and are still in use, such as using a plastic coating to hold the component in place. However, the most commonly used technique is to crop and bend the leads simultaneously under the board in such a way that the component cannot escape.

One alternative technique for retaining the component is to leave the leads long and process the board through soldering, using what is termed a solder-cut-solder technique [42].

The third step is in the process is where flux is applied from underneath the PCBA (not the component side) by a suitable technique before the board is sent to wave or drag solder. Before the soldering takes place, the PCBA is preheated and then soldering takes place. This procedure is followed by inspection, testing, and coating. If any defect is found after inspection, repairs are performed [42].

Manufacturing Assembly Defects and Rates

The definition of defect can be widely varied, such as a circuit unfit for the purpose for which it has been constructed, or one which departs significantly from the normal or intended condition [16]. Whatever the definition of this term is, it defines the quality that is unacceptable to contract manufacturers or to the end customer. PCBA defects [30] can be classified into three headings:

Dual-in-line   149-pin array package (PGA)

Multiple lead sockets

FIGURE 6.4   Multi-leaded TH components.

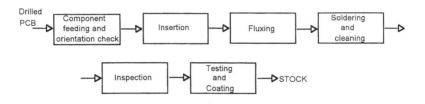

FIGURE 6.5   Typical through-hole assembly technology.

- Defects of bare PCBs (related to bare board manufacturing)
- Electronic component and connector defects (related to component manufacturing)
- Process defects

Printed circuit board assembly may fail the inspection and tests due to the following reasons:

1. The circuit functions, but does not satisfy the specifications required by the customer or manufacturer, i.e., excessive flux residue, solder balling, lack of solder, etc. (USA MIL-SPEC standards — Aerospace and Military requirements are particularly exacting.)
2. The circuit [34] fails to function due to a faulty component or components.
3. The circuit [15,41] fails to function due to a faulty solder connection, i.e., faulty solder joint, and/or faulty component positioning.
4. The circuit fails to function due to the PCB itself being defective, i.e., damaged tracks, lifted lands, through-hole vias, etc. [34].

By corrective action or other means, rework takes place on boards or components. Board rework includes repairing tracks, plating through-holes, or any repair job that is performed to recover the functionality of the PCB itself. Component rework includes removal or replacement of component, solder, and sometimes both to recover the whole circuit's functionality rather than replacing the PCB. Corrective action rework can also be classified as either board rework or component rework.

In the following sections, emphasis is only on component rework because the main subject of this study is component rework rather than board rework. From now on, PCBA rework is used to express component rework.

Despite improvements in printed circuit board assembly (PCBA) techniques, advanced control methodologies, and sophisticated inspection–test systems, defects are still common. Defect rates fluctuate between 50 to 10,000 per package million (ppm) depending on the process steps for SMT boards (99.995 to 99.00 yield in percent) [35]. It is generally recognized that defects on a PCBA cannot be completely eliminated with the current technology, and, typically the average claimed for the number of defective PCBAs varies between 10 and 35% [33].

6.3   Rework Technology and Assembly Robots

PCBA Rework Requirement

Since the 1950s, electronic component assembly equipment has been improved with advanced control methodologies, and assembly lines are equipped with sophisticated inspection–test systems, but defects are still common and defective PCBAs are reworked manually.

The PCBA rework requirement is well established at present, and this is unlikely to change, because:

- The high cost of scrapping faulty PCBAs is prohibitive.
- Only a fixed number of bare-PCBs are usually available to meet an order.
- Subcontractors must deliver a predefined number of working PCBAs from a limited number of bare-PCBs and components.

As is evident from the above scenario, rework of PCBAs is inevitable in PCBA manufacturing.

Unfortunately, there has not been any significant improvement in rework equipment. Rework equipment is still very basic and its development has not kept pace. Rework has traditionally been carried out by a group of skilled operators, with the help of various disassembly and assembly aids.

Rework has been performed by a group of skilled operators, using manual and semiautomated equipment mostly designed to remove and replace the components rather than performing the complete component rework process automatically. There is a growing need for automated rework because:

- Manual component rework is a difficult operation that requires well-trained technicians.
- Rework stations currently available cannot perform rework without people.
- Rework is carried out by people who may make mistakes and who experience interruptions, both of which reduce yield and quality.
- Increasing use of fine-pitch components makes even manual or machine-assisted rework difficult.
- The continuous move toward smaller components and smaller pitch sizes dictates that more accurate, reliable methods of rework are needed, which can only be achieved with automated systems.
- Rework requires some level of automation to keep the quality consistent since the high and costly quality standards of the assembly manufacturing process cannot be compromised by manual rework methods of lower quality.
- There is a need for a rework system that can perform rework on the basis of a batch size of one.

The rework process is still viewed as a side issue even though manufacturers consider that the quality of the reworked board is as good as the original board. Primarily, the techniques used for rework are manually assisted, and usually rely on the rework operator's judgement in the rework process.

Manual rework often introduces many problems; good joints are repaired because of inadequate inspecting, good components are damaged while repairs take place, long rework cycle times make the process expensive, and talented rework personnel cannot always be recruited. Most important, however, is that as the component pitch size on the board becomes smaller, the ability of humans to perform any rework at all is questionable. This phenomenon is driving rework equipment manufacturers to provide fine wheel-controlled X-Y positioning tables and PC controllers for existing equipment.

Industrial Rework Equipment

The equipment used in manual rework is not highly accurate. The tools are mainly hand tools. They include a selection of soldering irons, hot air guns and vacuum desoldering tips, pairs of tweezers, soldering wicks and wires, cleaning solvent and brushes, magnifying glasses, and fluxer and solder paste dispensers or other solder application tools. These are the basic types of rework tooling that have been traditionally used in rework since the existence of PCBA. Hand tools can be used to repair some of the SM components except pin grid arrays, PLCCs, and LCCCs.

The appearance of semiautomated equipment on the market has recently grown. With semiautomated equipment there is more control over the rework process compared to manual tools. The degree of automation depends on the manufacturer of the equipment. Component placement is easier with visual aid units, but the rework is still operator-dependent [9,13,43].

Two programmable machines are commercially available for the removal and replacement of SM components, but not for complete rework [5,12]. Both programmable systems are designed to handle batch production of SMD boards that require repetitive identical corrective action. Both machines are capable of removing and replacing defective items, but require an operator to set up the machine for the reworked component (time of heating, amount of heat, etc.), clean the pads from excessive solder and contamination, dispense the solder cream, and place the component in the second case.

The requirement for an automated rework system has been specified by Driels and Klegka [19] as an immediately useful device for electronic manufacturing. Since then, two reports of research have been published on the development of automated component rework [21,48], but results of actual rework success have not yet been reported.

Robots and Automated PCBA Rework

There has been a lack of international effort in the development of fully automated rework machines despite current defect rates being as high as 35% [33,35]. This is understandable since the cost of a fully automated rework cell may be as high as several hundred thousand pounds, but that is not the only option available. In the work described here, a rework cell has been developed which commercially could be an extension to a robotic PCB assembly cell, and for which much of the equipment for rework would already be available; this will make the process more economical, so it should be easier to justify the investment.

Robot applications in PCBA are increasing rapidly due to the need for small batches and quick changeover [27,49]. Consequently, the increasing use of robots and vision systems resulting from high accuracy and flexibility requirements have made them more acceptable, and the gradual reduction in their cost makes their use more feasible. Considering the robot's multifunctional ability, it is suggested that a well-designed cell could be economically deployed for the assembly, inspection, and rework of defective components.

At present, robotic assembly cells are not often utilized continually, and it would be possible to use these cells for rework as well. Even if a cell were being utilized, reducing production to carry out rework would be cost-effective because of the high cost of rework. Furthermore, since the PCBAs being repaired would probably have been assembled by the same cell, cost and technical problems associated with component and PCBA feeding, jigs, sensory requirements, grippers, etc. would not usually occur, and information about the layout of the board already stored by the cell controller could also be made available to the rework system.

6.4 Overall Development Planning

Since no attempt has been made to automate rework, and while the available technology is sufficient, the methods, tooling, and procedures suitable for automation are not known.

When considering automating a manual process using an industrial robot, choosing appropriate hardware and suitable methods is very important. Doing this effectively requires the knowledge of existing manual rework procedures, methods and tools, and available technology on automation including robots, sensors, control equipment, robots, etc. Therefore, future studies have to be performed on:

- The general electronic component rework procedure
- Rework problems and considerations
- Factors affecting rework
- The tools and systems used in rework
- Important rework steps that affect automation

Based on the findings obtained from the above studies, alternative automated rework systems can be proposed to choose the best technique so that all requirements might be fulfilled. At this stage, proposed systems can be designed in a top-down manner in sketch form and the most suitable one determined.

The determination of a suitable rework technique results in the study of core automated rework procedures, and solutions can be proposed for other rework tool requirements. After making proposals for automating each rework step, available tools shoule be analyzed to obtain the required tools for the development, because the proposed rework methods can only be achieved if proper tools and equipment are chosen.

The overall system architecture is known, and detailed automated rework procedures may be developed. These procedures provide detailed information about the activities of all rework tool devices and control systems.

After these studies, in order to produce well-engineered CIM systems, among other things, a combination of powerful system development methods and tools which allow the development of complex rework systems are needed, because the cost and complexity of the development does not permit any room for error in the system the design. In addition, the development requires considerable efforts focused on software and process development and total system integration.

The use of system analysis tools aids the development of suitable software and defines functional requirements for each rework step.

The last part of the study involves the design, manufacturing, integration, and interfacing of each piece of equipment to achieve total system integration. The steps mentioned above are followed in later sections of this chapter.

6.5   Detailed Studies of Rework

Manual Rework Procedure

The general electronic component rework procedure is given below for both SM and TH components: [40]

1. Prepare PCBA for rework: clean PCBA from dust and contamination.
2. Identify component(s) to be removed and remove obstructions and conformal coating if present. TH: Declinch the leads of the faulty component.
3. Flux the target area.
4. Preheat the local target area to below the reflow temperature to prevent thermal shock.
5. Reflow the leads/terminations of the faulty component with a suitable heater.
6. Remove the defective component with tweezers or a vacuum suction tool.
7. Clean the pads or holes to remove excess solder.
8. Clean flux from the board using a solvent cleaner.
9. SM: Pre-tin and flux or dispense solder cream to the pads. TH: Preform leads on the new TH component.
10. Pick up a new component and insert, or insert the new component.
11. Apply flux and preheat the target area.
12. SM: Reflow solder paste. TH: Solder component leads.

Factors Affecting Rework

Various parameters affect the quality and effectiveness of rework; some of these are listed below:

- Heat application: Excessive heat causes delamination of the board and damage to the component and PCBA.
- Placement accuracy: For standard SMDs (50 mil-pitch), the allowed minimum offset in X and Y is 0.15 mm, and this decreases with the rotational offset which is 0.1 to 1°.
- Gentle handling: Components have delicate legs.
- Static protection: Any static electricity damages the components.
- Seating force: Excessive force between the SM component and the solder cream-dispensed pad may cause bridging in SM rework or may bend the legs.
- Accurate solder dispensing for SMDs: Small and different sized pads require very accurate solder dispensing and the right amount of solder cream.
- Proper soldering and desoldering of THDs: Multilayer boards make proper desoldering–resoldering extremely difficult due to thick boards, different heat requirements, and multileaded components.
- Declinching of TH components: During removal, components have to be declinched properly; otherwise they may damage plated through-holes.
- Type of adhesive and heat application in the case of SMDs: Adhesive must be softened, which can only be achieved by heating the whole component during desoldering; otherwise removal of the component is difficult.
- Accessibility to defective component: Some good components may obstruct the defective item and may not allow removal.
- Flexibility of picking devices: Capability for dealing with different-sized components is necessary because some odd or nonstandard components always exist.

Thermal Considerations and Problems

The removal of components from a PCBA is principally based on the remelting of its solder joints. Thus, the most critical aspects of component removal from a PCBA are the application and control of the applied heat, the selection of a suitable method, skilled and qualified personnel in the case of manual rework, and good control of the machine and tool elements. The heat application in rework takes places in two stages, preheating and heating.

Preheating

Thermal shock is just as damaging during the repair process as it is during initial assembly. Rapid heating causes a variety of problems, including separation of lead frames from plastic bodies, broken device wire bonds, and cracked solder joints. Fracturing of ceramic capacitors is a well-documented failure mode [26,40]. The problem is even worse on multilayer boards due to the various mismatched thermal coefficients of expansion of different layers.

The problem is compounded by the fact that, during repair, heating is usually localized. This causes much larger differential stresses within the substrate than are seen during the production of soldering. Ceramic substrates and components are particularly prone to fracturing, but ordinary epoxy glass materials can also be damaged especially if they are multilayer boards. Preheating is also necessary to activate the flux.

As in the wave and reflow soldering processes, the solution to these problems is to preheat the assembly immediately prior to soldering. Ideally, the preheater should gradually raise the temperature of the assembly to within about 100°C of the solder's melting temperature. For tin–lead alloys, this suggests a preheating temperature of about 125 to 150°C. Increasing preheating temperature over this range degrades the solder joint integrity of adjacent components [36]. The rate of temperature increase for the preheating temperature should be held around 2°C/s maximum. Some SM ceramic components do not allow heat increases of more than 1°C/s [31,32].

Heating

Excessive heat can damage the component or PCB structure. Excessive heat can also destroy the adhesive bond between the resin system and the conductive patterns (delamination) consisting of the lands, conductors, and plated through-holes in the base material.

Insufficient heat will not adequately or rapidly melt the solder, thereby affecting component removal. Conductive patterns (lands, conductors, and plated through-holes) can be pulled from the PCB structure with the component by not adequately melting the solder off all the solder joint(s).

As a consequence of the above, the amount of heat needed to remelt a solder joint is critical to the component removal process. There are several heat factor considerations that need to be understood in order to perform optimum component removal and replacement. One of these factors is the substrate (PCB) itself. PCB substrates are constructed from many different materials, from epoxy glass to ceramic, and differ in their number of layers, ground planes, and quantity and distribution of components. Other factors are the mass of the solder joint, the mass of the component, the mass and thermal characteristic of the heat source, and the thermal coupling or linkage between the solder joint and the heat source. These differences produce different thermal loadings, which must be considered when developing an automated rework cell [28,46].

Most of the SM components that are fabricated from alternating layers of conducting material and dielectric are designed to withstand soldering temperatures of 260°C for up to 10 s, so the maximum heating temperature should not go above 260°C [20]. The maximum rise of the temperature of the component body should not exceed 5°C/s during heating [1], and TAB components should not be taken above 220°C [40].

6.6 Determination of Reflow Methods and Automated Rework Techniques

The Effect of Reflow Methods on Automation

Studies of manual rework tools and activities have indicated that:

- The general rework procedures for TH and SM components are totally different.
- TH and SM component desoldering and resoldering require different tools.
- Desoldering and resoldering tools and methods determine the rework procedures for each component technology.
- There are many desoldering and resoldering techniques that can be used, but none is perfectly suited to each component technology (see Tables 6.1 and 6.2 for a comparison of the current removal and replacement methods for SM and TH).

All the above information is important for the development of a fully automatic robotic rework cell, but the different tools required merit special attention because the different component rework technologies do not allow the use of one heating (reflow) method for component removal. As a consequence, any development for automating rework must consider how effective heating can be achieved using as much common equipment as possible so that economic requirements might be fulfilled.

Because the heating methods chosen completely condition the layout, the procedures, the methods of rework, the form of the automation, and the design and manufacturing of the rework cell, the determination of appropriate desoldering and resoldering methods is not only important for the quality and reliability of rework; it also conditions the development of the whole rework cell.

An added complication is that studies have been restricted to considering existing heating methods and techniques and adapting them to suit automation. This is inhibiting because the insufficiency of

TABLE 6.1 Comparison of the Present Removal and Replacement Methods for SM Component Rework

	Soldering Iron (SI)	SI with Reflow Block	Heated Tweezers	Hot Gas Pencil	Directed Hot Gas/Air	Hot Bar	Focused IR	Unfocused IR	Laser Soldering	Hot Plate/Belt	Vapor Phase	Infrared Reflow Oven
Resoldering	√	No	No	√	√	√	√	√	√	√	√	√
Desoldering	√	√	√	√	√	√	√	√	No	√	√	No
Recommended items to repair	Group 1	Groups 1, 2, 3, and 5	Groups 3 and 4	Groups 1, 2, and 3	Groups 3, 4, 5, and 6	Groups 3, 4, 5, and 6	Groups 4, 5, and 6	Groups 1, 2, 3, 4, 5, and 6	Groups 1, 2, 3, 4, 5, and 6	Groups 1, 2, and 3	Groups 1, 2, and 3	Groups 1, 2, 3, and 4
Equipment cost for full set	£500–1500	£1000–2500	£1500–3000	£500–1000	£5000–30,000	£10,000–40,000	£10000–	£5000–	£100,000–	£500–	–	–
Speed of reflow	Very slow	Fast	Very fast	Fast	Fast	Very fast	Fast	Fast	Very slow	Fast	Fast	Fast
Appearance of rework	Acceptable	Good	Good	Good	Very good	Very good	Very good	Very good	Very good	Good	—	—
Geometric limit	√	√	√	None	√	√	√	No	No	No	No	No
Needs other equipment for solder cleaning	√	√	√	√	√	√	√	No	√	√	√	√
Dedicated tooling	Some tips	√	√	No	√	√	√	Four only	No	No	No	No
Tool mark	√	√	√	No	No	√	No	No	No	No	No	No
Degradation of all components by intermetallic growth	No	No	No	Some nearby	No	No	No	No	No	√	√	√
Adjacent component reflow	No	No	No	Possible	No	No	No	Possible	No	√	√	√
Needs desoldering/resoldering equipment	Except group 1	Resoldering	Resoldering	None	None	None	None	None	Desoldering	None	None	Desoldering
Double-sided PCBA repair	√	√	None	√	√	√	√	√	√	No	No	No
Adjacent item disturbance	None	None	None	Possible	Possible	None	None	None	None	√	√	√
Suitability for future SMDs	Not for multi-lead	Not all types	Not all types	Not for multi-leaded	With extra tool	With extra tool	With extra tool	√	√	May be some	May be some	May be some

Group 1: Chip capacitors, resistors, etc;; **Group 2:** Transistors, diodes, etc;; **Group 3:** SOICS; **Group 4:** PLCCs; **Group 5:** QFPs; **Group 6:** LCCCs.
Source: Reference 24. With permission.

TABLE 6.2 Comparison of the Present Removal and Replacement Methods for TH Component Rework

	Soldering Iron and Braids	Solder Extractor [Vacuum Desoldering Iron (V.D.I.)]	Solder Flow Bath	Hot Clamp	Heater Block
Resoldering	√	No	√	No	No
Desoldering	√	√	√	√	√
Recommended items to repair	Group A	Groups A and B	Groups A, B, C, and D	Groups A, B, and C	Group A, B, C, and D
Equipment cost for full Set	£50–500	£1000–2000	£3000–8000	—	—
Speed of reflow	Very slow	Slow	Very fast	Fast	Fast
Appearence of rework	Acceptable	Good	Good	Good	Good
Geometric limit	√	√	No	√	√
Needs other equipment for hole cleaning	No (but only suitable single-sided PCBAs)	No	No	√	√
Dedicated tooling	No	No	No	√	√
Tool mark	√	√	No	No	√
Degradation of all components by intermetallic growth	No	No	No	No	No
Adjacent component reflow	No	No	Possible	No	Possible
Needs desoldering resoldering equipment	No	Resoldering	No	Resoldering	Resoldering
Double and multilayer PCBA repair	No	√	√	√	√
Effects to SMDs on double-sided PCBAs	No	No	√ (some)	No	√
Adjacent item disturbance	No	No	No	No	No
Replacement and adjustment of tooling heads	No	No	Nozzle	Special grippers	Heater tool change

Group A: Axial, radial components; **Group B:** Small DIPs-SIPs; **Group C:** Large DIPs-SIPs; **Group D:** PGAs.
Source: Reference 24. With permission.

existing heating methods may limit the rate of component rework and result in uneconomic conditions, which might be overcome by new methods.

Reflow Method Selection

Manufacturers of current equipment have tried to extend the applicability of various heating methods to the reflowing of solder. These range from the use of simple soldering irons to infrared (IR), hot gas, solder fountains, and sophisticated laser systems. Unfortunately, there is still no single method that is appropriate for all devices, and many of the techniques are not applicable to some components due to the nature of the components and the associated PCBs (Tables 6.1 and 6.2).

Directed hot air/gas and unfocused and iris-focused IR methods were determined those mainly used in SM manual rework (Table 6.1), while the solder fountain was found most appropriate for use in TH rework (Table 6.2). Vacuum desoldering iron and soldering iron couples can also be used for TH rework since the soldering irons have been automated [50].

Because there are no other suitable methods for reflowing solder joints, it is worth considering the applicability of the main reflow methods to automation. This approach allows the use and adaptation of existing rework tools which have been developed over many years, thus reducing the time necessary to develop the fully automatic rework cell. It also allows current methods to be improved by automation, because the limitations of current equipment are well known.

The suitability of each reflow method for automation is dependent on many factors; the most important ones are listed below:

- Cost of removal and replacement equipment
- Suitability (technically and economically) for robotic rework
- Ability to perform both desoldering and resoldering operations with the same heat source for all components
- Reliability of rework
- Possibility for temperature gradient control
- Amount of tool, head, or nozzle changes needed
- Potential for disturbance of adjacent components
- Ability to remove and replace components by using assembly grippers
- Speed of rework

Alternative Proposals for Automation

The suitability of appropriate reflow methods for automation has been studied based on the above criteria and some alternative proposals are made in the following sections.

Hot Air/Gas-Based Surface Mount Component Rework

Hot air/gas is one of the primary methods used for removing and replacing SM components in manual rework. Hot air/gas units are not expensive, so they can be economically justified for use in automated rework.

There are various possibilities for integrating this method into a robotic assembly cell in order to perform removal and replacement of components, but the most suitable is one designed in such a way that the hot air/gas unit is attached to the robot's end effector (Figure 6.6). In this method, the robot's end-of-arm tools may be equipped with an interchangeable nozzle head assembly, which may incorporate elements such as suction tools. During rework, the robot's end-of-arm tools are located above the defective component, and the component may be heated and removed. Replacement may be carried out in reverse order. This method is technically feasible, but the number of nozzles required represents a major investment. The hot air/gas method can require as many as 60 nozzle heads to cover all SM components and new package designs may also require new nozzles. For every nozzle, a tool parking facility is needed;

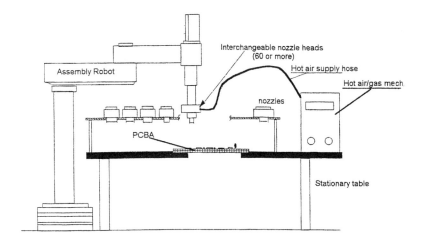

FIGURE 6.6   Hot air/gas method incorporated to robot's end effector. (Reference 24. With permission.)

automatically changing nozzles can also be very costly due to the high cost of tool changers (usually $300 or more each) despite the relatively low cost of the nozzles. Furthermore and very significantly, a large number of tool changers will occupy a large portion of the robot's envelope which is then not available for other tasks.

Iris-Focused Infrared-Based Surface Mount Component Rework

The use of IR is increasing in manual rework because of its many advantages. Usually, a 150 watt halogen light bulb is used to develop 1 to 1.2 μm short-wave IR light. The light is focused through lenses and the spot size of the heat source is manually adjusted through an iris ring. To cover different sizes of SM components, the system requires four lenses, and continuous adjustment of the spot size is provided for each lens.

As a heat source, this method is less expensive than most other techniques (Table 6.1). It has good potential for automation. The most suitable method for use in automated rework is fixing the heat source in the work envelope of an X-Y positioning table and moving the PCBA into position under the source (this allows parallel activity, enabling the manipulator to do other work at the same time). The operation principle of this method is simple. The PCBA is located on an X-Y positioning table, and the table is driven to a location where the IR heat source is positioned above the table. The defective component is heated until all joints are desoldered and then the positioning table moves the defective component to a safe picking location where the robot can pick up the defective item (see Figure 6.7). Resoldering may be performed in reverse order. This technique is feasible in a technical sense, but the requirement and economics of an X-Y positioning table must be considered. The iris-focused IR unit is capable of performing both desoldering and resoldering operations for almost all SM components. Closed-loop control is possible, and the temperature rise may be closely controlled, which should eliminate component and PCBA damage.

Soldering Iron–Vacuum Desoldering Iron-Based Through-Hole Rework

Soldering irons (SI) and vacuum desoldering irons (VDI), are popular in manual rework, but they do not satisfy the needs of automated rework. However, if these methods are combined and an automated version of a soldering iron designed for robotic soldering applications is used, most TH component rework can be accomplished. This method relies on the use of a vacuum desoldering iron to desolder each leg and then suck molten solder sequentially from each joint, and a soldering iron and flux-cored

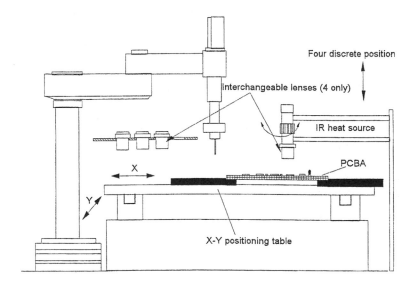

FIGURE 6.7   Iris-focused IR rework method. (Reference 24. With permission.)

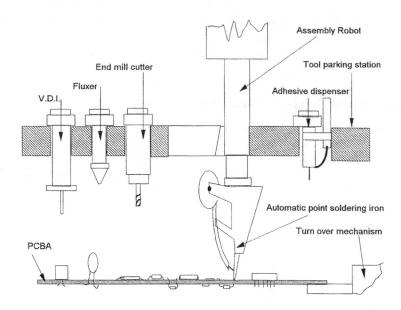

FIGURE 6.8 Soldering iron vacuum desoldering iron-based TH rework. (Reference 24. With permission.)

solder wire (self-fluxing soldering) which is fed continuously to the end tip of the soldering iron to form a solder joint. Continuous solder wire fed soldering irons have been used for a long time to solder odd-form components [47] and in other robotic soldering applications. Furthermore, they can be operated by computer control and carried by most industrial robots [10].

To automate TH rework using an SI–VDI couple, first the SI, the VDI, and other tools such as the fluxer, the adhesive dispenser, and the leg cutter must be attached to the robot's end effector by some means to enable tool exchange. Second, this method requires a fixture with automatic turn over facilities to turn the PCBA upside-down (Figure 6.8).

This method is both technically and economically feasible, but other criteria must also be considered. This method is prone to errors because total removal of solder from each hole is not guaranteed. If a sweat joints occurs, removal of the component results in damage to the plated through-holes or possibly to the PCB itself. Most SIs and VDIs are equipped with closed loop temperature control facilities, but this is not enough to determine whether the joints are desoldered or resoldered.

Solder Fountain-Based Through-Hole Rework

The solder fountain is the only method that can perform both resoldering and desoldering for most TH components using one heat source; this method is successfully used in manual TH rework. The solder fountain is based on the wave soldering principle and incorporates a set of nozzles that can be easily raised and lowered through which molten solder is pumped. The wave generated has the shape of a fountain and is controlled by the pump speed, the nozzle height, and the nozzle shape. This method is a good candidate for automated rework and a possible methodology is given below:

1. The PCBA is fixed to the table in a position with the components facing up.
2. The positioning table aligns the legs of the defective component above the leg cutter and cuts them.
3. The target area is located above the fluxer and flux is applied by spraying.
4. The positioning table aligns the defective component above the nozzles of the solder fountain and when molten solder melts the joints, the manipulator removes the component from the PCBA.

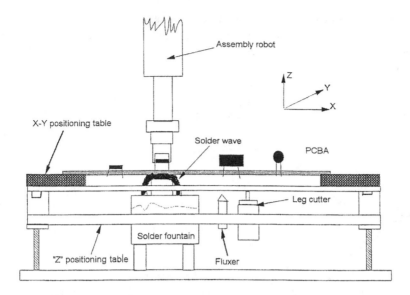

FIGURE 6.9   Solder fountain-based TH rework. (Reference 24. With permission.)

5. Replacement takes place immediately after component removal and the manipulator discards the defective component and picks up a new one.
6. The positioning table aligns the target area above the fluxer and flux is applied.
7. The target point is located above the solder fountain and the solder fountain is operated to melt the solder that is left in the holes after the defective component has been removed.
8. The manipulator places a new component into the holes when the solder in the holes has melted.

The schematic of this method is shown in Figure 6.9; this method is technically possible but the cost of an X-Y-Z positioning table must be justified.

Comparison of Alternative Reflow Methods

A comparison of hot air/gas-based and IR-based SM rework methods showed that the IR-based rework method was found to be more suitable than the hot air method for the reasons stated below [24]:

1. The IR method requires only four interchangeable parts (lenses) whereas the hot air/gas method requires 60 nozzles. The nozzles also need interchangeable tools which facilitates mounting them on the manipulator's end effector. This arrangement would cost approximately $30,000. Despite the requirement of an X-Y table, which costs around $12,000, for the IR method, the hot air/gas method will cost much more.
2. The IR method uses the robot envelope more economically, due to the elimination of nozzles and nozzle parking facilities.
3. The IR method desolders and resolders all SM components; the hot air/gas method blows method small components away.
4. The IR method allows closed-loop control of component temperature because there is a sufficient gap through which remote temperature measurements can be taken. This may eliminate one of the most important concerns: heat damage to components and PCBs. Furthermore, closed-loop control eliminates the need for experimental studies to determine heating characteristics of boards, components, etc.
5. There is no potential for the displacement of surrounding components which may create further rework problems.

6. Since there is no moving gas, there is no built up of static.
7. The method allows the use of the assembly robot's gripper and vacuum suction heads, which can be considered a major economic savings compared to the use of nozzles, which are all equipped with vacuums.
8. Despite the requirements of spot size adjustment and lens changing, the IR method eliminates nozzle head design and manufacturing problems which are currently concerns for hot air/gas systems.
9. Because there is closed-loop temperature control, the reliability of reworked joints or PCBAs should be better.
10. Noncontact heating eliminates tool marks.

The disadvantages of this method are that reflowing of nearby joints on closely packed PCBAs is possible, and reflow may occur at higher temperatures due to the use of short-wave IR.

The solder fountain and VDI–SI couple-based reflow methods have been compared. The study revealed that:

1. Solder fountain-based TH component rework is more effective because solder joints using a solder fountain are reliable joints, and removal can be performed more safely due to the elimination of sweated joints. Furthermore, exclusion of sweated joints also eliminates the need for secondary heating, which might degrade the PCB and take extra time.
2. Even though the solder fountain method requires an X-Y-Z positioning table, the requirement of an X-Y positioning table for all SM rework leaves only the need to justify a Z positioning table. This can indeed be justified, because the VDI–SI method requires a board reorientation facility.
3. The use of a solder fountain and X-Y-Z positioning table eliminates the scattering of metal parts on the PCBAs during leg cutting (removal).
4. The simultaneous desoldering and resoldering of the joints of the defective component decreases the rework cycle time compared with sequential removal and resoldering of each leg.
5. No adhesive dispensing is necessary. This also eliminates adhesive curing and dispensing facilities.
6. A solder fountain is less expensive than VDI–SI tooling.

If double-sided mixed technology PCBAs are reworked, the method also has some disadvantages. If any insulated wire jumper exists in the vicinity of or just below a defective TH component, it may melt and cause shorts. If any SM components, such as micro-tabs, bobbins, or micro switches, which are not resistant or suitable for wave soldering, are located in vicinity of or just below a defective TH component, these can be damaged during removal. Therefore, these components and jumper wires would need to be removed during the preparation of the PCBAs for rework.

6.7   Determination of Other Rework Requirements

Once decisions are made on the reflow methods for TH and SM components, analysis and decision making based on rework requirements can be carried out to specify the other rework tooling necessary for automation, and other related problems could be solved.

Rework Tools

To achieve fully automated rework, the following needs to be considered [25].

Preparation of PCBAs for Rework

The cleaning of PCBAs and removal of conformal coatings are not necessary because the cell is designed to rework assembly defects directly after manufacture when they are not covered by dust or other contamination and conformal coatings (PCBAs are coated after the final test of the finished goods) and are not subject to field defects. The other problem when preparing a PCBA for rework is that of removing obstructions (i.e., heat sinks, jumper wires, threaded fasteners, other components, etc.). The first three must be removed manually because of their method of attachment to a PCBA, but they are relatively easy to remove and they represent only a small proportion of board components. Other obstructions can be removed by the rework cell and replaced after the defective component has been repaired.

Underside Heating

To protect the PCBA from localized heat shock in order to reduce delamination and activate the flux, underside heating is essential in rework. A hot air device was found to be appropriate for this task.

Cleaning Solder from Pads

After a defective SM component has been removed, the pads of SM components must be cleaned of excess solder so that an accurate amount of solder can be put on each pad before a new component is positioned. Uneven distribution of solder endangers placement accuracy and the beads and spikes that are often left on the pads when a component is removed may cause tilting of the new component. Pad cleaning can be achieved using a vacuum desoldering iron (VDI) that can either be attached to the assembly robot's arm using interchangeable adapters or a located in a stationary position; the cleaning is performed by the X-Y-Z positioning table.

Solder Cream Dispensing

Soldering of SM components requires solder to be dispensed as paste. Solder pastes are dispensed in several ways including syringes, pressure-fed reservoirs, or guns. Since the robot must be capable of handling more than one tool during rework, the assembly robot can be equipped with an interchangeable tool to handle the solder dispenser and dispense solder cream.

Component Fluxing

Fluxing of the defective component and its location area is essential during rework and must be applied to solder joints of a defective component to improve heat transfer, dissolve the oxide on the joints, and allow the molten solder to wet the chemically clean surface. Since rework requires local area fluxing and effective control of the amount of flux dispensed, spray-type fluxers were found to be the most suitable for the cell.

Declinching of Through-Hole Legs

The protruding clinched legs of defective TH components must be cut off to remove TH components from boards. This process can be achieved using an end mill cutter.

Removal of SM Components

Because adhesives are not used for single-sided PCBAs, surface mount components may be directly removed using suction tools when the solder is molten.

Post-Desoldering and Resoldering Cleaning

After the removal and replacement of components, the defective area is usually cleaned with a decreaser and brush to remove flux and other contaminants. This step can be eliminated from consideration because when rework is being carried out in conjunction with assembly, this facility will already have been provided.

Heat Control

A very important consideration in rework is the safe application of heat when reflowing solder. Two programmable proportional integral derivative (PID) temperature controllers have been used for closed-loop control of IR top heating and hot air bottom heating, and this has proved to be effective regardless of heat load. Figure 6.10 illustrates this arrangement. The bottom heater control system is also used for TH rework.

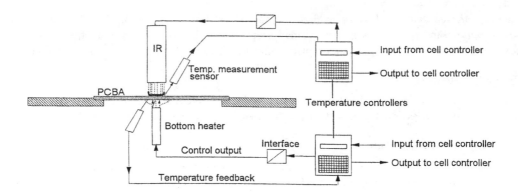

FIGURE 6.10   Schematic of IR and bottom heater control mechanisms. (Reference 25. With permission.)

TH solder joint desoldering/resoldering has different requirements than that of SMs. Although pre-heating requirements can be met using closed-loop bottom heating control, there is a requirement for the detection of the reflowing of solder joints. This is an added complication because the reflow of TH solder joints cannot be detected with sensory tooling. The only option considered viable was to perform experimental work to determine the time needed to reflow each type of joint so that this information could be passed to the cell controller which could then monitor and control the process.

Temperature Measurement

Two temperature measurement sensor requirements have been identified: one for bottom heating control and one for IR heating control. For bottom heating control, a contact-method temperature measurement sensor is used, while the IR top heating temperature measurement is achieved using a noncontact temperature sensor. Therefore, the IR heating process is not affected by a probe of a contact-type measurement sensor.

Sensory Control System

A rework cell should be equipped with assembly sensors for component verification, placement force sensing, etc. Additionally, since PCBA rework exhibits much more process variability than PCB assembly, factors such as board warping due to the weight of assembled components or heat distortion during soldering, etc. need additional sensors to ensure that:

- Fiducial marks on PCBAs are registered; this is more important here than with component assembly because of the involvement of post-desoldered PCBAs which may be deformed or suffer dimensional changes due to heating during desoldering.
- Defective components are located and checked for the type of fault, access to the component, and obstructions like heat sinks.
- After SM component removal, the post-desoldered area is free of missing pads, detached component legs, etc.
- After all SM pads have been cleaned of excess solder, pads are free of solder beads, etc. which may endanger component placement.
- Solder cream flows.
- After solder cream has been dispensed, all the pads have solder cream.
- New SM components are registered relative to the manipulator and the target site. This also includes checking component presence and coplanarity of the legs if it is a small outline integrated circuit (SOIC) or other multileg component.
- The end quality is checked for bridging, misplacement, and other defects.

Because of the intrinsically unstructured nature of the PCBA rework environment, in-process inspections are essential to provide regular information updates. At various points in the rework process, the cell controller has to make decisions on the basis of the uncertain results of the previous actions, e.g., whether all the solder has been removed from the pads, what the exact locations of the defective component and tools are, etc. An inspection system with three cameras to check both sides of the PCBAs and to register new components on the robot's end-of-arm tooling is necessary [23].

Other sensors are needed to

- Monitor the forces exerted in shear (for glued SMDs on double-sided PCBAs) and axially (through-hole) so that these may be compared with preset threshold values. This is especially important for TH component removal; otherwise, during removal, the PCBA may be damaged if there is a joint which has not been desoldered on the component being removed.
- Detect the local deflections of a PCBA which may not be flat due to warpage or the weight of components. This knowledge is required for leg cutting, solder dispensing, and solder fountain operations.

Supervisory System

Examination of the rework process has revealed that although most of the activities are carried out in a predefined sequential manner, rework often requires the simultaneous activation of several pieces of equipment and involves the use of process knowledge either stored in the system or acquired during rework. A knowledge-based rework process planning system is necessary to retrieve process information and to invoke task-oriented rework and inspection routines. A major feature of the planning system should be its ability to operate dynamically in real time prior to and throughout the rework process.

In order to cope with the proposed batch size of one, the application software should be generic and capable of PCBA rework without human involvement. This may require extensive data support which may come from:

- Automatic testing equipment (ATE) to determine and identify defective components
- CAD/CAM data specific for reworked PCBAs identifying and giving the positions of the defective and surrounding components
- In-process sensory devices that may supply various pieces of information such as type of defect, obstructions, location of post-desoldered defective component, etc.
- Preexisting PCBA data pools that may supply various data requirements such as rework procedures and detailed data concerning tool locations, etc.

The cell controller has to adjust the system so that it can accommodate changing conditions automatically, and the target planning kernel needs to be dynamic based. In addition to the above, the rework cell requires a PC and input/output (I/O) board in order to perform supervisory tasks and control rewoork tools remotely [23].

6.8   Development of Core Automated Rework Procedures

Based on the knowledge of manual rework activities, available tools, and requirements of automation, automated core rework procedures were developed for both TH and SM components.

Two rework procedures were developed because there are two types of component technologies, and each technology requires completely different tools and rework methods. Figure 6.11 shows the predefined core rework procedures from which automated rework methods may be developed. As evident from Figure 6.11, each time inspection and sensory input is received, the cell controller makes the appropriate decision to carry out rework. Therefore, the rework is not totally performed in sequential manner. Some stages may be repeated and some eliminated depending on the rework requirement of each defect and the particular situation during rework.

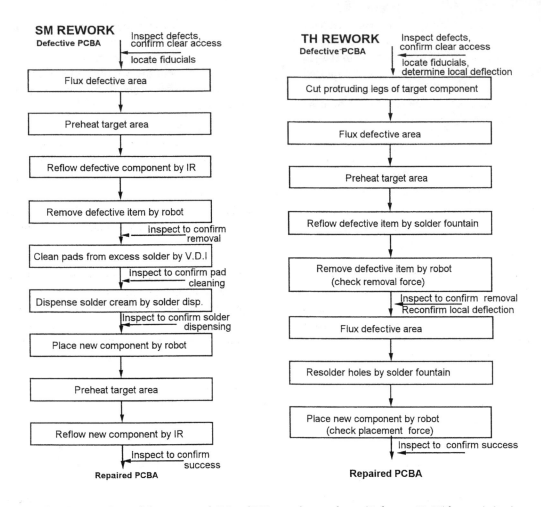

FIGURE 6.11   Core of the automated SM and TH rework procedures. (Reference 25. With permission.)

6.9   Equipment and Rework Tooling Selection

The proposed automated rework methods can only be achieved if proper tools and equipment are chosen or manufactured to suit the specific needs of each method, because the equipment is as important as the methods themselves. Therefore, the available tools should be analyzed to show the availability of the required tools, to illustrate what needs to be developed, and to determine further automation requirements [23].

Manipulating Devices and Controllers

An Adept-One robot is used as an industrial assembly robot. It is a four-axis, direct drive, *SCARA* (selective compliance arm for robotic assembly) configuration robot. It covers a large circular envelope with a 9 inch (228 mm) inner radius, a 31.5 inch (800 mm) outer radius, and a 7.7 inch (195 mm) vertical stroke with a 13.2 pound (or 6.2 kg) nominal payload.

An open frame Z positioning table is designed and manufactured. Figure 6.12 presents it schematically. The elevation table is driven by four lead screws, with an accuracy of ±0.05 mm and a resolution of ±0.001 mm, that are located in each of the table's corners. A stepper motor manipulates the system.

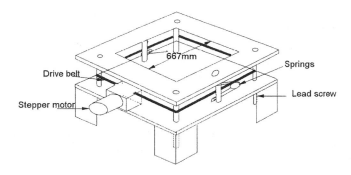

FIGURE 6.12 "Z" positioning table.

An open frame X-Y positioning table that meets the rework requirements is procured. It has a 660 × 660 mm wide open frame and a 305 mm travel (total of 610 mm travel on X and Y). It is equipped with the standard lead screw drive stages having accuracies of 4 μm/25 mm and resolutions of 1 μm. The table is equipped with DC motors.

Accuracy requirements of electronic component rework and assembly were considered for the selection of the manipulating devices.

Reflow and Resoldering Devices

Iris-Focused IR Heat Source

An iris-focused manual IR unit is the only option, and it is adapted for the cell. The infrared source/lens system is comprised of an infrared source, an upper lens assembly, an iris assembly, a lens attachment, and a control system. The 150 watt IR source is located in the lamp housing at the top of the unit. The source emits short wave IR radiation at 1 to 1.2 μm wavelengths depending on the power adjustment. The purpose of the upper lens assembly is to collimate the radiation from the IR source into a formative beam before projection through the iris assembly. The iris assembly is located between the upper lens assembly and the lens attachment and serves, in conjunction with the lens attachment, to vary the diameter of the IR beam (spot size) to suit the component to be removed. The lens attachment focuses the beam into a spot. The spot size is determined by the divergence angle of the lens attachment. Four interchangeable lens attachments are available to cater to different sizes of components. This feature also requires automatic lens attachment facility. Each of the lenses requires a different distance between the lens and the PCB [7].

Solder Fountain

An adjustable segmental nozzle type solder fountain can be used. The machine is based on the wave soldering principle and incorporates a set of nozzles through which molten solder is pumped. The wave generated has the shape of a fountain and is controlled by pump speed, nozzle height, and shape. Nozzles are raised and lowered with turning the screws located in each nozzle with a screwdriver. Although the mechanism looks crude, the design of the nozzle arrangement provides a constant solder wave height regardless of the number of nozzles raised or lowered. Variable nozzle areas range from 12 × 12 mm to 110 × 25 mm through nine nozzles [16].

Vacuum Desoldering Iron

Vacuum desoldering irons (VDI) use a combination of controlled heat and continuous vacuum in the same tip, which is hollow and heated. Some of the specifications are: closed-loop temperature control (idle temperature stability ±1.1°C), operating temperature range of 316 to 482°C, installed total power of 150 watts (0.65 A), 0.11 kg head weight, and a solder collection chamber [8].

Bottom Heater (Preheater)

A hot air blower device is chosen as a bottom heater to meet the underside heating requirements as determined. It consists of a built-in heater, an air pump, and built-in electronics. Some of the features are the operating temperature from 20 to 600°C, the air volume of 30 liter/min, and a 260 watt total performance with overall dimensions of 36 × 200 mm.

Reflow Control Devices

For the control of SM component heating processes, a PID type temperature controller (three term) is used. It has a 200 ms sampling period, 4 to 20 mA (DC) input and output, and two alarm functions (process variable high and low limit — two settings). An on/off type temperature controller with K-type thermocouple input (K-type: nickel chromium/nickel aluminium), relay output, and two set points is deployed for bottom heating control. In addition, a K-type thermocouple and noncontact temperature measurement sensors are obtained to meet the specifications of the controllers and to achieve closed-loop control. The noncontact temperature measurement sensor uses a spectral response of 8 to 14 μm to measure the surface temperature of SM components and provides analog output (4 to 20 mA DC).

Other Rework Tools

End Mill Cutter

A continually adjustable (0 to 18,000 RPM) electric motor with a mandrel attachment was used for cutting the protruding TH component legs.

Solder Dispenser

There are three methods for dispensing solder cream. These are the air-driven syringe, the peristaltic pump, and the positive displacement pump. The positive displacement method is a more reliable method and there are two different types of positive displacement possible: positive displacement pinch action dispense head with micrometer adjustment and rotary displacement pump.

Since the primary limitation of the positive displacement pump is its inability to easily change the volume of paste dispensed, a rotary displacement pump is procured. It dispenses solder cream with positive displacement action using a rotary feed screw principle. The solder cream is held in a feed reservoir under a positive head of air depending on the viscosity of the material. This positive air pressure, supplied by the air line, forces the fluid material out of the barrel, into the vertical feed shaft of the valve body, and then through the angled feed shaft to the feed screw chamber. Material flow from this point to the dispense point is controlled by the feed screw rotation in feed direction. The feed screw is driven by a DC motor (see Figure 6.15b). Applying a DC voltage signal to the DC motor rotates the feed screw, and the solder cream is forced out of the dispense tip [14]. The solder cream is dispensed to pads by providing a suitable gap and relying to the tackiness of the solder cream to stick to the pads.

Fluxer

Two normally closed, air-actuated, needle-type liquid spray valves are employed for bottom and top fluxer applications with an operating frequency of 400/minute. The area of coverage is determined by the nozzle size and the distance between the nozzle and the work surface. The spray area coverage ranges from 0.2 to 3.0 inches in diameter for the nozzle distance range of 1 to 12 inches.

Auxiliary Robot Tools

Quick-exchange end effector mechanisms are used to automatically exchange components and place heads and the solder dispenser on the robot arm. The choice of the end effector exchange mechanism determines the tool parking requirements for the robot. Tool parking facilities are identified and provided for tool parking of SM and TH grippers and suction tools, spare IR lenses, and of solder dispensers.

The vacuum suction heads are manufactured to remove and place SM components. Placement and removal of TH grippers require a parallel jaw gripper with pusher action. These are not directly available in the market; a few are made to meet specified TH component placement requirements by a specialist company. In addition, a special gripper combining the removal and replacement features in the same gripper-finger body is developed for axial TH components. Due to financial constraints, the components are supplied to the robot from component kitting magazines rather than feeders and magazines, and removed defective components are discarded into a container.

Control Equipment

Assembly Robot Controller

The Adept-One robot is operated using an MC controller. The current configuration enables one robot control task and up to 15 other process control tasks to be executed simultaneously. This greatly simplifies the integration and programming tasks. It also provides 32 input and output port [3].

The Positioning Controller

A unidex 14, PC bus-based, multiaxis, motion controller is chosen to manipulate the two DC motors of the X-Y positioning table and two stepper motors allocated for the Z positioning table and iris spot size adjustment of the IR heat source. The Unidex 14 gives PC/XT/AT, PS/2-30, and compatible PC users the ability to integrate motion control into the PC bus. It occupies one full slot on the PC bus and meets all IBM input/output (I/O) standards. Any language which has the ability to do I/O, such as Basic, C, and Pascal, can be used to program the Unidex 14 [2].

Sensory and Supervisory Control Devices

The vision system used for rework inspection is a Pi-030 68030/68882. It is based on a Motorola MC68030-20 CPU processor running at 20 MHz, with an MC68882-20/68003 floating-point maths coprocessor, 1 Mbyte of DRAM, and with a frame store capability of 1 Mbyte, which can be configured in several ways including four frames of 512×512 with 256 gray levels. The system is configured so as to cross-compile C programs written on an IBM PC compatible or a SUN, VAX, or similar workstation, and download the compiled code via an Ether Net connection. A quartz–halogen light projector with collimators and a Nikon lens are also procured for image processing.

For the cell controller activities, a 386 PC, with a 130 Mb hard disk and 16 Mb of RAM is obtained and used in the development.

A 24-channel digital input/output PC add-on board with a remote solid-state relay subsystem is obtained to perform the on/off switching of rework tools through various solenoids or relays that activate tools and to enable binary sensory inputs. The board occupies one of the PC extension slots in the cell controller PC, and it is configured to have 4 input and 20 output channels.

6.10 Subsidiary Automation of Tools

Since purchased rework tools are for manual rework, some require alterations to allow automation. In order to adjust the spot size of the IR heat source for a component reflow, the iris ring must be manipulated automatically to control the size of the spot. This has been done by employing a stepper motor, timing belt, and pulley mechanism (see Figure 6.13, side view). Control of the stepper motor is achieved by the fourth axis of Unidex 14S controller.

Z motion of the positioning table is used to set the focusing distance between the IR heat source and PCBA since each of the IR lenses requires a different focusing distance. The noncontact measurement sensor is also targeted at four discrete positions by using two micropneumatic actuators, as shown in Figure 6.13, front view.

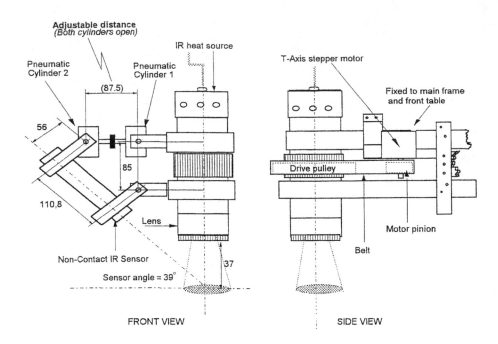

FIGURE 6.13   Angle adjustment and automatic spot size adjustment mechanism and dimensions.

Solder Fountain — The nozzles of the solder fountain are to be raised and lowered by turning the adjustment screws, which are located inside each nozzle. This can possibly be automated by following the manual adjustment pattern so that a pneumatically or electrically driven, torque-controlled screw-driver can be mounted to the robot's Z actuator using a rotary actuator, which would bring the screwdriver into action when required.

Fluxer — The distance between the fluxers and the workpiece is adjusted by manipulating the Z positioning table so that a continuous spot size increment from 6 to 76 mm in diameter is obtained.

Vacuum Desoldering Iron — The vacuum desoldering iron (VDI) is one of the devices that must be lowered and raised to clear the working space of the X-Y-Z positioning table. The most economic way of achieving this requirement is to use pneumatic manipulation. The iron system is attached to a non-rotating linear actuator together with a compensation mechanism designed to protect boards from being damaged by overpositioning the positioning table. Figure 6.14a illustrates the schematic of the mechanism.

Leg Cutter — There are four pieces of rework equipment which have to be located underneath the table: the solder fountain, the bottom fluxer, the bottom camera, and the leg cutter. The operation of the bottom fluxer and the camera are remotely achieved without any contact with the PCBA; therefore, they are positioned below the other two devices at the base Z coordinate (see Figure 6.14b). This leaves the requirement to manipulate one of the two remaining pieces of equipment, because only one piece of equipment can be in contact with the PCBA at a given time. Therefore, the leg cutter is moved up and down by using a sliding mechanism and a pneumatic linear actuator (see Figure 6.14b).

Bottom Heater — Since closed-loop control of bottom heating requires a suitable thermocouple mechanism which can be retracted and raised to provide contact on the bottom of PCBAs for closed-loop control whenever it is necessary, a single-acting miniature linear actuator is used for this process (see Figure 6.15a.)

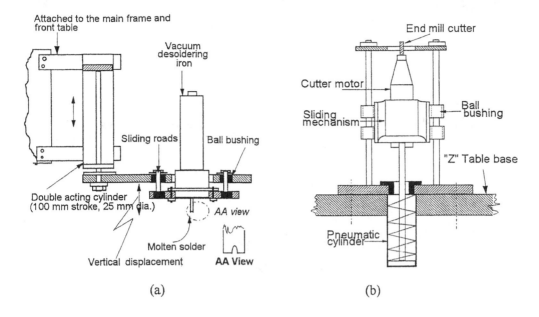

FIGURE 6.14 (a) Schematic of vacuum desoldering iron, and (b) Leg cutter raise-lower mechanism.

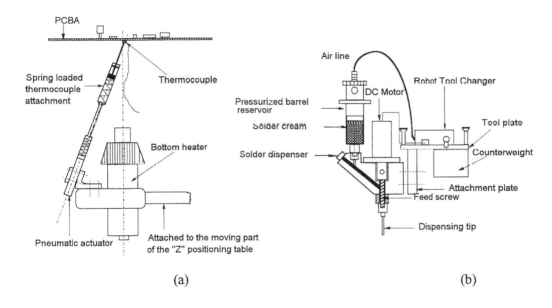

FIGURE 6.15 (a) Bottom heater thermocouple raise and lower mechanism, and (b) Attachment of solder dispenser to the plate-tool changer assembly.

The bottom heater is fixed to the moving frame of the Z positioning table so the gap between the bottom heater and the PCBA can be kept constant.

Solder Dispenser — Figure 6.15b also shows the integration of the solder dispenser to the robot's tool changer.

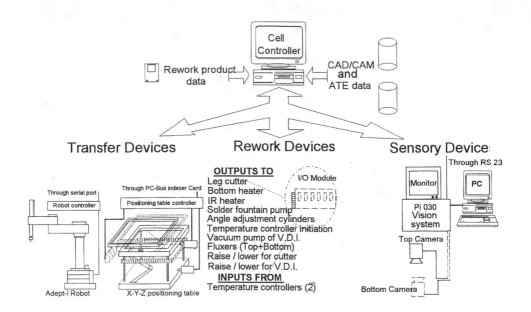

FIGURE 6.16   Control hierarchy of the rework cell.

6.11   Top-Down View of System Architecture

The studies carried out up to this level indicated that automated rework requires a supervisory control, the Adept-One robot, the X-Y-Z positioning table, the sensory devices, and various rework equipment to perform the rework automatically based on a batch size of one. Based on these findings, the cell's system control architecture has been determined. Figure 6.16 shows the control hierarchy of the rework cell at the top level. The rework cell is supervised by a 386 PC functioning as the cell controller. It is equipped with the input/output (I/O) module and the PC bus indexer card to control the I/O and positioning control requirements, respectively; all rework tools, the Adept-One robot, the X-Y-Z positioning table, and the sensory devices (PI030 vision system) are interfaced with the cell controller.

6.12   Detailed Automated Rework Procedures

When considering rework automation, which replaces human senses with remotely manipulated mechanical, electronic, and pneumatic devices, the core needs expansion in order to group the functional controller and task related activities in a procedural order. Detailed representation of rework procedures is quite difficult because of the unstructured nature of the PCBA rework environment and regular sensory information updates.

However, automated SM and TH rework procedures maybe simply illustrated as a flowchart in order to illustrate the primary logic and flow of rework based on predefined core rework procedures. Figures 6.17 and 6.18 present the SM and TH rework flow diagrams, respectively.

As seen from Figures 6.17 and 6.18, rework activities are carried out in a predetermined manner even though some steps are repeated if the first attempt fails. This feature provides a good opportunity for describing step-by-step rework procedures, assuming that all the decisions that take place in both types of rework are positive, and the defective component is not obstructed. Detailed rework procedures have been formulated and outlined in flowcharts. Table 6.3 presents part of these charts as an example illustrating the technique used. Rework, inspection, and cell controller activities are grouped in separate columns but listed in a sequential manner to illustrate their interdependent relationships. In addition, these procedures provide detailed information about activities of all rework tools, devices, and control systems.

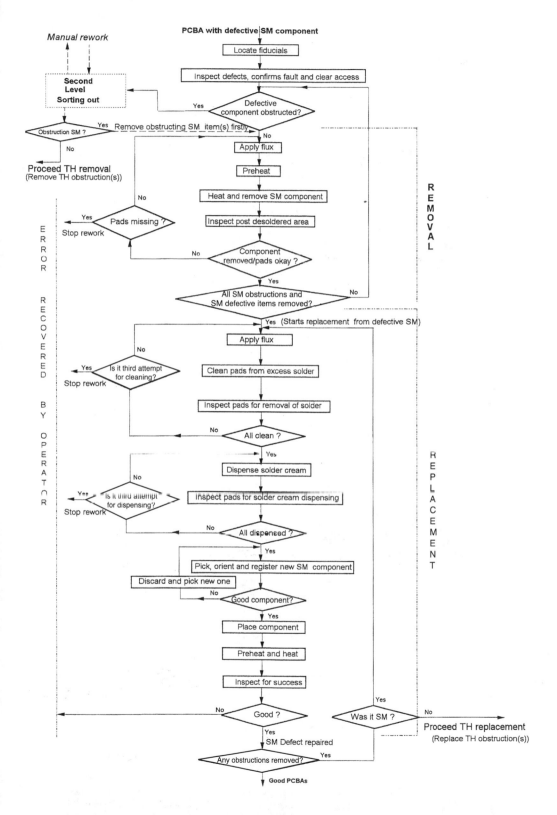

FIGURE 6.17 Flowchart of SM component rework procedure. (Reference 25. With permission.)

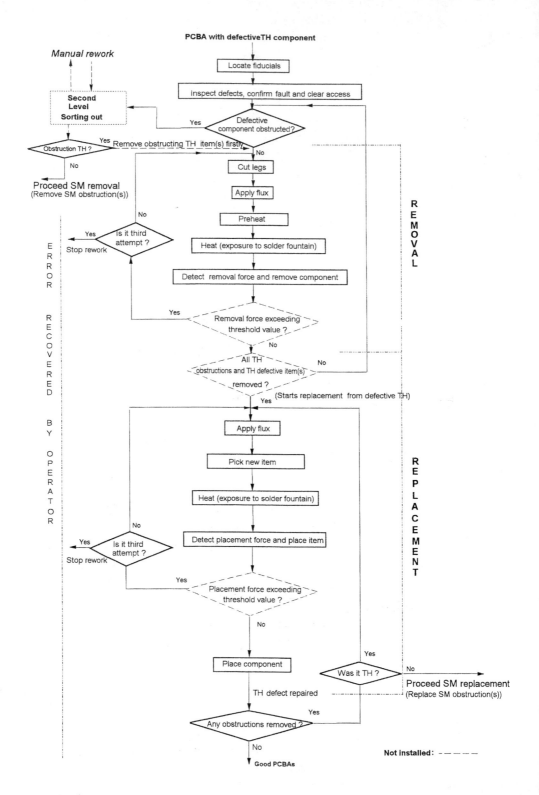

FIGURE 6.18   Flowchart of TH component rework procedure. (Reference 25. With permission.)

TABLE 6.3 Part of a Detailed Rework Chart for SM and TH Rework

Step	Rework	Vision	Cell Controller
1	Operator reads PCBA serial number and PCBA design ID		
2			Operator types in PCBA serial number and PCBA design ID
⋮	⋮	⋮	⋮
32	Robot/solder dispenser dispenses solder cream to dispensing point; positioning table moves to inspection area	Vision checks solder cream ball existence on the target pad	Cell controller formulates and controls solder dispensing on a target check point
33	Positioning table locates pads on solder dispensing area — robot/solder dispenser dispenses solder cream		Cell controller initiates solder dispensing and controls it
34	Positioning table locates the pads on inspection area for solder cream inspection	Vision inspects position and defects of solder cream	Cell controller formulates solder cream inspection
⋮	⋮	⋮	⋮
41	Positioning table locates the item on inspection area	Vision inspects post-resolder joint quality	Cell controller initiates post-resolder inspection

6.13   Total System Modeling Requirements

As seen from the nature of the problem, the development of a robotic rework cell requires a team of research engineers, each working on different parts of the development because the cell development is composed of three separate research areas: the cell controller, the intelligent sensors (vision and others), and hardware development and integration.

For the cell controller aspect of the project, a generic application software capable of attempting PCBA rework without human involvement is required. It relies on extensive support of data such as test results from the automatic PCBA testing equipment (ATE), CAD data to provide design data for the PCBAs, in process inspection data from installed sensory devices, and preexisting data that include a PCBA data pool which is generated by the cell and stored. The cell controller adjusts the system to automatically adapt to various conditions and activate the various tools when required.

For the in process inspection aspect, a generic application software is required for regular information updates of component location, orientation, solder paste volume, solder joint shape, and new component registration on the robot's end of arm equipment. These provide a sufficient amount of data for the cell controller to make decisions on the basis of the uncertain results of the previous actions at various points in the rework process.

In contrast to the previous two, the last part of the project is indeed manufacturing process development related but with complicated manufacturing steps and numerous tool controls. This is because the rework itself inherits the assembly process steps and the reverse of assembly due to the removal and replacement of defective electronic components of two different technologies (TH and SM). Basically, it is a miniature electronic manufacturing system with extra removal features, and therefore it necessitates a manufacturing system analysis tool.

The first and second parts are heavily based on software development, but are also interrelated to each other and the third part. The third part is based on the manufacturing process and requires a physical system modeling technique for the design. Briefly, the design of the rework cell requires an information system modeling technique in addition to physical system modeling. Therefore, the development requires considerable efforts focused on software and process developments and total system integration.

Various methodologies, techniques, and tools have been developed for the design and analysis of complex systems. The modeling techniques usually can be divided into information, decision, and physical system modeling. These three modeling methods and some techniques in each class are [18,29,52]:

1. Information system modeling (ISM): SADT, SSAD, IDEF1, MERISE, AXIAL
2. Decision system modeling (DSM): IDEF0, SSAD, GRAI
3. Physical system modeling (PSM): IDEFO, Petri-nets, GRAFCET, Simulation, GEMMA

These methods and techniques are reviewed elsewhere in great detail [37,45,52]. Each method has various advantages and disadvantages based on its particular application. The above classification and the requirements of the rework cell demonstrate that information and physical system modeling techniques are required for software and manufacturing (rework) process requirements, respectively, for the rework cell design and development.

If the problem of rework automation, its physical system design requirements, and nature are considered, neither ISM nor PSM methodologies are sufficient to deal with the complete design of the rework cell using a single technique in either of the methodologies to achieve successful total integration and development. Because the problem includes both software and process development, suitable analysis tools are required for each. If the design is considered separately, SADT (Structured Analysis and Design Technique) is much more suitable for the cell controller and in-process inspection parts of the development, since these particularly include software development [51]. The development of methods, and tool parts of the development require an analysis tool like IDEFO.

On the other hand, the use of different system design techniques at each part of the project might hinder progress, since each part of the development is interrelated and interwoven. Therefore, a single modeling technique must be used to define the boundaries between the work of each coresearcher and to determine input/output data, control, and other activities for each boundaries at the top level. Hence, in the first stage the automated rework system is analyzed in a top-down manner using the SADT package TEAMWORK which is an analysis and design tool from Cadre Technology in the United States. This has led to definition of the problem among researchers and clear limitations of each researcher. Figure 6.19 represents the top two level diagrams. As seen from the figure, the first level diagram consists of five bubbles as (1) identify rework tasks required, (2) locate fiducials on PCBA, (3) plan and coordinate rework, (4) perform rework, and (5) inspect PCBA.

A comprehensive dictionary feature of the TEAMWORK computer tool has also been used and complied with the activities and flows identified. These help to define the boundaries between each coresearcher. In addition, definitions of common vocabulary and terminology also minimized further communication and interfacing problems.

After first defining the problem using the SADT technique, the first two researchers have continued to use the SADT but the third researcher has requested an analysis tool suitable for manufacturing system design rather than ISM. Bubble four defines the boundaries of the third researcher. Although the first two researchers have used SADT technique, the details related to them is not given here. Instead, the details of analysis are given for the physical system design.

Physical System Modeling Requirements and Analysis

ISM and PSM techniques are based on modeling the problem as a collection of subsystems of inputs and outputs, and then modeling each of these subsystems as a further collection of subsystems is a widely recognized system analysis tool [17,44,57]. But PSM techniques have been developed to analyze manufacturing systems and they account for information and material flows as well as control and mechanism. Consequently, PSM techniques are the most suitable for the physical system design.

The square-labeled Perform Rework represents the activities that comprise the rework process. This is basically a manufacturing process and is more suitably analyzed using PSM techniques. Therefore, the second level activities of Perform Rework are presented as subsystems using the top-down approach of the decomposition technique such as used in IDEF0 and SADT, without incorporating the flow information as the trees representation (see Figure 6.20). These subsystems are exploded until each activity reaches the bottom level. To illustrate the method used, only one set of lower level diagrams, which is Perform SMD removal, is presented in Figure 6.21. The approach of determining the bottom level activities has been taken because the rework activities, being manufacturing facilities, are much easier to synthesize with a bottom-up functional integration approach since it is a manufacturing facility [51].

Each of the bottom level rework activities has been studied to determine the control mechanisms, the related tooling, input and output data formats, and electronics in order to represent functional requirements

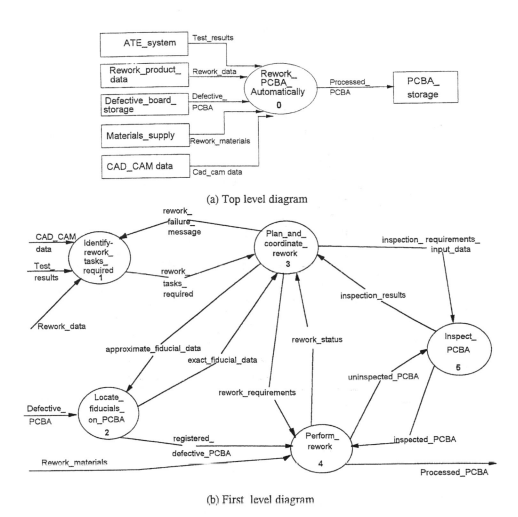

(a) Top level diagram

(b) First level diagram

FIGURE 6.19 Automated robotic rework cell system specification.

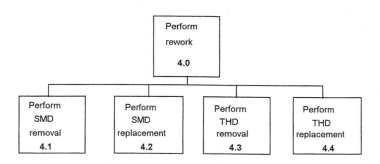

FIGURE 6.20 Subsystems of Perform Rework.

for performing each rework step. The notation used is given in Figure 6.22. As evident from Figure 6.22, the notation uses three different arrows: one for physical, one for data, and one for signal initiation for hardware tooling. This top-down decomposition and bottom-up synthesis approach has not only revealed the functional requirements for each of the steps in rework process, but has also eased the integration

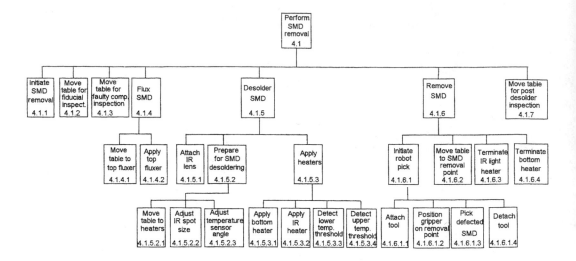

FIGURE 6.21   Lower level diagrams (perform SMD removal).

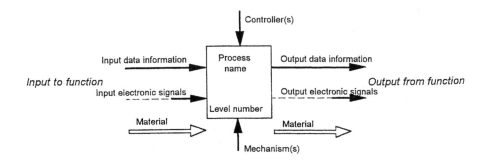

FIGURE 6.22   Bottom-up, functional integration syntax notation.

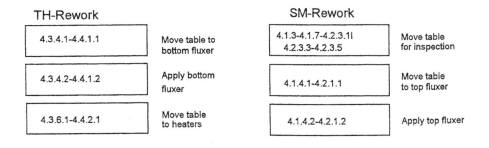

FIGURE 6.23   Some similar rework steps of surface mount and through-hole rework.

on both the mechanical and the software level. In addition, these studies have also revealed that some of the activities are similar to each other with regard to input-output function, controller, and mechanism requirements, although each of them are used to perform different activities. The similar rework activities have then been determined in order to eliminate repetition for both the development of the robot programs and the cell controller activities. For instance, Figures 6.23 and 6.24 represent some of the similarities found for each rework technology and between TH and SM rework, respectively, as an example. The results of the studies were shared and used in common by the all researchers.

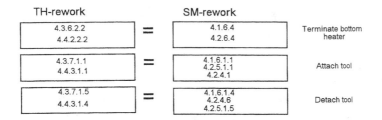

FIGURE 6.24   Some similarities found between TH and SM rework steps.

6.14   Robot and Vision System Software Development

Rework process analysis has revealed that various generic rework actions or steps are exactly the same. The determination of the similar actions drastically reduced the time spent for the development of the cell controller and the robot programs.

Unlike the Unidex 14 motion controller and digital I/O control card, which are integrated by plugging into the cell control computer, the robot and vision systems have their own remote controllers. The established RS-232 communication lines only provide information transferring ability and they require local software to manipulate their own controllers. Fortunately, the similar architecture can be used to model the software for both robot and the vision systems; for illustration purposes, only the robot program is discussed here.

The function of the robot is to remove and replace TH and SM components, to dispense solder cream, to position a new component above the vision system for inspection, and to discard the defective components. VAL II, the robot operating system and programming language, has been used to create robot programs. Since the rework environment is unstructured and the aim is to be able to deal with a batch size of one, flexibility of the rework cell plays an important role in the development of the robot software.

After consideration of the above, the following have been identified for the development of the robot programs:

- The cell controller commands the necessary robot task, and the robot executes the task accordingly.
- The rework data necessary to manipulate the robot must be obtainable from the supervisory cell controller. Sources of data may include sensory information, CAD/CAM data of PCBAs, and automatic test equipment (ATE) results (i.e., the robot does not contain any rework data).
- The robot programs are to be developed in a hierarchical manner with the communication software, which acts as the agent, at the highest level and supporting routines at a level(s) to perform the required functions.
- The robot is to inform the cell controller about the result of the executed robot routine. This information allows the cell controller to plan and coordinate the next action or carry out error recovery if a failure occurs.

On the basis of these decisions, robot software has been developed and the general structure is shown in Figure 6.25a. This consists of a housekeeping domain called MAIN for handling communications; it can execute any of the 14 subroutines and talks to the cell controller. Generally, all the robot rework routines have a similar structure to what is shown in Figure 6.25b. This way of organizing the robot software has some advantages:

- Good modularity/modification of any robot subroutines will not affect the others.
- Consistency in program: all subroutines follow a global generic structure.
- Ease of software modification.

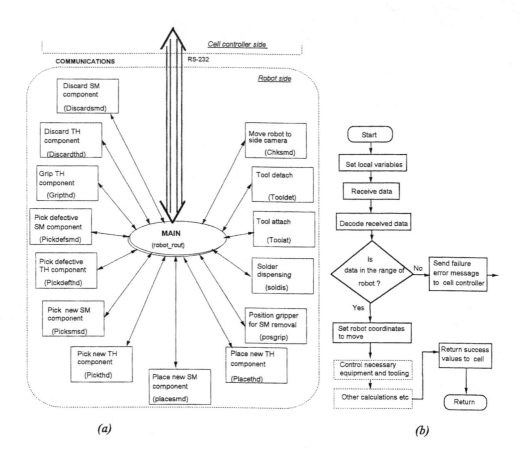

FIGURE 6.25 Robot software structure and general flowchart of the robot programs.

6.15 System Integration and Interfacing

Interfacing

The bottom level analysis data has also been used during hardware interfacing and integration. Ports, signal lines, and equipment are numbered or named as defined in the analysis. The use of common terminology both on paper and in real application has also reduced the problems that occur in the interfacing stage of development.

Figure 6.26 illustrates the interfacing between the robot and the cell controller. Two lines (X-Y axes) of direct current (DC) and one of the stepper motor (for the Z-axis) of the X-Y-Z positioning table are interfaced with the cell controller via the positioning controller drivers. The T-axis driver is used for IR spot size adjustment. The arrangement is shown in Figure 6.27. In order to be able to communicate with the vision system, the cell controller and the vision are interfaced using an RS-232C serial communication. The structure of the interfacing is basically illustrated in Figure 6.28.

Figure 6.29 illustrates the arrangement and control logic of the rework tools at the top level and gives corresponding signal numbers to control the rework equipment.

Infrared Heat Source and Closed-Loop Control System — The aim of the closed-loop control of IR heating is to monitor and control the temperature rise during reflow so that a reliable constant time/ temperature profile can be achieved regardless of thermal load. Therefore, a means of interfacing is necessary between the IR heat source and the temperature controller's output to adjust the power to the IR heat source. The IR heat unit's manual controller was discarded. Figure 6.30a presents a way of interfacing

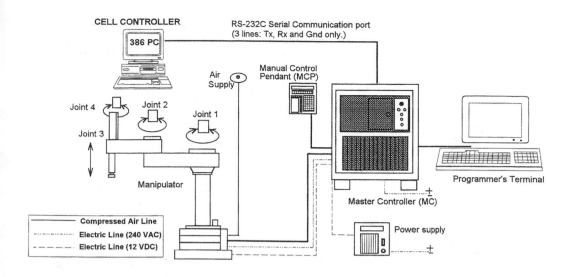

FIGURE 6.26   Interfacing between the robot and the rework cell controller.

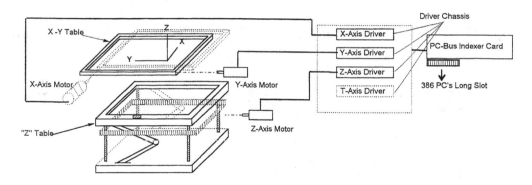

FIGURE 6.27   Interfacing between the positioning controller and the X-Y-Z positioning table.

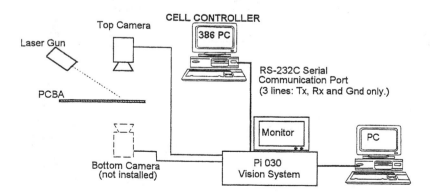

FIGURE 6.28   Interfacing between the cell controller and installed sensory devices.

the IR heating system and integrating the system to the cell controller. The event number is a setting, which defines when to trigger the input/output (I/O) module. With the current setting, the controller triggers the input module when the process temperature exceeds a certain level (205°C), whch is termed a process value (PV) event.

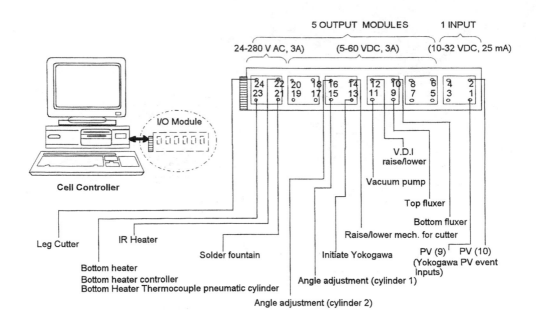

FIGURE 6.29   Interfacing between SM and TH rework devices to the corresponding signal lines at the I/O module.

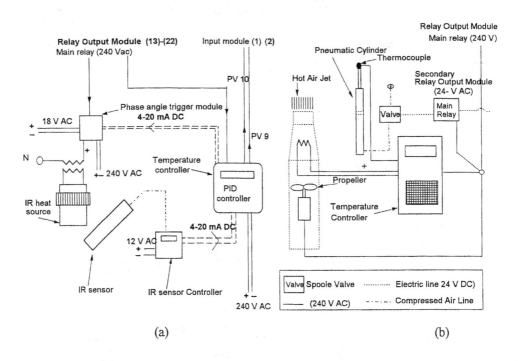

FIGURE 6.30   (a) Interfacing and control diagram of closed loop control of IR heating and (b) The bottom heater.

Bottom Heating Closed-Loop Control and Interfacing — The bottom heater unit is altered to control the heating process. This is done by separating the main power supply from the air fan and heat resistant unit so the heat resistant unit can be switched on/off by the bottom controller's relay depending on the thermal demand while the air fan is operated constantly during the bottom heating (see Figure 6.30b).

During rework, the bottom heater, the bottom heater controller, and the pneumatic cylinder, which raises and lowers the thermocouple attachment to the bottom of the PCBA, are triggered together. As seen in Figure 6.30, the bottom heater control is completely separated from the IR heating control mechanism, and is thus operated independently for both TH and SM rework.

Total System Integration and Cell Layout

The difficulty in designing a robotic assembly cell has three major aspects: economic, strategic, and technical. Economic analysis is concerned with choosing the most appropriate technical scenario at a reasonable cost. The strategic issues center on part presentation, flexibility, inspection, and throughput. The technical problems involve detailed equipment choice, its layout and operation, and confidence in acceptable performance. The above have been considered during the development of the cell, and all equipment has been installed. Figures 6.31 and 6.32 show the layout of the rework cell [25].

The structure of the cell may be better explained if it is divided into four regions as:

Rear Region — Close to the robot base column is a component feeding and tool parking zone where interchangeable tools such as grippers and lenses are stored. In addition to this, various solenoids used to control the rework tools have been placed underneath the back table (see Figures 6.31 and 6.32).

Underneath Center Region — Farther away from the base is the rework zone occupied by the X-Y-Z positioning table. The solder fountain, the leg cutter mechanism, the bottom fluxer, the bottom hot air heat source, and the bottom camera are positioned in a 250 × 210 mm envelope, which may be called the tool envelope (see Figures 6.31 and 6.32).

Upper Center Region — This is located above the underneath center region, where the IR reflow unit with a noncontact temperature measurement sensor and spot size adjustment mechanisms, the vacuum desoldering iron, the top fluxer, and the top camera are located in a 250 × 210 mm tool envelope (Figure 6.32)

Front Region — The outer zone consists of a supporting frame onto which stationary equipment such as the automated IR reflow unit, automated vacuum desoldering iron mechanism, and top fluxer are secured. The VDI and solder fountain controllers are also located in the front table (see Figure 6.31).

As seen in the Figure 6.31, the rework tools are placed as close as possible to each other to increase the size of the maximum reworkable PCBA. The X-Y-Z table is also positioned in the work envelope of the robot in such a way that the robot's envelope can be used economically for both rework and assembly. This design approach allows 400 × 400 mm sized PCBAs to be reworked while the rework tools occupy 54% of the robot's total working envelope of 15,370 mm^2.

A brief summary of the reworking of a defective component is given as follows. A PCBA with defective component(s) is placed on a fixture which is positioned in the center of the X-Y table. The X-Y-Z positioning table manipulates the target location of the PCBA to various locations where upper or lower rework tools are located to perform individual rework requirements such as cutting the TH component legs, solder dispensing, IR heating, pad cleaning, etc. The rework is carried out based on the procedure described above. In this arrangement, the function of the manipulator is to remove and replace the components, exchange the lenses of the IR heat source by the use of suitable grippers or end effectors, carry the solder dispenser and VDI to dispense solder, and clean pads, respectively. Detailed methods and procedures for performing automated rework are described by Geren and Redford [25].

6.16 Evaluation

Although the rework cell has been developed to deal with all possible standard components, the reworked components were limited to a few types to reduce the requirements for numerous grippers and suction tools. The components chosen were as follows.

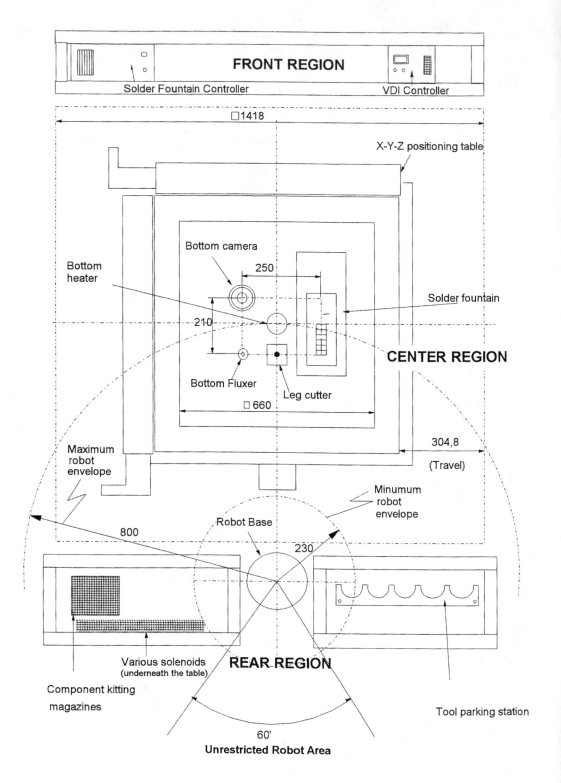

FIGURE 6.31 Illustration of *"underneath center and rear region"* equipment in the automated rework cell's layout.

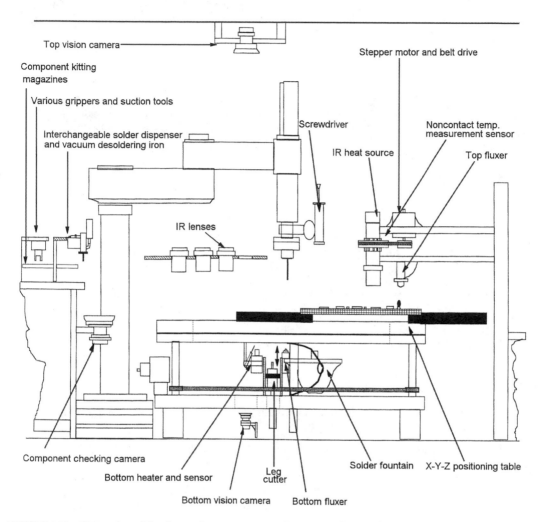

FIGURE 6.32 Illustration of "underneath-upper center and rear region" equipment in the automated rework cell's layout. (Reference 24. With permission.)

- Through-holes: DIP-20s and 5 pitch resistors or capacitors.
- Surface mounts: All types of rectangular chips (0805, 1206, 1812, 2220 etc.), SOTs, SOICs, and PLCCs.

Experiments were performed to validate the feasibility of the cell using the above components. It was observed that removal and replacement of both TH and SM components were successful. Components were removed and replaced up to three times, and no visual damage was observed; the reworked areas were indistinguishable to the naked eye.

6.17 Conclusions

An autonomous and fully automated PCBA robotic rework cell has been developed, implemented, and manufactured based on system modeling techniques. This cell is designed to be an integrated part of a PCBA cell and to perform assembly, rework, and in-process inspection of both TH and SM electronic components based on a batch size of one with a cell controller coordinating in-process inspection and the robot/tool operations. The use of system analysis tools led the designers through the design process since there are no guidelines.

The choice of desoldering and soldering tools is the one of the most important factors in the development of a fully automated rework cell. That choice determines the rework methods, procedures, and tools and affects the quality, reliability, and ability to deal with various component types. In addition, it affects the installation cost of the system.

Even though various rework methods and techniques could be applied to automated PCBA component rework as part of a robotic PCBA cell, IR and solder fountain desoldering/resoldering methods have been chosen for SM and TH component rework, respectively, because their technical superiority and appropriate rework procedures have been developed based on these methods. The procedures developed proved to work well and can be considered more effective than alternative rework methods because of their suitability to automation, effectiveness, reliability, and, most importantly, cost-effectiveness. Advantages of the chosen methods for SM rework are:

- All SM components are desoldered and resoldered without needing any auxiliary heads such as nozzles, heating heads, etc.
- The desoldering/resoldering method also allows utilization of current pick and place end effectors originally used for PCBA; i.e., for PCBA component rework, no special extra grippers are necessary.

Because of the above, the cost of a rwork cell is reduced.

- Heating of the whole defective component will enable adhesives to be softened if double-sided PCBAs are reworked, so it should be possible to rework double-sided PCBAs.
- Blowing or disturbing nearby components is eliminated when using an IR heat source, in contrast to the hot air/gas method.

Advantages of the chosen methods for TH rework are:

- TH components are desoldered and resoldered using the same source; no extra equipment is necessary.
- Both removal and replacement are fast.
- The reworked area looks good and is almost indistinguishable from the original.

Automation of rework requires extensive use of automation to achieve total process control. To achieve the aim, the tools and systems that are mostly available in the market require a substantial amount of modifications to make them suitable for automation. In addition, the electrical integration and interfacing these tools plays an important role in the automation process. The chosen tools and the developed architecture based on a supervisory control have confirmed that they are sufficient to achieve the primary aim. The capacity of the PCBA rework cell to repair a maximum of 400×400 mm sized PCBAs uses 54% of the robot's (Adept-1) total working envelope of 15,370 mm^2, leaving the rest for PCBA manufacturing activities. These results indicate that robotic rework is possible as an extension to a PCB assembly. This will make the economic justification of a rework cell much easier. The additional cost to extend an assembly cell to perform rework has been estimated at around $70,000.

References

1. Abbagnaro, L., Process development in SMT rework, *Circuits Manufacturing,* December, 50, 1988.
2. Aerotech, Ltd, *Motion Control Product Guide,* Aldermaston, Berkshire, 1989.
3. Adept-One, *Basic Software Training Notes for the Adept-One Robot with MC Controller,* San Jose, CA, 1989.
4. Techcon Systems Ltd, *TS 5000 Rotary Microvalve Positive Displacement Pump User's Manual,* Glasgow, Scotland, 1989.
5. Electrovert UK Ltd, *Light Beam Soldering Machine, LB-10,* Henley-on-Thames, Oxon.
6. Parkheath Ltd, *WLS 30 User's Catalog,* Cardiff, UK.
7. PDR Microelectronic Ltd, *1500 CA User's Manual,* Redhill, Surrey, UK.
8. Pacenter Electronic Production Equipment, *Pace Catalog,* Pagnell Bucks, UK.

9. *Hot Programs; Programmable Soldering Stations, Surface Mount International,* 5(5), 34, 1991.
10. R. A. Rodriguez Ltd, *Automatic Soldering Heads for Robotic Application,* Letchworth, Herts.
11. Siemens A.G., *An Introduction to Surface Mounting,* Munchen, Germany.
12. Eldon Industries, *Automatic Removal and Remounting of Surface Mount Components,* Compton, CA.
13. Bickerdyke, P., The heat of the matter, *Electron. Prod.,* 24(1), 20, 1995.
14. Coombs, C. F., *Printed Circuits Handbook,* 3rd ed., McGraw-Hill, New York, 1988.
15. Davy, J. G., A comprehensive list of wave solder defects and their probable causes, *Brazing Soldering,* 9, 50, 1985.
16. Davy, J. G., How should PWA solder defect inspection be performed, *Proceedings of the Technical Program: National Electronic Packing and Production Conference,* Nepcon East, Boston, 1988, 451.
17. Demarco, T., *Structured Analysis and System Analysis,* Yourden Press, Englewood Cliffs, NJ, 1979.
18. Doumeingts, G., Modeling techniques for CIM, Workshop on Computer Integrated Manufacturing Open Systems Architecture, Brussels, Belgium, March 1990.
19. Driels, M. R. and Klegka, J. S., An analysis of contemporary printed wiring board manufacturing environment in the U.S.A., *Int. J. Adv. Manufacturing Technol.,* 7(1), 29, 1992.
20. Elliott, D. A., Wave soldering concerns for surface mount assemblies, *Circuit World,* 15(4), 25, 1989.
21. Fidan, I., Merrick, M., Ruff, L. E., Millard, D., and Derby, S. J., Automation issues of SMD automated rework cell, *Proceedings of the 1995 American Control Conference,* Seattle, Washington, Vol. 3, 2174, 1995.
22. Geren, N., Chan, C., and Lo, E., Computer-integrated automatic PCBA rework, *Integrated Manufacturing Syst.,* 3(4), 38, 1992.
23. Geren, N., Design and Development of the Hardware for an Automated PCBA Inspection and Rework Cell, Ph.D. thesis, University of Salford, Salford, 1993.
24. Geren, N. and Redford, A., The significance of desoldering and resoldering methods in robotic automated rework, *J. Electron. Manufacturing,* 4, 41, 1994.
25. Geren, N. and Redford, A., Automated rework of printed circuit board assemblies — methods and procedures, *Int. J. Comput. Integrated Manufacturing,* 9(1), 48, 1996.
26. Hintch, S. W., *Handbook of Surface Mount Technology,* Longman Scientific and Technical, Horlow, 1988, UK.
27. Hollingum, J., Robot system builds customized printed circuit boards, *Assembly Automation,* 11(3), 21, 1991.
28. The Institute for Interconnecting and Packaging Electronic Circuits (IPC), *Suggested Guidelines for Modification, Rework and Repair of Printed Boards and Assemblies,* Revision C., Lincolnwood, IL, January 1988.
29. Kovacs George, L., Mezgar, I., and Nacsa, J., Concurrent engineering approach to support CIM reference model-based FMS design, *Comput. Integrated Manufacturing Syst.,* 7(1), 17, 1994.
30. Lea, C., *A Scientific Guide to Surface Mount Technology,* Electrochemical Publication, 1988, UK.
31. Lewis, A. D. and Martel, M. L., A new technology for surface mount rework, Technical Proceedings, EXPO SMT, 1988, 161.
32. Linman, D., Preheat improves VPS process, *Electron. Prod.,* February, 24, 1990.
33. Lo, E. K. and Goodall, A. J., Automatic in-process inspection during robotic PCBA rework, *J. Electron. Manufacturing,* 2, 55, 1992.
34. Mangin, C. H., Minimizing defects in surface mount assembly, *Electron. Packag. Prod.,* October, 66, 1987.
35. Mangin, C. H., Surface mount technology: prospects and pitfalls, *Assembly Automation,* February, 24, 1987.
36. Martel, M. L., The use of collimated infrared in surface mount rework, Proceedings of the Technical Program, Surface Mount 90, Boston, MA, August 1990, 867.
37. Mertins, K., Sussenguth, W., and Jochem, R., Integrated information modeling for CIM: an object-oriented method for integrated enterprise modeling, Computer Applications in Production and Engineering: Integration Aspects (CAPE 1991), Bordeaux, France, September 1991, 315.

38. Military Standards, MIL-C-28809, *Printed Wiring Assemblies,* Naval Publications and Forms Center, Philadelphia, PA, 1989.

39. Morris, B. S., *Surface Mounted Device Rework Techniques,* Advanced Rework Technology (A.R.T), Colchester, Essex, 1989.

40. Morris, B. S., The practicalities of surface mount rework, *Printed Circuit Assembly,* November, 22, 1989.

41. Nakagawa, Y., Automatic visual inspection of solder joints on printed circuit boards, *Robot Vision,* SPIE 336, 121, 1982.

42. Noble, P. J. W., *Printed Circuit Board Assembly,* Open University Press, U.K., 1989.

43. Prasad, R. P., *Surface Mount Technology — Principles and Practice,* Van Nostrand Reinhold, New York, 1989.

44. Ranky, P. G., *Flexible Manufacturing Cells and Systems,* CIMWARE Limited, Guildford, Surrey, 1990.

45. Ranky, P. G., *Manufacturing Database Management and Knowledge Based Expert Systems,* CIM-WARE Limited, Guildford, Surrey, 1990.

46. Siegel, E. S., Where is the heat? Part II, *Circuit World,* March, 58, 1991.

47. Schneider, M., Resoldering with the industrial robot, *Feinwerktech. Messtech. (WG),* 98(3), 89, 1990.

48. Strouth, T. and Orange, D., Automated repair of printed wiring assemblies, *Electron. Packag. Prod.,* November, 38, 1994.

49. Warnecke, H. J. and Wolf, H., Robotic insertion of odd components into printed circuit boards, *Assembly Automation,* November, 198, 1985.

50. Wassink, R. J. K., *Soldering in Electronics,* Van Nostrand Reinhold, New York, 1990.

51. Williams, D. J., *Manufacturing Systems,* Halsted Press, UK, 1988.

52. Wu, B., *Manufacturing System Design and Analysis,* Chapman and Hall, UK, 1994.

7

Model of Conceptual Design Phase and Its Applications in the Design of Mechanical Drive Units

Roman Žavbi
University of Ljubljana

Jože Duhovnik
University of Ljubljana

7.1 Introduction

Engineering design is a series of activities necessary for finding a satisfactory solution to a given task. Solutions are embodied in the resulting technical systems. A satisfactory solution is a solution that performs a given task within a certain time limit, with a certain reliability, for certain (production and operating) costs [1–3], and with a certain environmental impact.

The process of designing, as one of the key phases of product (technical system) development,[3-6] has, in recent years, received a great deal of attention from researchers in different areas of mechanical engineering, electrical engineering, computer engineering, civil engineering and architecture, psychology, philosophy, etc.

There are several reasons for such intense interest:

- Design is, in general, a result of human mental activity. With the development of artificial intelligence, genetic algorithms, and neural networks, man himself is becoming a subject of increasing interest in the natural science and engineering circles.[7,8]
- The process of designing comprises a search for appropriate working principles to provide a basis for the fulfillment of the required function, geometry, configuration, and dimensioning of elements which embody the working principles, selection of appropriate materials, and various analyses of elements.[3-5]
- An insufficiently described design process hinders the efficient development of computer-aided tools for the conceptual design of technical systems.
- 80% of costs in the life cycle of a technical system are determined in the phase of its design.[9]
- A trend of convergence toward the ability to manufacture technical systems is noticeable in the world, so that companies will be able to achieve and maintain their competitive advantages only by successful development of new technical systems. New in this case means those systems whose conceptual design is based on new physical laws that have not yet been applied in a certain field, or on new, as yet unused, combinations of already used physical laws.

The period of design research is marked by three phases:[9]

- Empirical: The first attempts to describe the process of design were mainly based on several years of experience of design engineers who analyzed and commented on various constructions and prepared guidelines for design and the evaluation of designs.
- Intellectual: Many researchers approached the modeling of the design process in a more disciplined manner, which led to more consistent theories and models. In this phase, questions were posed regarding the efficiency of the presented theories and methods for measuring the advantages of new methods.
- Experimental: In the past 15 years, much experimental work has been carried out, and there have also been advances in research methods. Experimental work in the field of design has specific characteristics due to the strong impact of the human factor and the fact that humans function in a certain cultural, political, sociological, and economic environment.

As already mentioned, the design process has received much attention in recent years, which is not to say that there was no interest in it before. The essential difference occurred mainly in the scientific disciplines from which researchers originated and in recognition of the significance of design as an independent branch of science. Researchers mainly originated from the ranks of design engineers, who drew attention to the large influence of design on the life cycles of products by analyzing the development of technical systems. The lack of any kind of process formalization inhibited the development of training for design engineers and prevented the establishment of engineering design as an independent technical science. The pioneers in the field of systematization and design methods in Europe are Hubka, Koller, Rodenacker, Roth, Pahl, Beitz, Zwicky, Pugh, Cross, French, Altšuller, and in the United States Asimow, Suh, and others.

The purpose of this review is to briefly present the work of researchers concerning the conceptual design phase of technical systems and the work of researchers who view design from entirely new angles.

The European school of design methodology is, in fact, German. We will examine only the main characteristics of the research results of its most prominent representatives, focusing on the field of conceptual design.

After determining the task, Koller determines the function structure, which implicitly predetermines the chain of physical laws, the type of physical laws, and their order. However, the index used to find the appropriate physical law is a function of the function structure and not, for example, of a physical variable. Establishing the function structure is entirely the task of the design engineer, who can rely only on his own experience. Koller himself[3] suggested the composition of function structure as the crucial activity in the conceptual design of technical systems.

The same approach was used in the methodologies of Pahl and Beitz[4] and Hubka.[5] Zwicky's morphological table is used in combining appropriate working principles to fulfill certain functions in the function structure. The combinations of working principles should be compatible, but they determine almost no rules for the evaluation of this compatibility. Again, authors explicitly emphasize the need for rich experience in composing the function structure and searching for compatible physical laws.

The main characteristic of both approaches is the fact that the function structure, in which physical laws fulfill certain functions, is composed in advance. The selection of physical laws is therefore subordinate to the function structure, which narrows the conceptual design of new technical systems.

On the basis of the study of patent literature, Altšuller in the former Soviet Union created a methodology, i.e., an algorithm for the conceptual design of technical systems, which is being constantly supplemented. The key element of the methodology is an assumption that all technical systems can be reduced to a so-called minimal technical system (MTS) which consists of three elements, one or two materials, and one or two fields. One or two elements of an MTS are always given in the problem formulation, while the third one should be added by the design engineer in order to create the starting scheme — a minimal technical system. It is possible that the problem formulation already determines all three elements or even more, but their interaction does not fulfill the task, and they therefore need to be changed and transformed into an effective MTS. The manipulation of basic elements is described in detail in the literature,[10] but it requires much experience on the part of the design engineer.[11]

Once an MTS has been designed, physical laws which link its elements together need to be found in tables following the algorithm. Physical laws are therefore sought again using a general scheme which needs to be composed in advance.

The above algorithm (or methodology), is not well known, probably primarily for the following reasons: The beginning of its development dates back to 1946, to the former Soviet Union, i.e., behind the "iron curtain." Furthermore, all early work was published exclusively in Russian. The first translation of a work from 1979 which describes this methodology in detail into German was published in 1986[10] in the former GDR, which did not strongly promote awareness of this methodology. A translation into English in 1988 did not generate much interest either. Altšuller's work should be mentioned, though, because of his different approach to these problems.

Altšuller's research results were used as a starting point for work by a research group of the Invention Machine Laboratory in Minsk, Belarus. This group, whose most active member is Sushkov, is attempting to produce a computer program for automatic conceptual design of technical systems, mainly at the level of redesign on the basis of Altšuller's methodology and in cooperation with the Knowledge-Based Systems Group from Twente University in the Netherlands.[12,13]

A computer program developed by the Invention Machine Corporation, co-founded by Tsourikov, who was also an associate of the Invention Machine Laboratory in Minsk, is based on the same principles. The available version of the computer program is essentially a mere hypertext version of Altšuller's work.[14]

Let us also briefly examine the approach of Japanese researchers, centered around Tomiyama at the University of Tokyo, to the conceptual design of technical systems. His approach to the beginning of conceptual design also assumes precomposition of the function structure; the difference lies in the way the function structure is supplemented. Supplements are recorded in the form of a state transition graph, which illustrates the desired functioning of the future technical system. On the basis of this graph, an automatic modeler based on the qualitative process theory (QPS)[15] selects appropriate working principles which can ensure state transitions recorded in the graph. In this case as well, the selection of appropriate working principles is subordinate to a predetermined function structure.

At MIT, Suh introduced his own approach to engineering design.[1] He defined design as mapping between functional and physical spaces (Figure 7.1). The functional space contains functional requirements (FR), i.e., the purpose of the designed technical system, while the physical embodiment of requirements

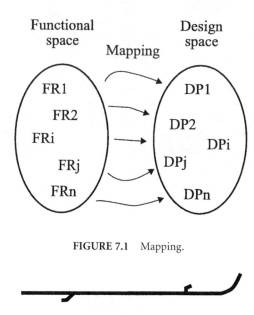

Functional space — Mapping — Design space

FR1 → DP1
FR2 → DP2
FRi →
FRj → DPi
FRn → DPj
DPn

FIGURE 7.1 Mapping.

FIGURE 7.2 Bottle and can opener. (Reference 1. With permission.)

expressed by design parameters (DP) is located in the physical space. However, mapping is not unary since there may be several embodiments (design parameters) for each functional requirement.

Suh wrote down the design axioms which make it possible for mapping to provide good (in technical sense) designs or technical systems. There are actually two axioms. The first axiom (the independence axiom) states that mapping must be such that changes to a design parameter only affect the relevant functional requirement. An example is provided for explanation (Figure 7.2): functional requirements are bottle opening (FR1) and can opening (FR2).

The bottle opener tongue and can opener cutter are design parameters DP1 and DP2. Under the first axiom, this design is good because DP1 exerts an influence only on FR1. In simple words, a bottle/can opener tongue/cutter can have any shape without affecting the quality of can/bottle opening.

The second axiom (the information axiom) states that, of all technical systems that fulfill the first axiom, the best one is that which contains the smallest amount of information. Information in this case is a measure of the amount of knowledge required to fulfill the functional requirement (FR). Examples of the application of the information axiom are discussed at length by Suh.[1]

Suh's approach is mainly to analyze predetermined technical systems, but he provides no guidelines (except general ones such as axioms) as to the conceptual design of technical systems; for that task, the design engineer must use other approaches.

Linde[16,17] of Coburg Technical University designed a strategy of innovation based on developmental contrasts (in German: *Widerspruchsorientierte Innovationsstrategie*, or WOIS), essentially characterized by the search for and design of problems whose content triggers innovative solutions. Here, the design engineer will use various principles, which are listed by the author on the basis of his own experience. In general, this strategy pays a great deal of attention to the formulation of problems.

The essence of the developmental contrast can best be presented by example: the efficiency of front-loading firearms depends on the accuracy of shooting and the speed of loading. In order to increase efficiency, it is therefore necessary to increase accuracy, which can be achieved by lengthening the barrel, and to shorten the loading time, which can be achieved by shortening the barrel. According to Linde, this contrast is of crucial importance for the formulation of the problem and its solution. The conflict is resolved by designing rear-loading firearms, whereby the requirement for an increase in accuracy (lengthening the barrel) will not have an influence on the speed of loading. Let us point out the congruence of this solution with the first axiom according to Suh.

Other parts of the strategy or conceptual design of technical systems consist of a group of methods introduced primarily by Altšuller (material-field system), Zwicky (morphological tables), and German researchers. Linde's strategy is actually the only method which quotes and uses Altšuller's results to a greater extent. From the macro-viewpoint, Linde's strategy is a tool for searching for new products.

These methods are general and are as such intended for the design of more or less arbitrary technical systems.

The LECAD Laboratory began the development of the first model of conceptual design of technical systems in 1989;[18] it was developed up to the prototype stage in 1992[19,20] and by 1995 it had been installed in a prototype design environment.[21] However, the fact that its conceptual design ability was limited was soon noticed. This limitation resulted from the fact that the model is based on the description of models of shape and on working principles derived from them. In addition, the selection of working principles is subordinate to the function structure.

Correction of this shortcoming will be described in the guidelines for further work.

Design as a Process of Conceptual Design of Products Leading to Manufacturing Design — Duhovnik's Perspective

The design process is generally defined by its characteristic phases. Individual authors define the phases or parts of this process differently, each presenting a different approach to design or design method[22-24]. It is, however, characteristic of almost all authors to try to define a generalized model of the design process. A few of them use characteristic product groups as the basis for their work,[25,26] while another group attempts to find confirmation for individual cases in a generalized model.[23,24]

As a rule, each description of the design process which is too definite in its individual steps process also predetermines the thought patterns leading to new ideas. In the conceptual design phase, i.e., in the initial phase of the process, human mental effort in finding new solutions is one of essential conditions for the successful designing of new products. Mental effort is the most important part of creating new products; only afterwards can the design process be framed by determining the conditions for the definition of realistic creations or solutions that are understandable to the environment. The determination of these conditions is more subject to the real environment. It is interesting that the degree of creativity of a certain solution is perceived quite differently in different environments. The reason for such difference of perception can be recognized in the actual potential of the evaluating environment. The lower the level of technical knowledge in the evaluating environment, the more the perception of creativity of technical solutions is limited. At the extremes of evaluation, the obtained results are at best equal to the knowledge of the evaluators in a specific environment.

Due to the above considerations, some authors recognized the design process as consisting of connected phases, but with special emphasis on the possibility of supplementing adopted solutions. During supplementing, requirements for new solutions are taken into account. These requirements include a condition that creative supplementing must be distinguished by quality, low price, and greater manufacturing flexibility. Greater flexibility in turn means shorter production cycles, and this enables greater reliability of supply or shorter delivery times.

Different models of the defined process were also created on the basis of the well-known product improvement curve (Figure 7.3).[27] This curve predetermines the demand for constant improvement of product function or shape. In theory, one can therefore speak of the iterativeness of the design process, which is intended for constant product improvement.

Products are defined by their function and shape. Function means a physical function that is presented with a working principle in the real technical world and taken into account in the conceptual design phase. In the manufacturing design phase, which also appears in the design process, function represents a specific function or its functionality. Shape has special significance. It is presented as a technical or an architectural characteristic. In conceptual design, it is usually presented in the form of a model of shape, which is replaced by a textual model in speech communication. In the manufacturing design phase, shape is presented by a graphical model, which nowadays is usually three-dimensional. For better presentation

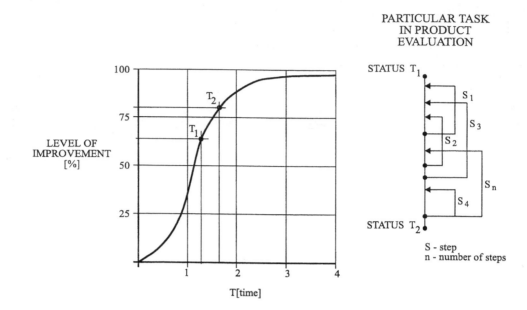

FIGURE 7.3   Product improvement curve.

and faster processing, three-dimensional models are usually presented as surface models in their use environment. This type of presentation is called visualization.

Manufacturing design is the phase of the design process in which all details are defined, all workshop documentation is made, technological conditions are determined, the prototype is made and tested, and the final decision on whether the product will be produced is made. Conceptual design is the phase of the design process in which the rough product shape which fulfills the required function is developed.

Manufacturing technology is an important criterion for the definition of shape. While the technology of use is important in determining function, conditions at the actual point of production are relevant for manufacturing technology. It is therefore understandable that the shapes of products change with the introduction of new technologies, while their functions are at least preserved, if not extended. A group of authors[7] presented such an iterative design process (Figure 7.4), which has the characteristics only of the execution of individual steps or, exceptionally, of phases of the design process, and is not involved in the determination of evaluation criteria. It is, however, essential that a certain type of processing is performed at each step. Processing can be performed at the level of function, model of shape, functionality, or technical shape.

If physical function is defined, product development can be supplemented either on the basis of manufacturing technology or due to changed individual functions which are represented by physically different elements or modules in the structure. A well-defined function structure and elementary product, or modular structures derived from it, determines the shape. For this reason, the definition of individual functions is especially important. Functions are then mapped into the real spatial world through models of shape in accordance with known working principles. In mapping, certain individual functions may be connected as soon as the working principles are defined. Another type of binding takes place when models of shape are defined. In this phase it may happen that the criteria determined for binding individual functions are not fulfilled, and the need for new or changed binding of individual functions may appear.

Iterativeness of the Design Process — Supplementing of Information

In the design process, it is hard to define accurately procedures for the development of new products because of the pronounced demand for constant supplementing of findings into new working principles and developed models for required functions. Due to this consideration, the description of the design

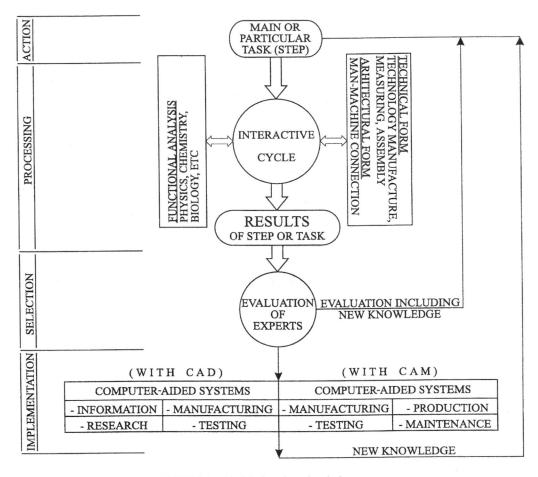

FIGURE 7.4   Model of an iterative design process.

process was supplemented as shown in Figure 7.5 with emphasis on the need for a stepwise procedure. It is important that the procedures themselves and their possible mathematical connections remain open in the description.

A clearly defined task is required for each step of the design process. Tasks may be the generation of an entire product or only its supplementing. In generating new products, the conceptual design phase is essential. Our approach to making a conceptual design model will be presented below in a concrete example. In the conceptual design phase it is important to define functions and the relations between them in order to enable the derivation of the required function of a certain product.

After the conceptual design phase, the same procedure of definition also appears in the manufacturing design process which, in our view, means a functionality analysis which is usually performed by calculating stress–strain, temperature, dilatation, and similar states. A functionality which is fulfilled through product dimensions is then checked with regard to its technical suitability for possible technological procedures. Limit dimensions of cross-sections are determined for critical points. The issue of accuracy of determination of critical points now appears for the first time. The technology to be used may confirm or disprove the obtained results for critical points; therefore intense interaction is required between the processes of determining the elements of functionality and technology.

The presentation of the results of work or an entire design phase can be evaluated by experts. However, this condition is omitted in some methods found in the literature,[8] even though it is, as a rule, the most

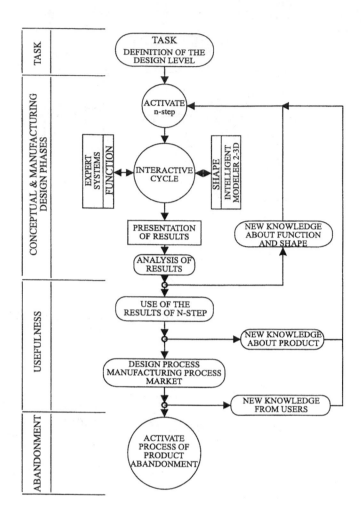

FIGURE 7.5 Stepwise presentation of the design process in its individual steps.

important one. Any mathematical model and product shape derived from it may become problematic if the evaluation of obtained results on the basis of criteria which are usually loosely defined (not described exactly and entirely) is not possible. Loose definitions can most often be recognized within economic criteria which change rapidly and may be one of the main stimulating or inhibiting factors for the continuation of the design process. After evaluation, work results may appear in a new feedback loop if there are many demands for their supplementing. In the opposite case, results can be applied directly.

Each product has a certain lifetime which varies among products. As a rule, the lifetime of products is programmed for a definite period of time. It is therefore important to recognize this period in advance and to determine the framework of supplements for each product. For very successful products the lifetime can be extended by improving them within a certain functionality or technical shape. In this case, the task is defined only for the third phase of the design process.

Why is it so important to emphasize this? Because it is asserted that, in the classical approach to the design process, the supplementing of products is not possible. In our case, because of the iterativeness and built-in requirement for supplementing, supplementing is possible.

If functional or other shortcomings are established during product monitoring, methodology may also include steps for product abandonment. Product abandonment may be built into the product itself, or come from the outside. In the former case, we distinguish between technical, technological, or economic abandonment, and various combinations of the three are possible.

Let us examine the following example. Product abandonment can be performed by defining new required functions or new element assemblies which may require several replacements during their maintenance and may become a vital part of our product. If individual functions are well described, if the binding function is suitable, and if the criteria for determining the vital assembly are determined in advance, the selection of bound elements can be derived mathematically. A selected assembly of bound elements and a clear determination of the product price range then determine the actual lifetime of the product.

It needs to be specially emphasized that for such a methodology, clearly defined principles are needed to present function, model of shape, functionality, and technical shape. Individual authors have carefully analyzed the principles for the presentation of models of shape and technical shape.[9,10] Recently, certain authors have attempted to examine the possibility of determining the principles of architectural shape by using procedures of pattern recognition. Recognized patterns are then mapped to new comparable models of shape.[11] This method is based on the establishment of characteristic functions on recorded samples. The prediction of new suitable architectural shapes is then performed using chaos theory.

Levels of Design

The definition of the design process by phases primarily depends on how function and shape are bound in the design process. Phases can be derived from the requirements for the binding method. In the conceptual design phase, basic or required functions are bound with the model of shape. Binding is performed through defined working principles in a defined morphological matrix.[12]

In the manufacturing design phase, functionality and technical shape and/or architectural shape are bound. This phase is easy to master in the form of individual steps; therefore, automated solutions exist. Decisions regarding the suitability of results obtained on the basis of such solutions are left to experts. Figure 7.6 presents the design process which, in general, consists of three phases and an *n* number of steps in each individual phase.

The exact number of steps is not determined which is very characteristic of iterative processes. The number of steps depends on product complexity or the depth of the developmental solutions used. The task, which is the first part of the design process, is intended to be a general description of requirements. As a rule, requirements are determined with regard to product use and its spatial influence. Usefulness

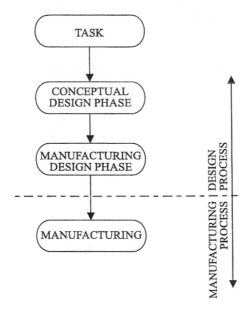

FIGURE 7.6 Three-phase design process.

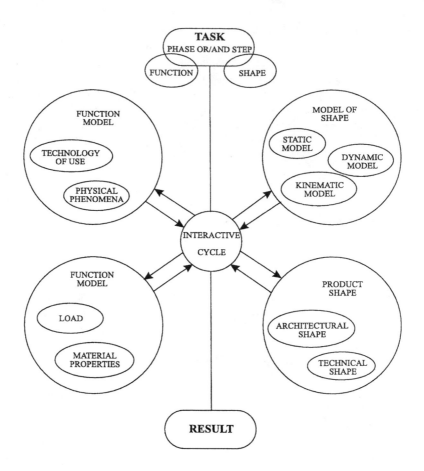

FIGURE 7.7   Connection between function and shape in a three-phase design process.

is roughly presented by the required function, and space by shape; it can therefore be said that the task should be a sensible way of presenting the connection between function and shape (see Figure 7.7).

If the task defines requirements for the entire process or only for one of its parts, it is therefore important to clearly define the level of designing. Part of the design process is considered to be the *n* number of steps in any phase, or the steps between the conceptual design and manufacturing design phases. It is not irrelevant whether product structure is designed completely from scratch or if only part of a product is reconstructed, and the relative difficulty of task definition also differs in these two cases. In the first case, a very thorough knowledge of the environment in which the product will appear is necessary, and the influence of the entire knowledge environment is taken into account. The knowledge environment, which needs to be defined especially broadly when designing from scratch, indirectly determines the final result of the design process. If the knowledge environment is poor, high-quality results of designing, i.e., products, cannot be expected.

Several approaches to determining the levels of designing are found in the literature.[1,3,12] However, the connection of design levels and characteristic analytical conditions has so far been recognized only by, Sushkov et al. [13]. If we limit our discussion only to the conceptual design and manufacturing design phases, it can be stated (taking into account the requirement for interaction between function and shape) that the design process contains the following levels (Figure 7.8).

- Designing from scratch
- Innovative design
- Variational design
- Adaptive design

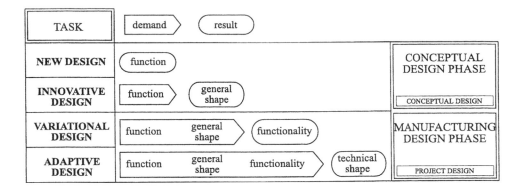

FIGURE 7.8 Levels of designing, connection between function and shape and required input data.

Designing from scratch means designing entirely new products, whereby a new working principle is determined for a new or known function. In the process of designing from scratch, one therefore needs to define the working principle, model of shape, functionality, and technical shape.

Innovative design means designing products by varying the working principles which fulfill the required function to the optimum degree. In innovative design, one needs to define the model of shape, functionality, and technical shape.

Variational design means designing products by varying loads, therefore comparable models of shape are obtained. In variational design, one needs to define the functionality and technical shape.

Adaptive design means designing products by adapting their dimensions to the technical and technological possibilities for their manufacture. In adaptive design, one needs to define the technical shape.

Modular structure at the level of function can be used in innovative design; we therefore usually speak of functional modular designing of machines and apparatuses. Modular structure at the level of shape can be used in variational design, in which case we speak of dimensional (shape) modular designing of machines and apparatuses.

It needs to be emphasized that there are different graphical presentations for the defined levels of design.[3,13]

With such methodology, the design process can easily be recognized as iterative, with *n* steps, three-phase, or multilevel. The interaction between function and shape is especially required in all types of processing. After each processing step in the design process, evaluation by several experts is necessary.

Functions must be clearly described in order to enable easier following of the design engineer's decisions along the decision-making tree. Below we will attempt to present a conceptual design model using an example of designing a mechanical drive unit. This example takes into account the methodological approaches for the determination of the basic, binding, and auxiliary functions based on the required function that a certain working principle must fulfill.

7.2 Model of Conceptual Design

In the conceptual design phase, a design engineer is faced with the problem of designing a technical system which fulfills a need or executes a required function. The design engineer combines the elements and more complex building blocks into assemblies (i.e., technical systems) that fulfill the required function, which arises from a need. The required functions can be simple, and therefore require simple technical systems, or complex, and therefore require complex technical systems.[19]

Qualitatively, our basic model can be described as follows: for every need or required function, there are a number of appropriate working principles that can fulfill the required function. Every working principle can be embodied using models of shape which execute the given basic functions, while all of them together execute the required function of the technical system. For each model of shape there is a binding function which functionally connects the models of shape into a technical system. Auxiliary functions describe characteristics of the working principles and models of shape. They are used as design constraints.

The basic building blocks of our model are the basic function structure, attribute vectors, and a rule for finding the successor of the models of shape.

In general, the selection of physical laws in the case of a predetermined function structure is predetermined, but many technical systems consist of existing components. The model of conceptual design is intended for the design of such technical systems.

Expressions Required to Understand the Model

A *function* is a task performed by an individual entity. This task should be expressed as neutrally as possible, independent of actual use. This generalization is intended to increase the applicability of individual entities (for example, transformation of mechanical engineering, transmission of mechanical energy, prevention of relative displacement, etc.).

- Required function: function determined by the design engineer to be fulfilled by the future technical system as a whole
- Basic function: function performed by the model of shape
- Binding function: function which needs to be fulfilled for the model of shape to fulfill its basic function. It enables the binding of models of shape into a flexible function structure
- Auxiliary function: function which defines the working principle or model of shape in more detail

A working principle is a verbal description of relationships or a physical law (for example, Gladstone–Dale law, exploitation of internal energy of oil derivatives, Joule's law, etc.).

A model of shape is an embodiment of a working principle, and basically does not contain any geometrical, material, or other details. These are expressed by auxiliary functions. The complexity of models of shape may vary (for example, timing belt and pulley, ball bearing, speed reducer, etc.).

A technical system is a group of relationships between models of shape and models of shape themselves. It performs the required function. The complexities of technical systems, like that of models of shape, vary. Technical systems of lower complexity compose those of higher complexity (for example, a bearing is a component of a gear train, which is a component of a speed reducer, which is a component of a mixer, and a mixer is a component of a paper production line. They are all technical systems, and each of them can be designed using a conceptual design model.)

Basic Function Structure

A predetermined function structure, which represents the generic model of the future technical system and at the same time also serves as the basis for setting up control procedures for binding model of shape, also served as the basis of our model. Naturally, each type of technical system requires its own basic function structure. In general, the basic function structure is the decomposition of a problem into subproblems, which enables simpler solutions, since the model was designed in such a way that the rule for finding successors can be reapplied during the inclusion of control procedures.

A design engineer composes the basic function structure of (any) technical system manually and in advance; the same applies to control procedures originating from it. The basic function structures composed in this manner can be stored in a special database, and they are accessed for the conceptual design of an appropriate technical system.

The function structure is presented as an AND/OR tree; its root is the required function, while other nodes are the characteristic functions of the technical system. By default, the final node is function *none*, with which the technical system, or a part thereof, is completed. The root and internal nodes can be type AND or type OR, depending on the technical system to be presented in general through its basic function structure.[19,21]

The function structure is called basic because it does not contain all functions of the future technical system, but only the most characteristic ones (the unavoidable ones).

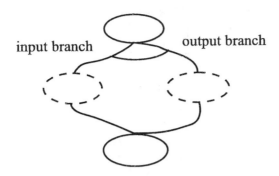

FIGURE 7.9   Basic function structure of a single-stage speed reducer.

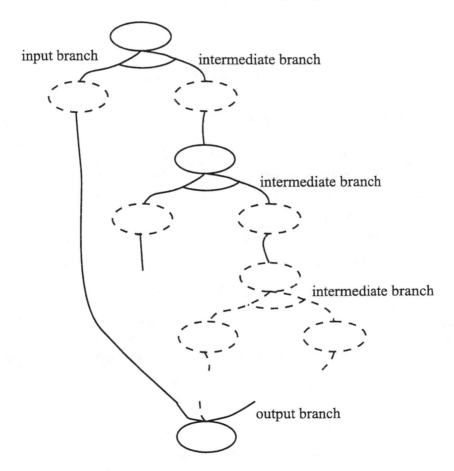

FIGURE 7.10   Basic function structure of a multiple-stage speed reducer.

Example

As an example, let us examine the basic function structure of single stage (Figure 7.9) and multiple stage mechanical drive units (Figure 7.10). The crucial part of their structure is the place where the function is performed (in a multiple stage mechanical drive unit, there are as many such places as there are stages). Depending on the working principle, this may be gears in mesh, friction contact, etc. Then the input and output branches of mechanical energy are recognized in the structure, as are intermediate branches, which are characteristic of double and multiple stage mechanical drive units. An important characteristic, which will be used to set up control procedures, is the frame of the unit, i.e., its casing, which connects all models of shape and provides output of reaction forces from the system.

The first control procedure which follows from the general structure of mechanical drive units depends on the number of stages of the mechanical drive unit; the second one depends on whether it is an input or output branch of mechanical energy, and the third one depends indirectly on the number of stages.

AND nodes in both figures indicate that the following two conditions need to be satisfied to fulfill the required function (in the case of mechanical drive units, the transformation of mechanical energy):

- Input of mechanical energy,
- Output of mechanical energy.

The selection of the basic function structure depends on the gear ratio. The function of component binding is represented by the last node in the basic function structure, and the conceptual design of a technical system, i.e., speed reducer in our case, can be completed with it.

Attribute Vector

The conceptual design model presented in this section is intended for composing technical systems from existing components. These can be represented entirely by attribute vectors:

$$\text{attribute vector} = (\text{fun}_i, \text{wp}_{ij}, \text{auxw}_{ijm}, \text{mos}_{ijk}, \text{auxm}_{ijn}, \text{bfun}_{ijkl})$$

$$\text{fun}_i \in \{\text{function}\}, \quad i = 1, 2, \ldots, F$$

$$\text{wp}_{ij} \in \{\text{working principles}\}, \quad j = 1, 2, \ldots, W$$

$$\text{auxw}_{ijm} \in \{\text{auxiliary functions of working principles}\}, \quad m = 1, 2, \ldots, A$$

$$\text{mos}_{ijk} \in \{\text{models of shape}\}, \quad k = 1, 2, \ldots, M$$

$$\text{auxm}_{ijn} \in \{\text{auxiliary functions of models of shape}\}, \quad n = 1, 2, \ldots, N$$

$$\text{bfun}_{ijkl} \in \{\text{binding functions}\}, \quad l = 1, 2, \ldots, B$$

Attribute vectors can be written on the basis of function trees (Figure 7.11), and these in turn are composed on the basis of existing models of shape.

In addition to functions, each description of a model of shape must also contain the following information for the model to function properly: working principles and their auxiliary functions which finally define the working principle, name of the model of shape, including the auxiliary functions which

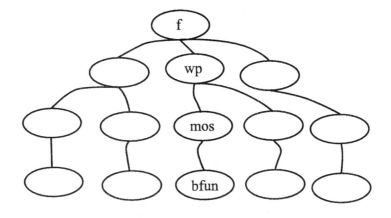

FIGURE 7.11   Function tree.

define the model of shape in detail and through which the design engineer expresses the design requirements, and the binding functions which serve to link models of shape into a technical system.

In addition to the requirements essential for the functioning of the model, the description may also contain information on the types of applied strength and other analyses and on individual special features of the model of shape (special features in the production, assembly, use and maintenance, etc.).

Graphical representation in the form of function trees is clear and appropriate for the presentation of models of shape which can fulfill individual functions, while attribute vectors are a more compact form, suitable for software production.

The content of the description of attribute vectors is of crucial importance for effective use of the model, and it should be produced by a person with considerable experience of design and the use of described models of shape. An inadequate description limits the applicability of models of shape. The description should therefore be neutral, i.e., independent of the examples of use.[1] One has to be aware that the more general the use of a model of shape, the more demanding its description.[35] Take the example of a gear, as a special machine part, and a washer, a general machine part with an extremely wide range of applications. It can be established in general that the use of neutral functions ensures the widest solution, space (Figure 7.12).

A rigid description of the functions of models of shape, which would depend on the type of use and would be wasteful regarding space, would look as follows:[19]

1. Description: increase of torque
2. Description: reduction of torque
3. Description: transformation of rotational into translational movement

Such a description would also prohibit new ways of using existing models of shape (Figure 7.13).

A general, neutral description would be as follows: transformation of mechanical energy. Transformation here does not denote the transformation of energy into other types of energy but a change of parameters (torque, rotation speed direction of movement, etc.) of mechanical energy.

The description of models of shape by attribute vectors is the most extensive part of preparations for the development of computer-aided design systems in the conceptual design phase, produced on the basis of a model presented in this section. Naturally, the set of available models of shape needs to be constantly supplemented.

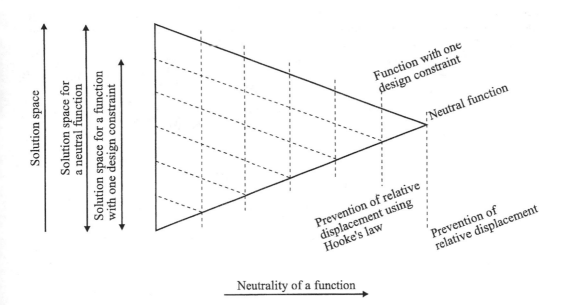

FIGURE 7.12 Function neutrality and solution space. (Adapted from Reference 5. With permission.)

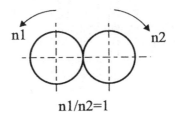

$$n1/n2=1$$

FIGURE 7.13   A new case of use prevented by a rigid description, unless given explicitly.

Example

To clarify, let us examine an attribute vector of a shrink-fit (Figure 7.14) which is used to prevent the relative rotation of components.

As mentioned above, the functional description must be neutral, i.e., independent of the modes of use of models of shape.

The basic function fun, which is fulfilled by shrink-fit, is recognized as prevention of relative rotation of components — the term *relative* is used here to denote the rotation of one part of the shrink-fit (in general, the ring) against the other (in general, the plug).

The working principle wp, which enables this function, is Hooke's law, which allows calculation of the contact pressure from the elasticity of material. In addition, friction also plays an important role — it is recognized as the appertaining working principle. The auxiliary function of the working principle, auxwp, is sliding upon overloading.

The name of the model of shape mos is shrink-fit, while the characteristic auxiliary functions of the model of shape auxm are the prevention of relative axial displacement, output of forces in one dimension (1D), and a low concentration of stress in shrink-fit elements due to geometrical features.

The binding function bfun of shrink-fit, which enables the placement of shrink-fit in a technical system, is the output of reaction forces and rotation.

The binding function may not be apparent at first sight, but we must be aware that the internal part of shrink-fit (plug, in general) is used for the transfer of mechanical energy in 1D, and that, therefore, no additional model of shape is required to perform this function. Upon defining the upper binding function, we find that a model of shape bearing corresponds to it, which serves for the output of reaction forces from the system and enables rotation.

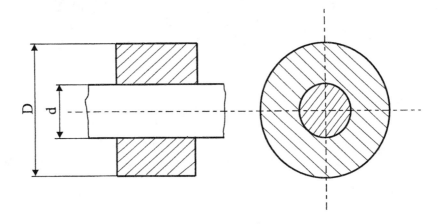

FIGURE 7.14   Shrink-fit.

The upper qualitative description of shrink-fit can then be translated into an attribute vector:

$$\text{fun}_3 = \text{prevention of relative rotation}$$
$$\text{wp}_{35} = \text{Hooke's law and friction}$$
$$\text{auxw}_{351} = \text{sliding upon overloading}$$
$$\text{mos}_{359} = \text{shrink-fit}$$
$$\text{auxm}_{351} = \text{prevention of relative axial displacement}$$
$$\text{auxm}_{352} = \text{output of force in 1D}$$
$$\text{auxm}_{353} = \text{low concentratin of stress}$$
$$\text{bfun}_{3591} = \text{output of reaction forces and rotation}$$

Rules for Binding Models of Shape

The essential rule is the rule for finding the successor of a preselected model of shape described by an attribute vector.

Qualitatively speaking, successors are found using the binding function of the current model of shape, since each model of shape can fulfill its partial function only if its binding function is fulfilled, and that can be achieved through its successor — a new model of shape. In other words, successors are those models of shape whose basic functions are identical to the binding functions of their predecessors: $\text{bfun}_{ijkl} \equiv \text{fun}_w$.

Since the binding function can be used to find several functionally appropriate models of shape, i.e., successors, these must somehow be collected and the best one selected. This is done using the analytic hierarchy process, which enables pairwise comparison of selection criteria and pairwise comparison of models of shape. The method and its use will be presented below.

Example

The following example explains the rule for finding successors.

A gear was selected as an example of a model of shape for which successors will be sought. The binding function of a gear is prevention of relative displacement, and we must find the models of shape which can fulfill it. On the basis of attribute vectors, we can establish that the successors can be the following models of shape, since their basic functions are identical to the binding function of a gear:

AV(prevention of relative displacement, wp, auxwp, square key, auxm, bfun)
AV(prevention of relative displacement, wp, auxwp, adhesive, auxm, bfun)
AV(prevention of relative displacement, wp, auxwp, splined shaft, auxm, bfun)

Flexible Function (Model) Structure

A flexible function structure is the result of using a design model, and it represents a group of models of shape (in which case it is named "model structure") or the required and basic (binding) functions (in which case it is named "function structure"). Why is it called flexible? Because it is not predetermined — it grows along with the design process, depending on the selected models of shape and their binding structures. In this manner, the risk of excluding individual models of shape in advance due to a rigid function structure with predetermined partial functions is avoided. Individual models of shape can simultaneously fulfill more or fewer function than anticipated by a rigid function structure. Such a function structure enables the variation of only those models of shape which fulfill the same partial function and have the same binding function as anticipated by the rigid function structure. This shortcoming was often overlooked in the past.[21]

Example

Let us examine the exclusion of a model of shape as a candidate for the conceptual design of a technical system as a consequence of using a rigid function structure in the following example.

Assuming that the function structure in Figure 7.15 is rigid, it can be fulfilled only with a combination of a gear, a square key, and a shaft, but it excludes the gear and shrink-fit and pinion combinations (Figures 7.16 and 7.17).

Our model enables the use of each of the three combinations, since it is not based on a predetermined rigid function structure. A flexible function structure grows with time, depending on the selected model of shape.[19]

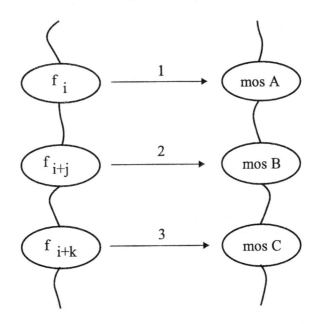

FIGURE 7.15   A rigid function structure.

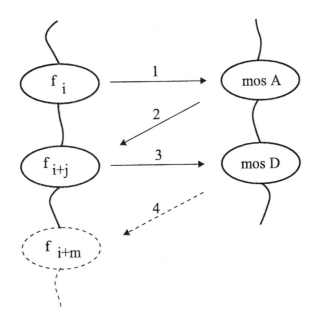

FIGURE 7.16   A flexible function structure I.

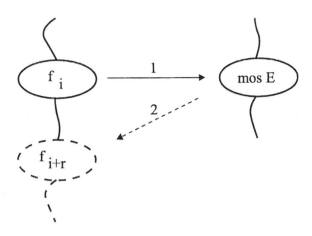

FIGURE 7.17   A flexible function structure II.

Synthesis of a Design Model

In the above subsections, we discussed the basic building blocks of a design model, so that a diagram of the progress of synthesizing a technical system can now be presented (Figure 7.18). With the selection of a model of shape, its binding function is also selected indirectly, and it is fulfilled by a new model of shape, its successor. In this manner, chaining (successive use of the rule to search for successors) is continued until all basic functions generated during the composing phase are fulfilled, and the result is a model/function structure of a technical system.

Let us summarize the functioning of a basic model: working principles which are appropriate for the required function given at the beginning are found first; models of shape with binding functions (which again need to be fulfilled) are then found for each selected working principle (from the model viewpoint, the binding function becomes the basic function). This is repeated until all basic functions generated during composition are fulfilled, and the result is a model/function structure of a technical system. We can say that the model consists of the basic model and control procedures. The setting up of such a model enables quite a deterministic approach to composing technical systems.

7.3   Implementation of the Model

The conceptual design model described in the previous sections will serve as the basis for execution. In describing the execution, we will bear in mind mechanical drive units, even though the conceptual design model is generally applicable. For example, Figures 7.9 and 7.10 present the basic structures of single- and multiple-stage mechanical-drive units, where solution trees of AND/OR graphs can be recognized:

- The root of the tree represents the required function. The root of the tree is an AND node if the required function is transformation of mechanical energy.
- Every AND node has two branches, which represent the input and output (or intermediate in the case of a multiple-stage mechanical drive unit) branch for mechanical energy.
- The OR node represents the characteristic basic function which has to be fulfilled.
- The final node corresponds to the function none, by default.

Let us again point out the fact that basic function structures are written down manually by the design engineer, and that they only contain the characteristic functions of the technical system.

In order to solve an AND node, that is, to fulfill the function of transformation of mechanical energy, all its successors must be solved, i.e., the input (intermediary) and output branch of mechanical energy. It is characteristic that the number of AND nodes is equal to the number of stages of the mechanical drive unit;

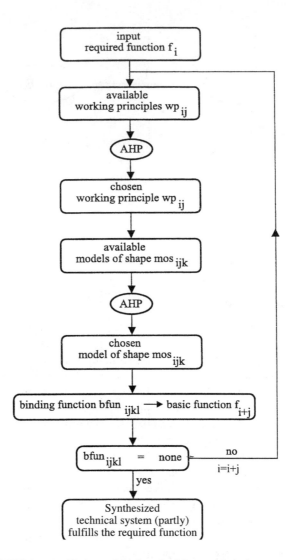

FIGURE 7.18 Basic model of synthesizing a technical system.

this feature is exploited to good effect in the conceptual design model for mechanical drive units. Such features must be found by design engineers in each type of technical system, since only in this way will they be able to compose the basic function structure and the control procedures deriving from it.

However, as already mentioned, Figures 7.9 and 7.10 present only the basic function structures, while the detailed structure is generated by the user through the use of the conceptual design model. Put more simply, this is seen as the generation of successors in the basic function structure, which has been indirectly mentioned several times in this section.

Formal Record of Certain Crucial Rules

Let us now examine a few rules in a conceptual design model written in PROLOG (Edinburgh syntax).[19] The basic rule for stringing models of shape — generation of successors:

```
select_function(Function):
        process_working_principle(Function, Working_principle),
        process_model(Function, Working_principle, Binding_function, Model),
        select_function(Binding_function).
```

The definition of the predicate process_model(function, binding_function, model):

Rule for the selection of a model of shape with regard to the function, working principle, and auxiliary functions:

```
process_model(Function, Working_principle, Binding_function):-
    attr_vect(Function,Working_principle, Auxw, Model, Auxm, Bind, Model_list)
nl, nl, write_on_screen(Model_list, 0),
express_aux_model(Function, Working_principle),
select_model(Function,Working_principle, Binding_function, Model_list, Model).
```

A rule which enables the conclusion of branch composition:

```
select_function(none):-
    nl, nl, tab(8), write('---Stop---'), !.
```

The binding function *none* is assigned to the frame, i.e., the casing, the task of which is to bind all components together. This marks the conclusion of composition of an individual branch of a mechanical drive unit.

With the basic conceptual design model, a user can bind only individual models of shape in individual branches: input (intermediary) and output. In order to generate the entire structure of a technical system (mechanical drive unit), control procedures also need to be taken into account (these depend on the type of each specific technical system), which are included in the basic model as additional conditions that must be fulfilled.

The rule which enables the continuation of composition in the intermediary or output branch of a mechanical drive unit:

```
select_function(none, Ends, Mesh, Branch, Status):-
    Ends_new is Ends-1,
    Ends_new > 0,
    select_function(transform_mech_energy, Ends_new, Mesh, Branch, Status).
```

Composition in the next branch is possible once composition in the previous one is concluded (function *none*), while the initial function for composition in a new branch is naturally the transformation of mechanical energy (AND node!).

The rule which enables the conclusion of composition of the entire technical system:

```
select_function(none, 1, Mesh, Branch, Status):-
    nl, nl, tab(8), write('---Stop synthesis---'), !.
```

It can be seen from the basic function structure of mechanical drive units that the new mesh (AND node!) is generated in even, i.e., output branches, which is used as the next rule:

```
select_function(output_forces_&_rotation, Ends, Mesh, Branch, Status):-
    Mesh > 1,
    Mesh_new is Mesh − 1,
    even(Branch),
    select_function(transform_mech_energy, Ends, Mesh, Branch, Status).
```

Not all necessary rules and predicates were presented, since we only wished to present the essential features of model execution.

According to an analogy with AND/OR graphs, we can see that the graph is analyzed for overall depth first, since first the input branch is composed in its entirety, and then the output (intermediary) branch of mechanical energy is composed — in the case of multiple stage mechanical drive units. During conceptual design, the growth of only one drive unit is monitored, but the system searches for all appertaining models of shape for each basic function. It can therefore be said that the system performs a local search, similar to breadth-first search (Figure 7.19).[19]

In general, one of the shortcomings of treating graphs using the depth-first search scheme is the risk of infinite loops.[36] In our case, there is no such risk, because the binding function of a model of shape can

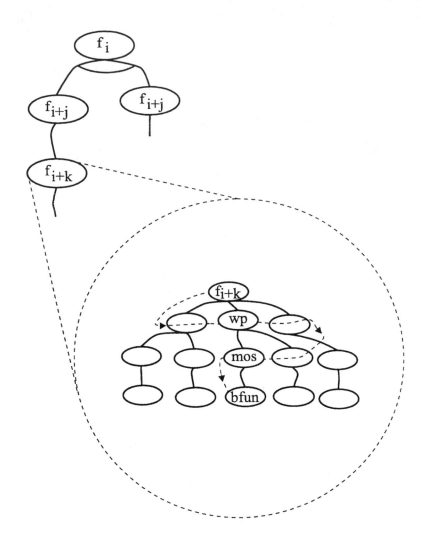

FIGURE 7.19   Overall depth-first search and local breadth-first search.

never be identical to the basic function that the same model of shape is performing. Another general disadvantage of both methods for graph search (depth-first and breadth-first search) is their functioning in the event of a combinatorial explosion.[36] In our case, this kind of risk does not exist either, because only the growth of one technical system is monitored at a time (mechanical drive unit).

7.4   Assessment of Appropriateness of Working Principles and Models of Shape

We must emphasize that no heuristic function, on the basis of which automatic composition or decisions regarding the selection of successors would be possible, is used to compose a technical system (analogy: extension of an AND/OR graph). Heuristic assessments depends on the problem field, and there is no general method for them.[36] In our case, the preparation of such an assessment would be too demanding,[19] and, in addition, full automation of the composition of a technical system would exclude the design engineer from this process, which is certainly not our original intention.[37]

Analytic Hierarchy Process

In the CAD Laboratory, we have decided to use the analytic hierarchy process method (AHP), the essence of which lies in pairwise comparisons of criteria, relevant for the decision, and pairwise comparisons of alternatives according to each individual criterion.[39,40] Pairwise comparisons are important because there are generally no measures for the evaluation of the functional appropriateness of components. Pairwise comparisons can implicitly also consider social, political, economic, and cultural contexts.[9,38] The other important characteristic of the method is decomposition of a complex problem and the setting up of a hierarchy.

The properties of the AHP method mentioned are precisely those which are characteristic of decision making in the conceptual design phase. It is of interest that in spite of the maturity of the method, it has not yet been used for the selection of the functionally most appropriate components in the phase of conceptual design of technical systems. The reason for this is probably a lack of trust in methods which require qualitative assessment as input data (9-degree scale in the case of the AHP method), which are then processed using mathematical procedures.

The AHP method can also serve as an implicit record of expert knowledge represented by criteria (i.e., design constraints, auxiliary functions) and pairwise comparison of alternatives (i.e., working principles, models of shape) according to each individual criterion.

Theoretical Basis

The selection of the functionally most appropriate alternatives (i.e., working principles, models of shape) using the AHP method can be divided into four phases:

- Decomposition of the problem (i.e., fulfillment of a required/basic function) to subproblems and the setting up of a hierarchy
- Pairwise comparison of criteria (i.e., design constraints) influencing the decision
- Pairwise comparison of the alternatives according to the chosen criteria
- Determination of the global priority of the alternatives

Hierarchy

The hierarchy enables pairwise comparisons of alternatives based on to an individual criterion, independently of other criteria.

In the application which will be presented in a later section we used only the three-level hierarchy (Figure 7.20):

- The basic function (with the working principle) as the goal
- Design constraints as criteria influencing the choice of the alternative
- Appropriate components as alternatives

The hierarchy itself presents the model of factors influencing the decision or the choice of the working principles/models of shape, and the problem which arises is the choice of the criteria important in the selection of the functionally most appropriate alternatives; the choice of criteria is also the most creative part of the selection process.[40–42] Through the criteria, the design engineer expresses the design constraints which should be taken into account and fulfilled by the alternative and, indirectly, by the technical system as a whole.

The first step in the selection process using the AHP method is therefore the setting up of the hierarchy: determination of the goal, criteria, and alternatives in general.

Pairwise Comparison of Criteria

The next step is the pairwise comparison of criteria (i.e., design constraints), which determines the relative importance of criteria in the choice between the alternatives.

Let us now look at matrix **A** of pairwise comparisons of criteria,

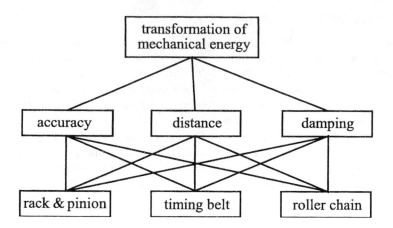

FIGURE 7.20   A hierarchy.

TABLE 7.1   Comparison Scale

Intensity of Importance on an Absolute Scale	Definition
1	Equal importance
3	Moderate importance of one over another
5	Essential or strong importance
7	Very strong importance
9	Extreme importance
2, 4, 6, 8	Intermediate values between the two adjacent judgments

$$
\mathbf{A} = \begin{array}{c|ccccc}
 & c_1 & c_2 & c_3 & \cdots & c_n \\
\hline
c_1 & a_{11} & a_{12} & a_{13} & \cdots & a_{1n} \\
c_2 & a_{21} & a_{22} & a_{23} & \cdots & a_{2n} \\
c_3 & a_{31} & a_{32} & a_{33} & \cdots & a_{3n} \\
\cdots & \cdots & \cdots & \cdots & \cdots & \cdots \\
c_n & a_{n1} & a_{n2} & a_{n3} & \cdots & a_{nn}
\end{array}
\tag{7.1}
$$

where c_i, $i = 1,\ldots\ldots, n$ denote the criteria, n is the number of criteria, and a_{ij} is the comparison of criterion c_i with criterion c_j. The values of comparisons are explained in Table 7.1. It has to be mentioned that [39]

- If we compare a criterion with itself:

$$
a_{ii} = 1
\tag{7.2}
$$

- And from the consistency of comparisons:

$$
a_{ij} = \frac{1}{a_{ij}}
\tag{7.3}
$$

The number of comparisons we have to make is $n(n-1)/2$.

Pairwise Comparison of Alternatives

In a similar way, we now compare the alternatives separately for each individual criterion; we obtain nn matrices $\mathbf{B}_i$, $i = 1,\ldots, nn$ (nn = number of criteria) of pairwise comparisons of the alternatives.

$$
\mathbf{B}_i =
\begin{array}{c|ccccc}
 & l_1 & l_2 & l_3 & \cdots & l_n \\
\hline
l_1 & b_{11} & b_{12} & b_{13} & \cdots & b_{1n} \\
l_2 & b_{21} & b_{22} & b_{23} & \cdots & b_{2n} \\
l_3 & b_{31} & b_{32} & b_{33} & \cdots & b_{3n} \\
\cdots & \cdots & \cdots & \cdots & \cdots & \cdots \\
l_n & b_{n1} & b_{n2} & b_{n3} & \cdots & b_{nn}
\end{array}
\tag{7.4}
$$

where l_i, $i = 1,\ldots, n$ denote the alternatives, n is the number of alternatives, and b_{ij} is the comparison of alternative l_i with alternative l_j according to criteria c_j.

The findings given in the above section also hold true in these comparisons.

Overall Ranking of Alternatives

The last step is determination of the global priority of the alternatives.

The eigenvector $\mathbf{W}$ (also priority vector) of matrix $\mathbf{A}$ of pairwise comparisons of criteria is calculated first:

$$
AW = \lambda_{max} W
\tag{7.5}
$$

where $W = (w_1, w_2, w_3,\ldots, w_n)^T$, and λ_{max} is the maximum eigenvalue.

The eigenvectors Y_i (also local priority vectors), $i = 1,\ldots, nn$ of the matrices $\mathbf{B}_i$ of pairwise comparisons of alternatives for individual criterion are calculated in the same way:

$$
\mathbf{B}_i Y_i = \lambda_{i\,max} Y_i
\tag{7.6}
$$

where $Y_i = (y_1, y_2, y_3,\ldots, y_n)^T$, and $\lambda_{i\,max}$ is the maximum eigenvalue.

The number of eigenvectors (i.e., local priority vectors) is therefore equal to the number of criteria.

The global value of the priority of the alternatives vector $\mathbf{R}$ is the product of matrix $\mathbf{E}$, where columns denote the eigenvectors Y_i of the matrices $\mathbf{B}_i$ of pairwise comparisons of alternatives and the eigenvector W of matrix $\mathbf{A}$ of pairwise comparisons of criteria:

$$
EW = R
\tag{7.7}
$$

$R = (r_1, r_2, r_3,\ldots, r_a)^T$, where r_a is the global value of priority of the alternative a. The alternative with the highest value is the best choice. Saty discusses this further.[39,40]

The Measure of Consistency of Pairwise Comparisons

Deviations from consistency are expressed by the following equation:[39]

$$
CI = \frac{\lambda_{max} - n}{n - 1}
\tag{7.8}
$$

and the measure of consistency is named the consistency index (CI).

The next measure is the random index (RI), which represents the consistency index of randomly generated reciprocal matrices (size $n \times n$) of values on the 1 to 9 scale:

n	1	2	3	4	5	6	7	8	9	10
RI	0.00	0.00	0.58	0.90	1.12	1.24	1.32	1.41	1.45	1.49

The third ratio is the consistency ratio (CR), which is used to estimate directly the consistency of pairwise comparisons:

$$CR = \frac{CI}{RI} \qquad (7.9)$$

CR values lower than 0.1 (CR < 0.1) are acceptable.[39,40] If the ratio is higher, it is advisable to correct the pairwise comparisons in order to preserve the consistency of estimation.

The control of consistency of pairwise comparisons also serves to prevent unjustified forcing of certain solutions.

Comparison Scale
The applicability and appropriateness of the comparison scale (Table 7.1) have been evaluated and confirmed by theoretical comparisons of many other scales, as well as in many practical examples of use.[39,40]

7.5 Application

Let us now examine the conceptual design of a printer head drive, which can be classified into the family of single-stage mechanical drive units (the basic function structure for such technical systems is presented in Figure 7.9), whereby due to the limited space the use of the AHP method will be presented only in the selection of models of shape for the required function and for the selected working principle.

The recognized required function of the printer head drive is transformation of mechanical energy. Selecting this function generates the root of the tree (i.e., future flexible function structure) which is an AND node (Figure 7.21).

In the next step, the computer program proposes working principles, which can fulfill the required function:

1. friction
2. form-fit
3. friction and form, etc.

On the basis of design requirements and using the AHP method, the design engineer selects form-fit as the most appropriate working principle which can fulfill the required function. After this selection, the computer system retrieves and displays stored models of shape which materialize the above working principle:

1. rack and pinion
2. timing belt
3. roller chain, etc.

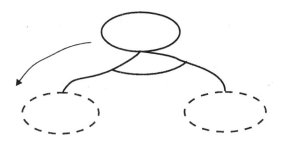

FIGURE 7.21 Generation of an AND node of the flexible function structure.

TABLE 7.2   Pairwise Comparison Matrix **A** of Chosen Criteria

	Accuracy	Distance	Damping
Accuracy	—	1	3
Distance	—	—	3
Damping	—	—	—

TABLE 7.3   Pairwise Comparison Matrix **B**$_1$ According to Accuracy

	Rack and Pinion	Timing Belt	Roller Chain
Rack and pinion	—	1/3	1
Timing belt	—	—	3
Roller chain	—	—	—

TABLE 7.4   Local Priorities

Timing belt	0.6000
Roller chain	0.2000
Rack and pinion	0.2000

In the next few pages we will examine the use of the AHP method for the selection of the most appropriate model of shape to fulfill the required function of transformation of mechanical energy more closely. We use a computer-aided evaluation system called SAATY, [41,43].

The design engineer (hypothetically) selects the criteria considered crucial for the choice of the most appropriate model of shape, for example:

• Accuracy of transformation
• Long center distance
• Damping (during transformation of mechanical energy)

In this way, the design engineer generates the design constraints which have to be satisfied by a technical system. The design engineer then has to perform the pairwise comparison of the above criteria, which is used to indirectly express their importance (Table 7.2).

In the pairwise comparison matrix **A**, the accuracy/distance comparison, for example, means that the accuracy of transformation is of equal importance ($a_{12} = 1$, see Table 7.1) to the long center distance.

On the basis of this pairwise comparison of criteria, the priority vector **W** is calculated:

$$\mathbf{W} = \begin{vmatrix} 0.4286 \\ 0.4286 \\ 0.1429 \end{vmatrix} \quad \text{CI: } 0.0000 \quad \text{CR: } 0.0000$$

The design engineer then has to perform a pairwise comparison of the models of shape proposed by the model for each chosen criterion (Tables 7.3, 7.5, and 7.7).

In the pairwise comparison matrix **B**$_1$, the rack and pinion/timing belt comparison, for example, means that the accuracy of transformation of the timing belt is moderately better ($a_{12} = 1/3$, see Tables 7.1 and 7.3) than the accuracy of the rack and pinion.

Local priority vector **Y**$_1$, local priorities (Table 7.4), and consistency according to the above comparisons are

$$Y_1 = \begin{vmatrix} 0.2000 \\ 0.6000 \\ 0.2000 \end{vmatrix} \quad \text{CI} = 0.0000 \quad \text{CR} = 0.0000$$

TABLE 7.5   Pairwise Comparison Matrix $\mathbf{B}_2$ According to Distance

	Rack and Pinion	Timing Belt	Roller Chain
Rack and pinion	—	1/5	1/5
Timing belt	—	—	1
Roller chain	—	—	—

TABLE 7.6   Local Priorities

Timing belt	0.4545
Roller chain	0.4545
Rack and pinion	0.0909

TABLE 7.7   Pairwise Comparison Matrix $\mathbf{B}_3$ According to Damping

	Rack and Pinion	Timing Belt	Roller Chain
Rack and pinion	—	1/5	1
Timing belt	—	—	5
Roller chain	—	—	—

TABLE 7.8   Local Priorities

Timing belt	0.7143
Rack and pinion	0.1429
Roller chain	0.1429

In the pairwise comparison matrix $\mathbf{B}_2$, the rack and pinion/timing belt comparison, for example, means that the useful center distance of the timing belt is essentially longer ($a_{12} = 1/5$, see Tables 7.1 and 7.5) than that of the rack and pinion.

Local priority vector $\mathbf{Y}_2$, local priorities (Table 7.6), and consistency according to the above comparisons are

$$Y_2 = \begin{vmatrix} 0.0909 \\ 0.4545 \\ 0.4545 \end{vmatrix} \quad \mathrm{CI} = 0.0000 \quad \mathrm{CR} = 0.0000$$

In the pairwise comparison matrix $\mathbf{B}_3$, the timing belt/roller chain, comparison, for example, means that damping of the timing belt is essentially better ($a_{23} = 5$, see Tables 7.1 and 7.7) than the roller chain damping.

Local priority vector $\mathbf{Y}_3$, local priorities (Table 7.8), and consistency according to the above comparisons are

$$Y_3 = \begin{vmatrix} 0.1429 \\ 0.7143 \\ 0.1429 \end{vmatrix} \quad \mathrm{CI} = 0.0000 \quad \mathrm{CR} = 0.0000$$

All consistency ratios are CR < 0.1; therefore all the judgments are consistent.

Based on the pairwise comparison of the criteria and the components with respect to individual criteria, the overall ranking of the components (Table 7.9) is as follows (global priority vector R ($\mathbf{E}\,W = R$)):

$$\begin{vmatrix} 0.2000 & 0.0909 & 0.1429 \\ 0.6000 & 0.4545 & 0.7143 \\ 0.2000 & 0.4545 & 0.1429 \end{vmatrix} \cdot \begin{vmatrix} 0.4286 \\ 0.4286 \\ 0.1429 \end{vmatrix} = \begin{vmatrix} 0.1451 \\ 0.5540 \\ 0.3009 \end{vmatrix}$$

TABLE 7.9 Overall Ranking of the Models of Shape

Timing belt	0.5540
Roller chain	0.3009
Rack and pinion	0.1451

TABLE 7.10 Pairwise Comparison Matrix **A** of a New Set of Criteria

	Accuracy	Distance	Damping	Noise
Accuracy	—	1	3	3
Distance	—	—	3	3
Damping	—	—	—	1
Noise	—	—	—	—

TABLE 7.11 Pairwise Comparison Matrix **B**$_4$ According to Noise

	Rack and Pinion	Timing Belt	Roller Chain
Rack and pinion	—	1/3	1
Timing belt	—	—	3
Roller chain	—	—	—

Based on the results, the design engineer chooses the component with the highest value, i.e., timing belt. In this way, the design engineer can choose all working principles and models of shape for the entire mechanical drive unit (technical system, in general).

Inclusion and Exclusion of Criteria

Inclusion of Criteria

It also frequently happens that after the working principle and model of shape have been selected, new design requirements arise which may even change the final choice. The AHP method and its computer implementation may serve to include new criteria in a simple manner and calculate their overall rank. Suppose that a new criterion is operating noise. The design engineer now has to compare initial criteria to the new one. He also has to perform a pairwise comparison of the alternatives according to the new criterion. After this, the overall ranking of the components has to be recalculated (Table 7.10).

On the basis of this pairwise comparison of criteria, the priority vector W follows:

$$W = \begin{vmatrix} 0.3750 \\ 0.3750 \\ 0.1250 \\ 0.1250 \end{vmatrix} \quad CI : 0.0000 \quad CR : 0.0000$$

Based on the newly included criterion, a pairwise comparison of the models of shape has to be performed next (Table 7.11).

In the pairwise comparison matrix **B**$_4$, the timing belt/roller chain comparison, for example, means that the operating noise of the timing belt is slightly lower than the operating noise of the roller chain ($a_{23} = 3$, see Tables 7.1 and 7.11).

Local priority vector Y_5, local priorities (Table 7.12), and consistency according to the above comparisons are

$$Y_5 = \begin{vmatrix} 0.2000 \\ 0.6000 \\ 0.2000 \end{vmatrix} \quad CI = 0.0000 \quad CR = 0.0000$$

TABLE 7.12   Local Priorities

Timing belt	0.6000
Roller chain	0.2000
Rack and pinion	0.2000

TABLE 7.13   Overall Ranking of the Models of Shape

Timing belt	0.5597
Roller chain	0.2883
Rack and pinion	0.1519

TABLE 7.14   Pairwise Comparison Matrix **A** of Criteria

	Accuracy	Distance
Accuracy	—	1
Distance	—	—

All consistency ratios are CR < 0.1; therefore all the judgments are consistent. Applying the pairwise comparison of the criteria and the components with respect to individual criteria, we can calculate global priority vector **R** ($EW = R$) and new overall ranking (Table 7.13):

$$\begin{vmatrix} 0.2000 & 0.0909 & 0.1429 & 0.2000 \\ 0.6000 & 0.4545 & 0.7143 & 0.6000 \\ 0.2000 & 0.4545 & 0.1429 & 0.2000 \end{vmatrix} \cdot \begin{vmatrix} 0.3750 \\ 0.3750 \\ 0.1250 \\ 0.1250 \end{vmatrix} = \begin{vmatrix} 0.1519 \\ 0.5597 \\ 0.2883 \end{vmatrix}$$

It follows from the overall ranking that even after the inclusion of a new design constraint, the timing belt remains the most appropriate model of shape.

Exclusion of Criteria

It may happen during design that the design requirements are reduced, which may alter the selection. Again, new requirements can be treated elegantly by executing the AHP method on a computer.

Let us now assume that the requirement for damping during the transformation of mechanical energy is removed and that only the following two requirements remain:

- Accuracy of transformation
- Long center distance

The new comparison matrix **A** follows (Table 7.14).

Priority vector **W** and consistency according the above comparisons are

$$\mathbf{W} = \begin{vmatrix} 0.5000 \\ 0.5000 \end{vmatrix} \qquad \text{CI: } 0.0000 \qquad \text{CR: } 0.0000$$

The consistency ratio is CR < 0.1; therefore, the judgment is consistent. Considering the pairwise comparison of the criteria and the components with respect to individual criteria, we can again calculate global priority vector **R** ($E\,W = R$) and new overall ranking (Table 7.15):

$$\begin{vmatrix} 0.2000 & 0.0909 \\ 0.6000 & 0.4545 \\ 0.2000 & 0.4545 \end{vmatrix} \cdot \begin{vmatrix} 0.5000 \\ 0.5000 \end{vmatrix} = \begin{vmatrix} 0.1455 \\ 0.5273 \\ 0.3273 \end{vmatrix}$$

TABLE 7.15 Overall Ranking of the Models of Shape

Tming belt	0.5273
Roller chain	0.3273
Rack and pinion	0.1455

FIGURE 7.22 Growth of a flexible function structure — generation of the first successor.

Even after reduction of design requirements determined before the selection of a model of shape, the order of appropriateness of models of shape remains the same. The timing belt remains the most appropriate. All models of shape and working principles can be evaluated in the described manner.

Let us now continue composing our technical system. Selection generates the first successor of the AND node in the input branch of mechanical energy (Figure 7.22).

The selection of a timing belt (with pulley) also implies the selection of its binding function, i.e., prevention of relative rotation. This function means that the model of shape which will fulfill it must prevent relative rotation of the timing belt on a shaft, for example.

According to the design model, the binding function becomes the new (basic) function, which needs to be fulfilled, and it therefore means another cycle of use of the basic conceptual design model.

The computer system finds appropriate working principles for its function:

1. Hooke's law and friction
2. Form-fit
3. Material connection (adhesive, solder, weld), etc.

Using the AHP method, the design engineer selects, e.g., a material connection, for which appropriate models of shape are:

1. Adhesive
2. Solder
3. Weld, etc.

The design engineer again applies the AHP method and selects, e.g., adhesive, the binding function of which is transmission of forces/torques, for which appropriate working principles are as follows:

1. Hooke's law
2. Viscoelasticity
3. Magnetic field, etc.

The design engineer selects Hooke's law, while the available models of shape are shaft, etc.

Once shaft is selected as the appropriate model of shape, the computer can again find appropriate working principles for its binding function, output of reaction forces and rotation:

1. Rolling
2. Oil wedge formation
3. Air cushion formation
4. Electromagnetic field formation, etc.

For such a function, the following design criteria can be assumed:

1. Low friction
2. Necessity of lubrication, etc.

Using the AHP method, the design engineer selects, e.g., oil wedge formation, for which appropriate models of shape are sought according to the conceptual design model:

1. Bush
2. Special journal bearing, etc.

By selecting bush, the design engineer also selects attachment as the binding function. Appropriate working principles are:

1. Hooke's law
2. Viscoelasticity, etc.

Using the AHP method, the design engineer selects viscoelasticity, for which the following models of shape are appropriate: plastic housings of different implementations, etc.

The binding function of plastic housings is *none*, with which the conceptual design of the input branch of mechanical energy of the developing technical system can be concluded according to the basic design model (Figure 7.23).

Since the performance site of the required function is the AND node, the design engineer must still compose the output branch of mechanical energy. Such continuation is dictated by the control procedure, which is determined by the basic function structure of a single stage mechanical drive unit (Figure 7.9 and the continuation of composition rule).

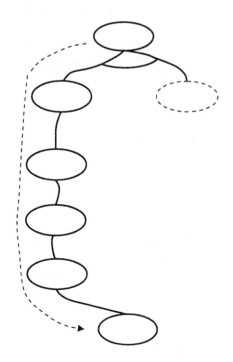

FIGURE 7.23   End of composition of the input branch of the technical system.

Since, as we have said, the performance site of the required function is the AND node, all direct successors of the same AND node perform the same function, i.e., the transformation of mechanical energy. In accordance with this, the user can now choose only among models of shape of the already selected working principle, and even these were determined at the selection of the input branch, since the combination timing belt–pulley requires (in general, at least) two pulleys. The only remaining choice is therefore:

1. Pulley
2. Pulley with integrated shaft
3. Pulley with integrated hinge, etc.

Here the design engineer decides on pulley with integrated hinge, which requires attachment as a binding function, and that is performed by the plastic housing, which was already seen during the composition of the input branch.

In this manner, the design engineer can design technical systems (in our case a printer head drive, a member of the family of single stage mechanical drive units in general) by using a computer system based on the conceptual design model (Figure 7.24).

Let us briefly repeat the composition of our technical system: the recognized required function is the transformation of mechanical energy, for which the computer system finds working principles which can fulfill the function. After its selection, a search for an appropriate model of shape follows, and with its selection, the relevant binding function is also selected. This function represents a new cycle of using the basic conceptual design model, which is repeated until the design engineer selects a model of shape, the binding function of which is *none*, which concludes the composition of the input branch of our technical system.

It can be seen from the basic function structure that the root of the tree is an AND node; therefore all the remaining successors, i.e., in our case the output branch of mechanical energy, also need to be fulfilled. The basic conceptual design model is also used to compose the output branch, and this is determined by a control procedure written on the basis of the basic function structure. Composition is repeated in a manner

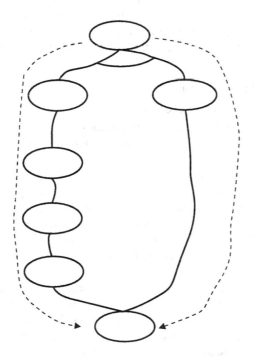

FIGURE 7.24 Designed technical system.

analogous to the one used for the composition of the input branch. The selection of working principles and models of shape is done using the AHP method.

7.6   Assessment of a CAD System for the Conceptual Design of Mechanical Drive Units

Assessment of the Conceptual Design Model

A conceptual design model consists of the basic model and control procedures derived from the basic function structure. The basic conceptual design model is simple and identical for all types of technical systems, and our experience to date suggests no problems.

The difficult phases of preparation of a computer-aided system are setting up the basic function structure, derivation of control procedures separately for each type of technical system, and the recording of the content of attribute vectors. In fact, all phases are crucial for the proper functioning of the model and its computer implementation, and they need to be produced manually by experts. A good approach to the production of basic function structures of complex technical systems is to create the basic function structures of individual functional subassemblies (in an automobile, these are the engine, gear box, suspension, etc.), which makes mastering the entire technical system easier. The derived control procedures then call the basic conceptual design model.

The difficulty of describing models of shape is shown above all in finding neutral basic and binding functions, which, in fact, provide for a variety of uses of the available models of shape. With the design engineer's cooperation, a computer-aided system for conceptual design of mechanical drive units is capable of designing flexible function structures (and, naturally, model ones) of single-stage and multiple-stage mechanical drive units with one input and output of mechanical energy, and of finding the relevant working principles and models of shape for each function. In multiple-stage mechanical drive units, it is possible to use different working principles for individual stages. The design engineer's cooperation means assigning the required function and design requirements and selecting the most suitable working principles and models of shape from among the appropriate ones using the AHP method.

The result of conceptual design is a flexible function structure and the most appropriate models of shape assigned to functions and described together with their characteristics.

Assessment of the AHP Method

The AHP method is suitable for the selection of the most functionally appropriate working principles and models of shape in the conceptual design phase, because it is characterized by the qualitative character of available information; no standard criteria or absolute measures are available for the assessment of functional appropriateness.

Using the AHP method, the above problem is solved by pairwise comparisons of both design requirements and appropriate working principles and models of shape (alternatives in general). By mutual comparison of design requirements, the design engineer indirectly determines their importance, which is naturally reflected in determining the final ranking of alternatives. The design engineer can assess the functional appropriateness of alternatives through pairwise comparisons, since, as said, he does not have any absolute criteria available, but he knows that a certain alternative is more or less suitable than the other for individual design requirements.

The application section presents a comprehensive assessment method, which could, in fact, deter users from using it. For everyday use, a more suitable method for the preparation of such a computer-aided system may perhaps be a method in which the user performs only mutual comparisons of criteria, while the remaining pairwise comparisons and relevant design requirements have already been determined by an expert and are ready to use. Naturally, the system needs to be constantly upgraded with new models of shape, and all mutual comparisons need to be updated.

The main obstacles to the introduction of the AHP method could be resistance to the form of expressing design requirements as required by the method, and mental fixation which would cause constant use of the same alternatives, even though there are others that are more suitable, although less common.

We would also like to draw your attention to the fact that humans are capable of simultaneously distinguishing between 7 (± 2) alternatives. Comparing more alternatives causes deterioration of harmonization to such a degree that the obtained results are no longer useful. The AHP method solves this problem in such a way that, prior to the assessment of alternatives, the alternatives need to be classified into characteristic classes. The classes of alternatives are assessed first. One of the possibilities is to use exclusion criteria which irrevocably exclude certain alternatives prior to comparing, and reduce their number to 7 (± 2). So far, this option has proven appropriate.[43]

It may also happen that a certain working principle is assessed as very promising, but its implementation is difficult, or, as yet, no appropriate model of shape is available. This could be avoided by immediately selecting the most appropriate model of shape for each basic function. Naturally, this approach strongly increases the number of appropriate alternatives, but these can be reduced in the manner described in the previous paragraph. This approach can also serve as stimulation for the development of new models of shape.

Since the individual steps of this method are clearly defined, the method is suitable for indirect recording of expert knowledge regarding the appropriateness of individual alternatives, and also for the comparison of individual alternatives on the basis of feedback information that design engineers receive from users of technical systems.

7.7 How Should We Continue?

The limitation of the described model arises from the fact that it is based on the description of existing models of shape and working principles derived from them. Limitations or even uselessness are naturally shown in creative design, which is the noblest form of design. Is it possible to find a model for such a design?

In designing, one can distinguish between at least four sources which contribute considerably to the degree of innovation of technical systems: applied physical laws, and the configuration, shapes, and materials of elements, which embody these physical laws.

- Physical laws (e.g., in the majority of cases, the laser has replaced the moving magnet and moving coil in turntables).
- Geometry (e.g., automobile body shapes have become more aerodynamic).
- Configuration (e.g., presentation of split systems in the design of air-conditioning devices).
- Materials (e.g., the use of PET instead of glass in the production of packaging for drinks).

Physical laws are the largest source (Figure 7.25), since no technical system operates contrary to them. The results of creative design are therefore always within the range of validity of physical laws.[44] They can explain the operation of the existing technical systems, therefore why not use them explicitly for conceptual designing?

The above idea served as the basis for the following hypothesis: from the required function, a characteristic binding variable needs to be extracted and used to find the physical laws that contain it. But since one physical law is usually not sufficient for the conceptual design of a technical system, others need to be found by chaining, which is performed using binding variables (Figure 7.26). Binding variables are analogous to binding functions, but they are even more general, while chaining is analogous to searching for attribute vector successors.

Let us examine a short example. We are faced with a task of torque transmission to a mixer. One of the possibilities is to use friction. If this possibility is to be exploited, a normal force needs to be provided, which is possible because of the buoyancy of the mixer blades. It is interesting that here, the environment in which the designed technical system operates actually represents parts of the technical system (Figure 7.27).

physical law

geometry

materials

configuration

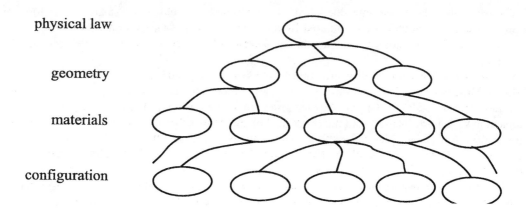

FIGURE 7.25   Tree with physical law in the highest position, followed by shapes, materials, and configurations.

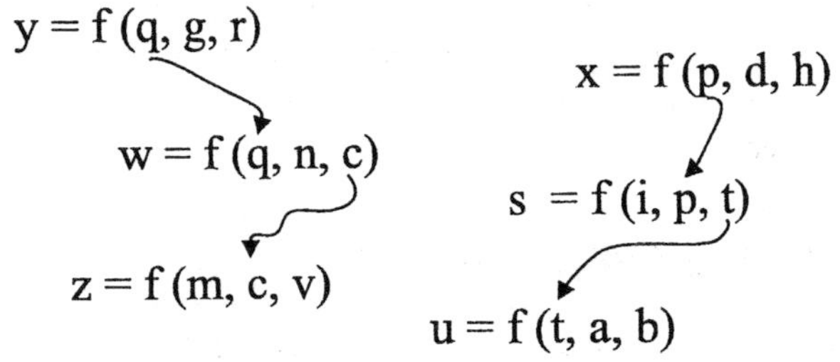

$$y = f(q, g, r)$$

$$x = f(p, d, h)$$

$$w = f(q, n, c)$$

$$s = f(i, p, t)$$

$$z = f(m, c, v)$$

$$u = f(t, a, b)$$

FIGURE 7.26   Idea of chaining.

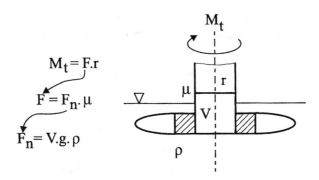

$$M_t = F \cdot r$$

$$F = F_n \cdot \mu$$

$$F_n = V \cdot g \cdot \rho$$

FIGURE 7.27   Mixer.

Confirmation of the above hypothesis and production of a new conceptual design model would mean an upgrading of the existing model and the transition to a new quality, since creative design would be possible on the basis of physical laws (see also Figure 7.25).

7.8 Conclusions

As stated in the introduction, there is great interest in design, and many different views of it have developed. Research results have been collected in various guidelines, methods, strategies, and models, but these are somehow lacking in final evaluation by their users. Only users are capable of assessing the usefulness of final results; it would therefore be advisable to encourage such assessments, since high-quality feedback would enable the improvement and perfecting of basically good methods and models.

The purpose of designing this conceptual design model is to systematize conceptual design and its simple implementation into a computer-aided system. A computer-aided system for the conceptual design of technical systems is meant to be an auxiliary tool and was not developed with the purpose of excluding the design engineer from the process of design. His role is primarily to present the variability of working principles and models of shape based on them, and the possibility of finding new ways of connecting models of shape into a functioning technical system.

From the viewpoint of composition, single-stage and multiple-stage mechanical drive units are classified as relatively simple technical systems which consist of quite special models of shape. These two facts make the designing of the basic function structures, derivation of control procedures, and production of attribute vectors much easier.

References

1. Suh, N. P., *The Principles of Design*, Oxford University Press, New York, 1990.
2. French, M. J., *Conceptual Design for Engineers*, 2nd ed., The Design Council and Springer-Verlag, Berlin, 1985.
3. Koller, R., *Konstruktionslehre fuer den Maschinenbau*, 3rd ed., Springer-Verlag, Berlin, 1994 (in German).
4. Pahl, G. P. and Beitz, W., *Konstruktionslehre*, 3rd ed., Springer-Verlag, Berlin, 1993 (in German).
5. Hubka, V., *Theorie der Konstruktions-Prozesse: Analyse der Konstruktionstätigkeit*, Springer-Verlag, Berlin, 1976 (in German).
6. Pugh, S., *Total Design: Integrated Methods for Successful Product Engineering*, Addison-Wesley, Wokingham, England, 1991.
7. Peruš, M., All in one, one in all (brain and mind in analysis and synthesis), DZS, Ljubljana, 1995 (in Slovenian).
8. Smithers, T., On knowledge level theories of design process, *Proceedings of AID 1996 Conference, Stanford, CA*, Kluwer Academic Publishers, Dordrecht, 1996, 561.
9. Wallace, K., Product development and design research, Keynote paper, International Conference on Engineering Design (ICED 97), Tampere, 1997.
10. Altšuller, G. S., *Erfinden: Wege zur Loesung technischer Probleme*, VEB Verlag, Berlin, 1986.
11. Sushkov, V. V., Mars, N. J. I., and Wognum, P. M., *Introduction to TIPS: A Theory for Creative Design, Artificial Intelligence in Engineering*, Vol. 9(3), Elsevier Science Publishers, Amsterdam, 1995, 177.
12. Killander, A. J. and Sushkov, V. V., Conflict-oriented model of creative design: focusing on basic principles, Third International Round-Table Conference on Computational Models of Creative Design, Heron Island, Australia, Dec. 1995.
13. Sushkov, V. V., Alberts, L. K., and Mars, N. J. I., Innovative design based on sharable physical knowledge, *Proceedings of AID 96 Conference, Stanford, CA*, Kluwer Academic Publishers, Dordrecht, 1996.
14. Invention Machine Demo Version, CD-ROM, Invention Machine Corporation, Cambridge, MA, 1996.
15. Ishii, M. and Tomiyama, T., A synthetic reasoning method based on a physical phenomenon knowledge base, in *AI System Support for Conceptual Design, Proceedings of the 1995 Lancaster International Workshop on Engineering Design*, Sharpe, J., Ed., Springer-Verlag, London, 1996, 109.
16. Linde, H., Gesetzmaessigkeiten der Technikevolution und ihre orientierende Nutzung in praktischen Innovationsprozessen mit WOIS, in *Proceeedings of the 10th International Conference on Engineering Design ICED 1995*, Prague, Vol. 1, Hubka, V., Ed., Heurista, Zurich, 1995, 152.
17. Linde, H. and Hill, B., *Erfolgreich Erfinden*, Hoppenstedt, Darmstadt, 1993.
18. Duhovnik, J. and Žavbi, R., Expert systems in machine design, in *Proceeedings of the 7th International Conference on Engineering Design ICED 90*, Dubrovnik, Vol. 2, Hubka, V. and Kostelić, A., Eds., Yudeko, Zagreb and Heurista, Zurich, 1990, 1038.
19. Žavbi, R., Expert System for Conceptual Design of Mechanical Drive Units, Master's thesis, University of Ljubljana, Ljubljana, 1992 (in Slovenian).

20. Duhovnik, J. and Žavbi, R., Expert systems in conceptual phase of mechanical engineering design, *Artif. Intelligence Eng.*, 7(1), 37, 1992.

21. Žavbi, R. and Duhovnik, J., Design environment for the design of mechanical drive units, *Comput. Aided Design*, 27(10), 769, 1995.

22. Encarnacao, J. and Schlechtendahl, E. G., *Computer Aided Design*, Springer-Verlag, Berlin, 1983.

23. Tomiyama, T., Umeda, Y., and Yoshikawa, H., A CAD for functional design, *Ann. CIRP*, 41(1), 143, 1993.

24. Duhovnik, J., Systematic design in intelligent CAD system, in *Intelligent CAD Systems I*, ten Hagen, P. J. W. and Tomiyama, T., Eds., Springer-Verlag, Berlin, 1987.

25. Suzuki, H., Kimura, F., Moser, B., and Yamada, T., Modeling information in design background for product development support, *Ann. CIRP*, 45(1), 141, 1996.

26. Russell, R. B., Youngsup, J., and Inyong, H., Feedback of manufacturing experience for DFM design rules, *Ann. CIRP*, 45(1), 115, 1996.

27. Dussauge, P., *Strategic Technology Management*, Prentice-Hall, Englewood Cliffs, NJ, 1991.

28. Duhovnik, J., Kimura, F., and Sata, T., Methodic in CAD, *Proceedings of Eurographics 1983*, North-Holland, Amsterdam, 1983.

29. Ulrich, K. T. and Eppinger, S. D., *Product Design and Development*, McGraw-Hill, New York, 1995.

30. Peklenik, J., An analysis of the manufacturing process topology and the process features as basis for the computer aided process planning (CAPP), 22nd CIRP International Seminar on Manufacturing Systems, Enschede, 1990.

31. Sluga, A. and Peklenik, J., The structure of the data and knowledge, CIRP — Seminar on CA-Design, Ljubljana, December 4–5, 1990.

32. Grabec, I. and Mandelj, S., Continuation of chaotic fields by RBFNN, in *Lecture Notes in Computer Science 1240*, Mira, J., Morcno-Diaz, R., and Cabestany, J. Eds., Springer-Verlag, Berlin, 1997, 597.

33. Hubka, V., Andreasen, M. M., and Eder, E. W. *Practical Studies in Systematic Design*, Butterworth, London, 1988.

34. Duhovnik, J., Tavčar, J., and Koporec, J., Project manager with quality assurance, *Comput. Aided Design*, 25(5), 311, 1993.

35. De Kleer, J. and Brown, J. S., A qualitative physics based on confluences, *Artif. Intelligence*, Vol. 24, Elsevier Science Publishers, North-Holland, Amsterdam, 1984, 7.

36. Bratko, I., *PROLOG Programming for Artificial Intelligence*, 2nd ed., Addison-Wesley, Reading, MA, 1990.

37. Fischer, G. and Nakakoji, K., Beyond the macho approach of artificial intelligence: empower human designers — do not replace them, *Knowledge–Based Syst.*, 5, 515, 1992.

38. Coyne, R., *Logic Models of Design*, Pitman Publishing, London, 1988.

39. Saaty, T. L., *Multicriteria Decision Making — The Analytic Hierarchy Process*, 2nd ed., University of Pittsburgh Press, Pittsburgh, PA, 1988.

40. Saaty, T. L., How to make a decision: the analytic hierarchy process, *Eur. J. Oper. Res.*, 48, 9, 1992.

41. Mrvar, A., Saaty's Multicriteria Decision Making, Diploma thesis, University of Ljubljana, Ljubljana, 1992 (in Slovenian).

42. Reedy, R. P. and Mistree, F., Modeling uncertainty in selection using exact interval arithmetic, *Design Theory and Methodology*, DE-42, ASME, New York, 1992, 193.

43. Žavbi, R. and Duhovnik, J., The analytic hierarchy process and functional appropriateness of components of technical systems, *J. Eng. Design*, 7(3), 313, 1996.

44. Žavbi, R. and Duhovnik, J., Prescriptive model with explicit use of physical laws, in *Proceedings of the 11th International Conference on Engineering Design (ICED 1997)*, Vol. 2, Ritahuhta, A., Ed., Tampere University of Technology, 1997, 37.

8

Computer Assembly Planners in Manufacturing Systems and Their Applications in Aircraft Frame Assemblies

T. Warren Liao
Louisiana State University

8.1 Introduction

Most products around us are assemblages of components. Some are relatively simple, consisting of a few different parts joined by a selected means. For example, some kitchen knives have wooden handles that are attached to the knife blade with metal fasteners. Others, such as automobiles and airplanes, are more complex. Complicated products normally use a greater number of parts of different sizes, shapes, and materials. All of these parts must be assembled using one or several joining methods such as welding, brazing, soldering, adhesive bonding, or mechanical fastening. One or more of these joining methods is consistently found in a manufacturing system built to produce a specific product. For example, welding is the major joining method used for shipbuilding, oil platform, and pipeline fabrications. Soldering is critical to electronic manufacturing. On the other hand, a high percentage of aircraft assembly operations require riveting.

Planning of production processes, whether they are forming, machining, or joining processes, is an important engineering activity in a manufacturing system. The objectives are not only to meet the quality and specification requirements, but also to reduce the production cost. The term *process planning* can be used generally for any production process. In a broad sense, process planning can be defined as the act of preparing detailed work instructions to produce a part or to assemble a product. It establishes the sequence of the operations to convert a piece part from its initial form to the final form or an intermediate product into the final product, as specified in the product design. For machining, the operation could be turning, milling, drilling, etc. For assembly, the operation involves welding, brazing, soldering, adhesive bonding, and mechanical fastening to combine the parts, subassemblies, and major assemblies into the final product. The term *assembly planning* is often used for the planning of assembly processes. The process (or assembly) plan as a result of planning provides an important input for design evaluation, production planning, and scheduling.

The process planning (or assembly planning) activity is usually performed by a process engineer (or assembly engineer). In order to prepare a process plan, the engineer must have the following knowledge [6]:

- Ability to interpret engineering drawings
- Familiarity with production processes and practice
- Familiarity with tooling and fixtures
- Knowledge of what resources are available in the shop
- Knowledge of how to use reference books, such as machinability data handbooks
- Ability to do computations on operation time and cost
- Familiarity with raw materials
- Knowledge of the relative costs of processes, tooling, and raw materials

Traditionally, the process planning task has been manually performed. Process planning follows an iterative procedure that generally has the following steps:

- Understand the drawings and design specifications.
- Identify datum surfaces and use this information to determine the setups.
- Select machines for each setup.
- Determine the sequence of operations in each setup.
- Select tools for each operation.
- Select or design fixtures for each setup.
- Select operating parameters.
- Prepare the final process plan.

Past efforts to automate process planning focused mainly on machining processes. These efforts can be roughly classified into four phases. The first phase consisted primarily of building computer-assisted systems for report generation, storage, and retrieval of plans. The second phase successfully implemented the so-called variant type CAPP (Computer-Aided Process Planning) systems. Previously prepared process plans are stored in a database. When a new component is planned, a process plan for a similar component is retrieved and subsequently modified by a process engineer to satisfy special requirements. The third phase focused on eliminating the process engineer from the entire planning function. Decision making logics required for process planning are implemented using artificial intelligence (AI) techniques to generate process plans automatically for new components without referring to existing plans. This type of CAPP system is generative. The fourth phase took another step farther by trying to automate the drawing interpretation process via the development of a feature extraction module, which served as the CAD interface. As CAD (computer-aided design) and CAM (computer-aided manufacturing) systems become popular, CAPP (computer-aided process planning) has been considered the critical link to realizing an important part of the CIM (computer-integrated manufacturing) concept. This was the primary factor behind the third and fourth phases of process planning automation. Despite the fact that much progress has been made, the goal of eliminating the process engineer from the entire planning function unfortunately has not yet been achieved. The major difficulties come from the incompleteness of design models for process planning, the vastness of process planning knowledge, and the uniqueness of each manufacturing system.

Compared with machining processes, the progress made in automating the assembly process planning lags far behind. This chapter intends to provide a review of assembly planners developed in the past and techniques used in their development, followed by a discussion of the application of some of these techniques in aircraft frame assemblies.

8.2 Techniques of Computer Assembly Planners

Assembly planning is needed to determine how to put together the components in the assembly including the assembly sequence, tools, fixtures, etc. Most of the studies and the computer assembly planners developed are restricted to the determination of assembly sequence. There are generally two ways to determine the assembly sequence(s). One starts from the assembly and ends when all components are disassembled. The other starts from individual components and ends when they are all assembled. The former is based on disassembly planning, which is also called decomposition, top-down, backward, or indirect planning. On the other hand, the latter is based on assembly planning, which is also called construction, bottom-up, forward, or direct planning. The majority of planners use disassembly planning to take advantage of its efficiency. The sequence of assembly is then obtained as the reverse of the sequence of disassembly. If the only concern is the geometric feasibility of the assembly (i.e., that parts do not intersect) then this approach has the important advantage that the planner never has to backtrack. This advantage vanishes when the goal is to optimize with respect to some reasonably complex goodness function. In addition, disassembly planning is not applicable in some cases, e.g., when there are snap-together parts. The sequence generated by disassembly planning might also incur problems during the task execution stage.

Both assembly and disassembly planning requires some knowledge and reasoning method to generate assembly sequences. The knowledge could include interpreting the assembly drawing, process and resource characteristics, and manufacturing heuristics accumulated based on past experience or even past assembly plans. The knowledge could be implicit in the assembly model, acquired from human experts in the development stage, or captured via an interactive program. The heuristics used typically depend upon the assembly and the type of operation that is involved. The reasoning method could be based on geometric reasoning, constrained search, rule-based, or case-based reasoning. The reasoning method could be implemented algorithmically. If so, it must be applied in a specific order. Various constraints (geometrical or non-geometrical) are considered during the reasoning process. When planning for automatic assembly, it is necessary to consider not only the constraints imposed by the product itself but also those imposed by the automatic or robotic system. Note that some planners generate all possible assembly sequences while others generate only one. Some consider subassemblies; others do not. Generating all possible sequences enables optimization later and offers alternatives for scheduling. Considering subassemblies allows concurrent operations, which means shorter in-process time.

Techniques for assembly sequence generation can be roughly classified into the following categories: the algorithmic approach, the integrated approach (usually consisting of algorithms and interactive programs), the knowledge-based reasoning approach, and the case-based reasoning (CBR) approach. Each of these approaches is described below.

Algorithmic Approach

The algorithmic approach organizes planning knowledge in the form of algorithms, which are, in turn, implemented using a procedural language like C or a functional language like CommonLisp. This approach has been utilized in both assembly and disassembly planning. In either case, geometric reasoning is often applied to verify the ability to assemble or disassemble based on various criteria such as accessibility, interference, blocking, structural stability, etc.

Disassembly Planning

Sekiguchi et al. [35] used this approach to automatically determine the disassembly sequence of machine units from the assembly drawings. The connective relations between the parts were extracted by analyzing the engineering drawings and represented as matrices. Four types of rules were created for generating the disassembling sequence: the first is to classify a part into the part group, the second is to select the part to be disassembled from outside to the inside, the third is to determine the direction to be disassembled, and the fourth is to determine which part is to be moved for the disassembly after carrying out a geometrical interference check. The sequence of disassembly was given in a list.

Assuming that an assembly task can be obtained by reversing a disassembly task, Khosla and Mattikali [19] developed a reasoning system that performs a disassembly planning of the model of mechanical systems and assemblies (MSA). The MSA is obtained by applying a topological merge operation on the set of components of the assembly. The locations and orientations of the components in the assembly are specified interactively by the designer by using a set of movement and orientation operators. A graph called the component graph is created to explicitly represent the mating between components. The mating between components is inferred automatically from the geometric model. The heart of the reasoning system consists of a module named "split" that accepts the given subassembly consisting of n components and finds a suitable division of these components in two groups that represent the two subassemblies or components that are formed after the action of one disassembly operation. The disassembly procedure (and possible alternatives) is recorded in the form of an AND/OR graph representing a hierarchy of subassemblies and components.

The algorithmic approach was taken by Lee and Shin [25] and Lee [24] to perform disassembly planning by establishing the ability to disassemble parts and subassemblies and by extracting preferred subassemblies. An assembly is represented by an attributed liaison graph in which a node represents a part and an edge specifies a relationship between two parts. The information of parts and edges are stored in part frames and liaison frames, respectively. Geometric reasoning is applied to facilitate an automatic deduction of the ability to disassemble a given part based on the accessibility and manipulability criteria. The system first decomposes the given assembly into clusters of mutually inseparable parts. An abstract liaison graph is then generated, on which each cluster of mutually inseparable parts is represented as a supernode. A set of subassemblies is then generated by decomposing the abstract liaison graph into subgraphs, verifying the ability to disassemble individual subgraphs, and applying the criteria for selecting preferred subassemblies to those subgraphs. The recursive application of this process to the selected subassemblies results in a hierarchical partial-order graph (HPOG).

Homem de Mello and Sanderson [16] presented an algorithm for the generation of all mechanical assembly sequences for a given product based on disassembly planning. The algorithm employs a relational model of an assembly represented as a quintuple $\langle P, C, A, R, a\text{-function}\rangle$, where P is a set of parts, C is a set of contacts between surfaces of two parts of the assembly, A is a set of attachments that act on a set of contacts, R is a set of relationships with each between pairs of elements of $P \cup C \cup A$, and a-function is a set of attribute functions whose domains are subsets of $P \cup C \cup A \cup R$. The basic idea underlying the decomposition approach taken is to enumerate the decompositions of the assembly and to select those decompositions that are feasible. A decomposition of an assembly $\langle P, C, A, R, a\text{-function}\rangle$ is a pairs of subassemblies $\langle P_{S1}, C_{S1}, A_{S1}, R_{S1}, a\text{-function}_{S1}\rangle$ and $\langle P_{S2}, C_{S2}, A_{S2}, R_{S2}, a\text{-function}_{S2}\rangle$. The set $C_{S1-S2} = C - (C_{S1} - C_{S2})$ is referred to as the contacts of the decomposition, which defines a cut-set. Conversely, a cut-set in the graph of connections of an assembly defines a decomposition of that assembly. The decompositions are enumerated by detailing the cut-sets of the assembly's graph of connections. Feasibility tests including task-feasibility and subassembly-stability are then applied to check each decomposition. The algorithm returns the AND/OR graph representation of assembly sequences.

For expressing an assembly as a hierarchical relation graph consisting of surface-level layer, part-level layer, and subassembly-level layer, Yokota and Brough [41] developed a program to determine precedence automatically from surface contact relations and labels classifying permitted movements. If the absence of a part gives more freedom of movement to another part, the former has precedence over the latter. The precedence relation is local to the parts concerned, and signifies the partial order of disassembly. Another method was used to find reduced precedence graphs (or subset graphs) by resolving loops, branches, and merges in the precedence graph. From the partial orders represented in the precedence graph, valid disassembly sequences for the assembly as a whole can be found. A procedure was also developed to identify groups of possible subassemblies from the precedence graph. Blocking between noncontact parts was not considered.

Dini and Santochi [12] presented an assembly planning software system called FLAPS (flexible assembly planning system) that implemented a disassembly procedure for the selection of the subassemblies and the assembly sequences of a product. The procedure is based on a mathematical model of the product, obtained by defining three matrices: the interference matrix, the contact matrix, and the connection matrix. Each matrix is evaluated along the x, y, and z directions of the Cartesian coordinate system of the CAD space. Consequently, the total number of matrices is 9. The possible subassemblies are automatically detected by satisfying some mathematical conditions applicable to these matrices. For each subassembly and for the whole product, all the possible assembly sequences are generated.

Wilson and Latombe [40] investigated the planning of assembly algorithms (sequences), which specify (dis)assembly operations on the components of a product and the ordering of these operations, from CAD data. The central concept is that of a non-directional blocking graph (NDBG) conceived to represent the internal structure of an assembly product. The construction of an NDBG is based on the identification of physical criticalities to decompose a continuous set into a finite number of regions that are treated as single entities. Its construction derives from the observation that infinite families of motions can be partitioned into finite collections of subsets so that the interferences among the parts are constant over every subset. Different from generate-and-test assembly sequencers, this representation describes the combinatorial set of parts interactions in polynomial space.

Laperrière and ElMaraghy [21] developed a generative assembly process planner (GAPP) that takes as input a solid model of the product to be assembled and outputs its feasible assembly sequences. A geometric reasoning method was developed to build the internal computer representation of the graph model automatically from the information contained in the B-rep file resulting from the product's solid model, using ICEM/DDN commercial hybrid solid modeler. Three types of mating relations are considered: contact, blocking, and free. For every contact surface pair, a 3×4 matrix called the freedom matrix is built automatically. An algorithmic engine is used next to generate assembly sequences by constrained expansion, which is accomplished through the computation of the cut-sets. The process starts by putting the product's graph model as the root node of a search graph. The process stops when a node has been generated where all edges have been removed. Three feasibility constraints are used to ensure that the generated assembly sequences are indeed feasible. They are geometric interference constraints, stability constraints, and accessibility constraints. Four criteria are considered to determine the relative goodness of the feasible alternatives: reorientations, parallelism, stability, and clustering. The final result is presented in an ordered list.

Considering the case where parts have toleranced geometry, Latombe et al. [22] developed two procedures for assembly sequence generation. The first is an efficient procedure that decides if a product admits an assembly sequence with infinite translations (i.e., translations that can be extended arbitrarily far along a fixed direction), which is feasible for all possible instances of the components within the specified tolerances. If the product admits one such sequence, the procedure can also generate it. For the cases where no such assembly sequence exists, another procedure is proposed which generates assembly sequences that are feasible only for some values of the toleranced dimensions. If both procedures fail to produce any sequence, then no instance of the product can be assembled. The NDBG concept presented by Wilson and Latombe [40] was extended to introduce the strong NDBG and the weak NDBG for representing possible and necessary blocking interferences among parts of the assembly for infinite translations, respectively. Though assumptions made do not reflect realistic imperfections of a manufacturing process, this work is definitely one step towards dealing with tolerances in assembly sequencing.

Assembly Planning

Ko and Lee [20] applied the algorithmic approach to derive the precedence graph of an assembly which is expressed as a graph showing the mating conditions between components. The assembly procedures are generated in two steps. First, each component in an assembly is located at a specific vertex of a hierarchical tree. Second, a precedence graph is generated from the hierarchical tree with the help of interference checking.

Roy et al. [32] proposed a design environment which acts as a preprocessing step before the actual assembly operation. The assembly is first modeled using the modified functional relationship graph (M-FRG) that captures the functional relationships between components or subassemblies. A procedure is then used for the generation of a second-order relationship graph (R^2), which specifies the temporal relationships among the relationships specified in M-FRG. Subsequently, a topological sorting algorithm is applied to the R^2 graph to give a topological order of mating relationships. A directed assembly tree is finally created by using the information available from the topological ordering of nodes in the R^2 graph and the R^2 graph itself.

Lin and Chang [27] developed a methodology for automatic generation of assembly plans for three-dimensional mechanical products. Matching and collision information is inferred from the assembly solid models. The information of part mating and collision is stored in three graphs: the part connectivity graph, the mating direction graph, and the spatial constraint graph. A frame-based representation scheme is also used to explicitly represent the assembly nongeometric information. To facilitate the manipulation of the symbolic information, a decision tree is designed to convert nongeometric frame models into a set of sequence constraints. Assembly plans are generated by a two-stage planning scheme. In the initial planning, assembly plans are generated based solely on product geometry considerations. This is achieved by a three-level planning strategy which analyzes the part connectivity relationships in a hierarchical manner and ensures collision-free insertion of individual parts. The initial assembly plans are further revised in the second stage under sequence constraints derived from nongeometric properties. Both high-level assembly sequence information and low-level assembly process details are included in the final assembly plans, allowing feasibility of applications in different manufacturing environments. Assembly precedence diagrams are used to represent feasible assembly sequences.

Ben-Arieh and Kramer [2] developed a two-stage methodology to consistently generate all feasible assembly sequences in consideration of the various combinations of subassembly operations. In the first stage, all feasible part introduction sequences are generated. The inputs to the first stage include a list of components, a list of contacts for each component, and a list of precedence constraints (geometric and technological). In the second stage, each feasible sequence is expanded to generate all feasible combinations of subassemblies. The algorithms were implemented in LISP. An assembly sequence is represented implicitly by nested lists of components.

Rajan and Nof [31] developed a new precedence constraint representation, called minimal precedence constraint, that minimally constrains the planning process. Procedures were also developed for its generation for components and liaisons from the CAD description of the assembly. The minimal precedence constraints for a component (liaison) i represents the alternate combinations of other assembly components whose presence in the current assembly state will preclude the assembly of the component (liaison). The minimal precedence constraints for component i, denoted as PR_i, are the set of all minimal subassembly states that will prevent its assembly. The minimal precedence constraints for a liaison between components i and j, denoted as PR_{ij}, are given by

$$PR_{ij} = (PR_i \land j) \lor (PR_j \land i)$$

The interactions of the minimal precedence constraints with cell constraints are analyzed to determine the existence of a global execution plan. In other words, the assembly and execution planning processes are integrated by simultaneously considering the satisfaction of assembly and cell constraints.

Chakrabarty and Wolter [4] implemented a simple planner (HAP, or hierarchical assembly planner) which views an assembly as a hierarchical collection of standard structures. HAP plans bottom-up in the structure hierarchy; first plans are generated for the lowest-level structures, and then they are merged to get plans for higher-level structures. A possible plan is represented as a hierarchy (tree) of partially ordered insertion sequences (POIS). Each node of the POIS represents an operation inserting a part or subassembly along some trajectory. HAP recursively generates a user-specified number of best plans for each part set, and then performs a best-first search for the best way to combine them. Combining plans for two different part sets to create a plan for the union of the part sets is called merging. Two kinds of merges are possible: direct and indirect. Indirect merges involve introducing a new subassembly to the plan, while direct merges do not. Each plan is evaluated based on a cost function, which is defined as a weighted sum of ratings of four criteria: directionality, manipulability, attachment, and similarity. The relative weights are selected by the user.

Integrated Interactive/Algorithmic Approach

The integrated approach normally uses a combination of interactive and algorithmic programs. The interactive program is used to capture some constraints on assembly plans from users. It is argued that interaction with users is needed because some constraints on assembly plans are not well understood to encode in a computer program, and other constraints are too computationally expensive to evaluate under current methods. The information (relations) captured by the interactive program together with information about other constraints is subsequently utilized in planning to generate all valid assembly sequences. The planning knowledge is organized in the form of algorithms just like the algorithmic approach. This approach was taken in the studies described below, which mainly differ in the form and number of questions asked.

Bourjault's approach [3] began by using the information contained in a parts list and an assembly drawing to characterize the assembly by a network, wherein nodes represent parts and lines between nodes represent any of certain user-defined relations between parts called liaisons. Nodes and lines are subsequently named. The precedence rules that permit algorithmic generation of (only) valid number strings representing assembly sequences are derived from yes/no answers to a series of questions about individual mates. The questions are of two general forms (where L_i is to be read as the liaison numbered i):

1. Is it true that L_i cannot be done after L_j and L_k have been done?
2. Is it true that L_i cannot be done if L_j and L_k are still undone?

The group (L_j and L_k) is called the body of the question. Questions of both forms occur in two stages of the technique, called step one and step two; the latter step is organized into a plurality of modules. The questions of step one each involve pairs of liaisons (there is only one liaison in the body); thus there are $2l^2$ questions (l is the number of liaisons in the assembly) in step one. The number of questions (Q) associated with later modules of step two depends on the answer set to all prior questions as well as on l, but the total number of questions cannot exceed 12^l. Symbolically, for $l \geq 3$,

$$12^l > Q \geq 2(l^2 + l)$$

Assembly sequences are presented as branches of an inverted tree with the initial disassembled state (no liaisons established) at the top rank and a representation of each valid first liaison connected to and below the top rank entry.

De Fazio and Whitney [10] modified Bourjault's method by asking the following two questions for each liaison:

1. What liaisons must be done prior to doing liaison i?
2. What liaisons must be left to be done after doing liaison i?

Answers are expressed in the form of a precedence relationship between liaisons or between logical combinations of liaisons. There are exactly $2l$ such questions, two associated with each liaison. The presumed starting state is that of disassembly; no liaison is established. To generate liaison sequences, begin by scanning the liaison list and the answers for those liaisons which are not precedented. Any of these may serve as the first liaison to be established. Line up representations of each first possible state across a rank and connect each with the starting state by a line. For each possible first liaison, explore for all possible subsequent states by again scanning the liaison list, the precedence relations (answers), and any other constraints imposed on the assembly, thereby generating another rank. In this fashion, one proceeds algorithmically to the finished state where all liaisons have been established. Note that some errors or oversights in answering usually lead to variations of the generated assembly sequences.

Baldwin et al. [1] developed an integrated set of user-interactive computer programs that generates all feasible assembly sequences for a product and then aids the user in judging their value based on various criteria. Generating all possible assembly sequences is done using the cut-set method to find and represent all geometric and mechanical assembly constraints as precedence relations. First, topological information about an assembly is entered and represented as a liaison diagram. Next, all possible subassemblies are generated by testing the connectivity of the corresponding subgraphs. The assembly cut-sets are generated next. Assembly cut-sets are defined by two part sets (i.e., subassemblies) N_i, N_j and the liaison set S connecting N_i to N_j. The assembly cut-sets are used to generate all the queries needed to determine the precedence relations for the assembly. The queries take the form, "Can the subassembly formed by the parts of N_i be disassembled from the parts of N_j?" All queries are then analyzed to generate precedence relations. Once the precedence relations are generated, they are passed to a program, which converts them into the assembly sequence diagram or an assembly sequence network. Each box in the network is an assembly state, and each line is an assembly move.

Chen [7] introduced a novel approach based on the concept of the constrained traveling salesman problem (TSP) to acquire all the precedence knowledge. The assembly is expressed as a liaison diagram. For each liaison, two questions are asked: If liaison x is not done, what liaisons combined cannot be done and what liaisons combined can be done and if liaison x is not done, what liaisons combined cannot be done and what are the dominant tasks? Tasks are said to be dominant if those tasks combined with the tasks that cannot be done will make the rest of the assembly operations possible. Based on the precedence knowledge, the algorithm FASG (feasible assembly sequences generation) generates all the feasible assembly sequences by using a pattern-matching algorithm which correctly expands all the qualified nodes to satisfy precedence constraints. The results are represented in a tree.

Wilson [42] presented a geometric assembly sequencing testbed called GRASP that implemented a dual approach developed to reduce the human input required to construct assembly plans. First, most assembly operations are validated automatically from the CAD models using simple and fast techniques that check the local freedom in removing a subassembly and the interference of its straight-line motion with other parts. Second, with each query the user is allowed to identify a set of parts that constrain a subassembly; the system uses this information to answer future queries automatically. It was shown that identifying constraining sets resulted in far fewer queries than the other methods. Planning proceeds backward from the assembled state and the developing set of sequences is stored in an AND/OR graph.

Knowledge-Based Approach

This approach generally models actions in assembly processes by production rules in the form of IF ⟨condition⟩ THEN ⟨action⟩. The condition part may consist of a single fact or a conjunction of several facts in assembly planning, either supplied by the user or induced by the inference engine. They form conditions of the rule, and only when those conditions are satisfied can the rule be fired. In a similar way, the action part models the action taken during an assembly process. The actions can be a mating between two components, a change of fixture or tool, etc. A rule-based system could also include procedural knowledge as part of the knowledge.

Most of the electronic assembly planning systems reported in the open literature are knowledge-based. Assemblies of electronic products could be divided into three levels: component packaging, printed circuit board (PCB) assembly, and equipment assembly. The components in PCB are basically of two types: leaded or through-hole and surface-mounted. Maria and Srihari [29] reviewed the application of knowledge-based systems in PCB assembly. Other systems that have been reported include HICLASS [28], OPGEN [13],

PWA-planner [5], SAPIENT [18], AUTOGEN [36], etc. Note that almost all rule-based assembly planners use backward chaining to perform disassembly planning. Only the assembly planners developed for mechanical products are reviewed below.

Sanderson and Homem-de-Mello [34] modeled an assembly consisting of a set of rigid objects as a relational graph showing the contact relations between objects. The problem of assembly sequence planning was considered a goal-directed planning problem which looks for ways to decompose the relational graph representation in a sequence of feasible cuts. A rule-based non-linear planning approach was used to generate feasible assembly sequences based on the relational model. The resulting plans were consolidated into an AND/OR graph representation.

In Delchambre's system [11], the user provides a simplified geometric model of the final product as well as some non-geometric data for specific component and for the final assembly. From this information and with the help of a geometrical reasoning module, the product graph is built containing all the relational information among the parts in the final product. Some precedence orders are then generated in consideration of geometrical, mechanical, technological, and stability constraints. From these precedence orders, assembly plans in the form of precedence graphs are generated. All knowledge was expressed in Prolog.

Huang and Lee [17] developed a knowledge-based assembly planning system coded in C and Lisp for the automatic generation of assembly plans from the CAD model of an assembly. The knowledge of assembly structure, precedence constraints, and resource constraints, represented in predicate calculus, formed the static knowledge base. Three basic rules were implemented: disassembly, change-fixture, and change-tool. Assuming that all disassembly plans are reversible, assembly plans are generated by repeatedly applying the above rules. The results are given in a list.

Tonshoff et al. [44] developed an expert system for determining the optimum assembly sequence for assembly tasks with few components. The assembly tasks determined by analyzing the assembly drawings and the parts list are represented in the knowledge base by facts. All joining positions for the objects to be assembled are automatically identified from the contact relationships. For each connection of the joining positions, technological data important for the description of the assembly process are taken from the drawing. Rules on assembly technology including task-related rules, object-related rules, and organizatory rules were formulated by studying possible process sequences. The task-related rules are used to avoid cases where improper assembly sequences result in situations where some objects cannot be assembled. The properties of the objects that influence the assembly sequence are formulated in object-related rules. In addition, organizatory rules regarding productivity and efficiency are taken into account, too. Dependencies between joining methods are considered to derive a priority list which is used to limit the options when determining the assembly sequence. A further reduction of the number of possible assembly sequences is achieved by including the technical function of the connection made from the separate components. The system was coded in Prolog by using the backtracking method (disassembly).

Henrioud and Bourjault [14] developed a prototype software program LEGA (written in Prolog) based on a top-down algorithm. Each product P is represented by a 5-tuple $\langle C, \Gamma, \Sigma, \Delta, f \rangle$ where C = set of components of P, Γ = set of the liaisons in P, Σ = set of attachments of P, Δ = set of complementary features of P, and f defines the set of components and functional features concerned for each attachment or each complementary feature. Starting from the end product, all the valid operations are given, each defining one or two parts. This process is repeated for each part until the components are produced. An operation is valid when it agrees with the set of operative constraints of three types: geometric, material, and stability. The constraints are described by a database which includes a set of facts and a set of rules. A set of global constraints may be added to avoid the generation of inefficient solutions. LEGA produced a set of lists representing the assembly tree. A new concept of precedence hypergraph was introduced, allowing a unified representation of disjunctive precedence conditions.

Chen and Pao [8] developed a case associative assembly planning system (CAAPS) for general mechanical assemblies based on the integration of neural computing techniques and rule-based systems. An assembly is expressed using a set of pre-defined design features. A component of CAAPS called CLIPS-based assembly planning module (CBAPM) generates a task-level assembly plan automatically. The CBAPM consists of five submodules. The preprocessor submodule (1) performs a topological and geometric data consistency check of the assembly B-rep input data, (2) calculates the plane equation for each face in the assembly, and (3) determines the candidate assembly directions for each component in the assembly. The liaison detection submodule determines any of three types of liaisons (insertion, attachment, and contact) between any two parts. The output of this submodule defines the product graph. The obstruction detection submodule determines which components obstruct the removal of a particular component by graphically projecting the component in question onto a given plane. The projections of all other components on the same plane that have non-null intersections ascertain the component's obstruction. The output of this submodule is a list of obstructions. Construction of the assembly plan is performed by the plan formulation submodule via a disassembly approach. The adaptation-and-modification submodule is to adapt and modify assembly plans efficiently through a PCM (plan cluster memory) that stores previous assembly plans. The assembly sequence generated is represented as an assembly tree pattern (ATP), in which each internal node represents an assembly operation and each leaf node is a mechanical part. The assembly plan can be obtained when an ATP is traversed in preorder.

Seow and Devanathan [45] proposed a temporal logic framework to unify all the implicit assembly sequence representations, upon which they could be precisely described and therefore mathematically reasoned upon, so as to generate the desired assembly sequences. An implicit assembly sequence representation refers to the conditions or constraints that assembly sequences must satisfy. MASS-C was built based on the proposed framework using Quintus PROLOG. The implementation has been carried out using the generate and test paradigm. Each possible combination of assembly tasks (in ordered list) is systematically generated and then tested to see if it satisfies the integrated constraint (hard as well as soft constraints).

Case-Based Approach

Unlike the other approaches, which all attempt to generate assembly plans from scratch, the basic idea of CBR is to derive a solution to a new problem effectively and efficiently by referring to the solution of a similar previous case and then adapting its solution appropriately to fit the new problem. The CBR approach comprises four major steps: case representation, case retrieval, case adaptation, and learning. The representation of cases is essential because it determines what the case memory can be used for, how cases can be retrieved, and whether there is enough information for case adaptation. Given a new case, the aim of the case retrieval process is to identify the most similar cases. The retrieval technique depends on the structure of case memory and the associated index scheme. The index schemes makes the retrieval process more selective and reduces the effect of the memory size. Case adaptation involves taking a selected case, making changes to the case, and labeling the revised case as the new solution. Learning in CBR is achieved simply by storing an old case in the case base.

Pu and Reschberger [30] implemented a test-bed assembly planner, called CAB-assembler, based on the idea of case-based reasoning. The case base consists of assembly solutions for subproblems, each corresponding to one step in the entire assembly process. Each case has the following fields: (1) a case feature vector including both geometrical and spatial information about the parts of the product; (2) the feature weight for each feature; (3) the success and failure history for each feature; and (4) a plan or solution for the case. Case selection is based on the priority of a case, which is defined as the sum of the product of each feature priority and its weight. The priority of each feature is calculated as follows:

$$
\text{priority}(f) = \begin{cases} 0 & \text{if features do not match or \# successes} < \text{\#failures} \\ \dfrac{\text{\#successes} * (\text{\#successes} - \text{\#failures})}{\text{\#successes} + \text{\#failures}} & \text{otherwise} \end{cases}
$$

Zarley [42] implemented a case-based assembly planner to produce process plans for pulley bracket assemblies. A pulley bracket assembly includes one or more pulleys over which cable will be run to actuate airplane flaps, etc. Pulleys require cable guides to prevent the cable from derailing. Each case in the case library has three parts: a problem description (the design), a solution (the process plan), and an index set. A discrimination network is used to index a case by passing its problem description through the network. The retriever finds one or more cases in the library with the same set of indexes, a superset of the indexes, or a subset of the current design input indexes. If an exact match is not found, the case-based planner then modifies the retrieved case's solution to accomplish the current design by feature substitution and feature addition. The end result is an ordered sequence of process instances which makes up a process plan.

Swaminathan and Barber [37] developed a system called APE (assembly planner using experience) that generates assembly plans based on a plan reuse philosophy. Input is provided to the assembly planner in the form of a graph of part connections, a graph of mating directions, and graph of mating constraints. The APE system indexes the stored cases based on their connections, mating directions, and spatial constraints. The planner is composed of six modules. The evaluator module divides the assembly into a number of constituent configurations, which are called loops. A loop consists of a sequence of mating parts including the directions in which they mate with the parts following them in the loop. These loops act as subgoals in the search for a solution by the retrieval module. Plans retrieved for all the subgoals are fused by the modifier and compositor modules into a set of plans that are consistent with the dependency constraints implied by each plan. Application specific constraints on the assembly are explicitly handled in the final phase of planning by the postprocessor module. The maintainer module reasons about the storage of newly generated plans and failure causing features. Assembly precedence graphs are employed for plan representation.

HYCAPP (Hyundai computer aided process planning) is a CAPP system developed by Cho et al. [9] for block assembly in shipbuilding using the CBR approach. A relational graph called part assembly network (PAN) was proposed to represent a block assembly of a ship. PAN is a kind of semantic network of frames, where each part is represented by a single frame with the part's attributes such as the shape and part number stored in the corresponding slots. The contents of a case consist of a PAN representing the complete structure of a block, its assembly sequence, the primary parts, and restrictions, noting certain crucial structural relationships between certain parts which must be assembled in a particular order. The adaptation process consists of two steps: first, the correspondence between the parts of the case block and those of the new block is found, and then the assembly sequence for the new block is derived from the assembly sequence for the old case block. The assembly sequence generated is stored as an ordered list in one of the case attributes named assembly sequence.

8.3   Methods for Input/Output Representation

Each assembly planner reviewed above was developed for a particular product, such as mechanical, electronic, ship, and airplane, or for part of the concerned product. Regardless of the product or part it is developed for, a computer assembly planner typically comprises three major components: assembly model representation, assembly sequence generation, and assembly plan representation. The techniques for assembly sequence generation were described in the previous section. This section summarizes the input/output representation methods that have been used.

Representation of Assembly Model

The assembly must be represented in some way to capture the components, their relationships, and the associated information. For a complicated product, a hierarchy of assemblies could be constructed. A component of an assembly at the higher levels could be an assembly itself composed of components at the lower levels. Different representation methods have been proposed. They include graphs, matrices, predicate calculus, and lists. Note that information contained in graphs and matrices can be translated into predicate calculus sentences or lists.

To date, the assembly model needed for assembly planning is not directly available in the CAD systems. Current CAD systems are created mainly to capture, represent, and utilize geometric shape information. They usually do not provide support for representing assembly information. Several researchers have investigated the issues involved in developing CAD systems for capturing and representing assembly information. Wesley et al. [39] described a geometric modeling system that generates a graph-based database in which assemblies are represented as nodes, and relationships among objects are represented as edges. Lee and Gossard [23] used virtual links to represent the relationships between components of an assembly in a tree-like structure. A virtual link is defined as the complete set of information required to describe the relationship and the mating features between mating pairs. Roy and Liu [33] proposed a semantic association model based assembly database to organize assembly-specific semantic information in a hierarchical structure. Despite these efforts, a CAD system that fully supports assembly planning is yet to be developed.

Most computer assembly planners assumed that a CAD interface is available to generate the assembly model from a CAD system. However, the work in the CAD interface area is incomplete. Some computer assembly planning systems take CAD models as input and derive the relationships using geometric reasoning [21,27,31, 38,40,43]. Such systems now can only deal with geometric constraints. This implies that CAPP for assembly has not gone beyond the third phase of process planning automation, as mentioned in the introduction.

The representation methods of assembly models that have been proposed are described below. Note that both graph and list have been used in representing the contents of a case in a CBR system.

Graph

The majority of representation schemes are graph-based with nodes representing parts, and links (or arcs) representing relationships. Such a graph is usually directional to show the precedence relations and attributed to indicate the part data and the relation type. For different products, the type and amount of information associated with nodes and links might differ. Many different variations have been developed for this type of representation. They include a component graph used by Khosla and Mattikali [19], a liaison graph used by Bourjault [3], De Fazio and Whitney [10], Baldswin et al. [1], Chen [7], and Lee and Shin [25], a relational model used by Homem de Mello and Sanderson [16], a hierarchical relation graph used by Yokota and Brough [41], NDBG used by Wison and Latombe [40], a graph model used by Laperrière and ElMaraghy [21], M-FRG used by Roy et al. [33], multiple graphs used by Lin and Chang [27] and Swaminathan and Barber [37], PAN used by Cho et al. [9], etc. Refer to each individual paper for details.

Matrix

The connective relations of a part with other parts along a specific axis can be expressed by a matrix. Each entry in the matrix denotes the type of connection, interference, and contact. This method was used by Sekiguchi et al. [35] and Dini and Santochi [12].

Predicate Calculus

Predicate calculus is a subset of formal logic commonly used as a knowledge representation method in AI. An atomic sentence in predicate calculus is a predicate of "arity" n followed by n terms enclosed in parentheses and separated by commas. Predicate calculus sentences are delimited by a period. A predicate calculus term is either a constant, variable, or function expression. In predicate calculus, variables can be universally or existentially quantified. First order predicate calculus allows quantified variables to objects in the domain of discourse and not to predicates or functions. A Prolog program is a sequence of Prolog clauses, which are analogous to sentences in first order predicate calculus. Therefore, predicate calculus is the representation method if Prolog is chosen to implement a planning system. This representation method has been used by Delchambre [11], Tonshoff et al. [44], Seow and Devanathan [45], and others.

List

A list is a collection of atoms or other lists, enclosed by parentheses. A LISP program is structured in list form. Therefore, list is the representation method if LISP is chosen to implement a planning system. An example of this is provided by Huang and Lee [17]. A case in a CBR system could be defined using a feature vector comprising a list of features. This has been used by Pu and Reschberger [30].

Representation of Assembly Sequence(s)

Some assembly planners generate only one sequence (or plan). Others generate all possible assembly sequences. Some generate sequences showing parts only while others generate sequences showing subassemblies too. The output representation schemes that have been used include ordered lists, precedence diagrams (or graphs), reduced precedence graphs, AND/OR graphs, assembly tree pattern, hierarchical partial order graphs (HPOG), assembly sequence diagram (or network), etc. Ordered lists are often used if a planner produces only one sequence. Otherwise, all of the above could be used to represent all possible sequences. Note that reduced precedence graphs are similar to hierarchical partial order graphs. Though it has not been shown, conversion from one to the other might be possible. A conversion between ordered lists, assembly trees, and precedence graphs has been proven possible [14].

Ordered List

The assembly sequence generated by the CBR approach is usually represented as an ordered list in which the first part to be assembled is placed at the top of the list and so on. The list could be nested according to which subassemblies are shown. This representation of assembly sequence(s) has been used by many researchers [2,9,12,30,31,45].

Precedence Diagram

The precedence diagram is a graphic illustration similar to a network plan showing all the tasks (of assembling the parts or completing the liaisons) necessary for assembly. The tasks are shown as circles (or nodes), and their dependence on each other is indicated by joining lines. This representation method has been used in Ko and Lee [20], Delchambre [11], Lin and Chang [27], and others. Roy et al. [32] represents similar information in the so-called directed assembly tree. A variation of precedence graph called precedence hypergraph allows the representation of disjunctive precedence conditions [14]. A hyperarc is represented by a set of arcs linking the originating node with a number of destination nodes and an arc of a circle centered on the originating node. Reduced precedence graph is another variation of precedence graph. The reduced precedence graph is not equivalent to the precedence diagram. It is composed of part nodes, collapsed nodes indicating possible subassemblies, and precedence arcs. The destination node in a precedence arc has precedence over the originating node.

AND/OR Graph

An AND/OR graph represents the final assembly as the root node. The final assembly could be derived from a number of different combinations of parts and subassemblies. Each combination is "anded" together and each possible combination indicates an "or" link to the assembly. Likewise, each subassembly could be derived from a number of different combinations of parts. A useful feature of the AND/OR graph is that it encompasses all possible partial orderings of assembly operations. Moreover, each partial order corresponds to a solution tree from the node corresponding to the final (assembled) product. The AND/OR graph representation allows one to efficiently search the space of feasible plans, and therefore is well-suited to problems where dynamic scheduling is important. AND/OR graphs have been used to represent possible assembly sequences by Sanderson and Homem de Mello [34], Khosla and Mattikali [19], and Homem de Mello and Sanderson [16].

Assembly Tree Pattern

An assembly tree pattern is a non-null-rooted directed tree, with labeled nodes and ordered edges [8]. An assembly tree pattern (ATP) represents the assembly sequence for an assembly in which each internal node represents an assembly operation and each leaf node represents a part. The tree structure itself shows the assembly hierarchy, subassemblies, independent subassemblies, and sequences in the assembly. The assembly plan can be obtained when an ATP tree is traversed in preorder.

Hierarchical Partial Order Graph

A hierarchical partial-order graph (HPOG) is a unified representation scheme for the temporal and spatial relationships in assembly operations [24, 25]. An HPOG is composed of nodes, supernodes, and branches. A node represents either a part or subassembly. A subassembly node is linked to a supernode, which contains specific information about the node such as liaisons to be connected. Temporal and spatial relationships among parts or subassemblies are represented by the edges which connect nodes and supernodes.

Assembly Sequence Diagram

Each node in the sequence diagram is a large box representing an assembly state. Black cells in the box indicate completed liaisons. A line between states, an arc, represents a physically possible move. Each sequence is a path that starts at the top and ends at the bottom of the diagram. This method has been used by De Fazio and Whitney [10], Baldwin et al. [1], and Laperrière and ElMaraghy [21].

8.4   Applications in Aircraft Frame Assemblies

An airplane is an integrated assembly of several sections including the wings, body, tailcone, stabilizer, flap, etc. Each section consists of several frame assemblies and skin covers. Each frame assembly can be further divided into subassemblies. For example, a wing frame assembly consists of a number of spar and rib assemblies. The major components that form these subassemblies include web, profiles, angles, fasteners, etc. Figure 8.1 shows the hierarchical product structure of a wing frame. A typical spar subassembly drawing is shown in Figure 8.2. The associated bill of material for the spar assembly is provided in Table 8.1. Many joining methods are used in airplane assembly. Among them, those commonly used include riveting, adhesive bonding, spot welding, and mechanical fastening. The selection of joining methods largely depends on the type of structure and material used. Riveting and fastening are the only two joining operations required for the spar subassembly in Figure 8.2.

Unlike mechanical products, an aircraft frame assembly is mainly composed of compliant sheet metal and/or composite parts. The thickness of these components is very small compared to their length and

TABLE 8.1   Bill of Material of the Spar Subassembly

Part No.	Part Name	Part Code	Quantity	Material	Notes
1	Web	Sp03-1	1		
2	Up-profile	Sp03-2	1		
3	Low-profile	Sp03-3	1		
4	Angle	Sp03-4	1		
5	Angle	Sp03-5	1		
6	Angle	Sp03-6	1		
7	Angle	Sp03-7	1		
8	Fitting	Sp03-8	1		
9	Fitting	Sp03-9	1		
10	Fitting	Sp03-10	1		
11	Fitting	Sp03-11	1		
⋮					
21	Rivet-1	R-3.5 × 7	20		
22	Rivet-2	R-4 × 11	104		
23	Rivet 3	R-4 × 12	6		
⋮					
31	Bolt-1	B-6 × 30	8		
32	Bolt-2	B-6 × 32	2		
33	Washer	W-6 × 1	10		
34	Spacer	Spr-6 × 1	10		
35	Nut	N-6 × 0.5	10		

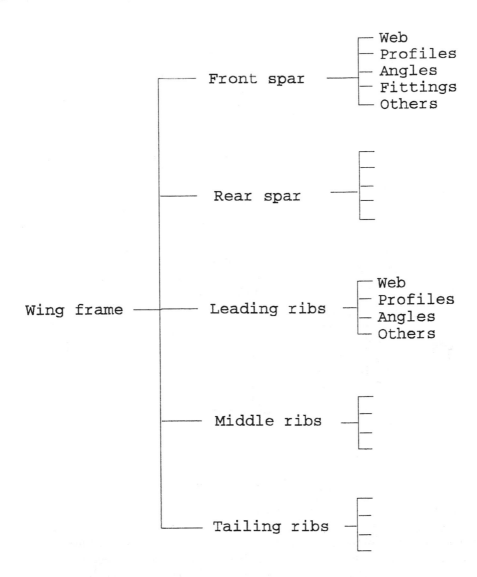

FIGURE 8.1 Hierarchical structure of an aircraft wing frame assembly.

width. The number of components in an aircraft frame assembly is also relatively large. Most importantly, each assembly is basically a collection of several layers of components joined together with many fasteners. These unique features of aircraft frame assemblies make their assembly different from the assembly of mechanical products in the following ways:

- Aircraft frame assemblies posses a fewer number of assembling directions (usually only two in the directions that are normal to the part surface.
- Aircraft frame assemblies require a high degree of fixturing to help the assembling operations (to ensure the rigidity and correct alignment of parts).
- A relatively large number of joints exists in the assembly. Among them, some are identical and others are different, connecting different number layers of parts together.
- Assembly planning tasks and assembly operations are not as easily automated because many technological constraints need to be considered.

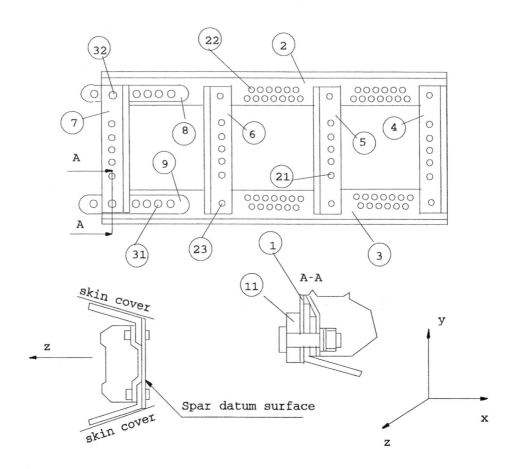

FIGURE 8.2   Spar subassembly.

Because of the above differences, the planning knowledge used in assembling mechanical products could not be directly used in planning aircraft frame assemblies. However, the general issues are the same. In developing an assembly planner for aircraft frame assemblies, we still need to address how the assembly model is represented, how the sequence of assembling operations is generated, and how the generated assembly plan is represented. With adequate planning knowledge, it is expected that both the knowledge- and case-based approaches could be used to develop planners for aircraft frame assemblies. There should be no exception for the algorithmic approach. No matter what approach is taken, the unique characteristics of aircraft frame assemblies should be considered in developing the assembly planner. Our assembly planner was developed based on the observation of the layer structure of aircraft frame assemblies [26]. Besides our work, we are not aware of other assembly planners developed for aircraft frame assemblies in the available literature.

In the following sections, our methods of assembly model representation, sequence generation, and plan representation for aircraft frame assemblies are discussed.

Assembly Model Representation

A new representation scheme called LADGA (layered acyclic directed graph with attributes) was developed for aircraft frame assemblies. The LADGA consists of nodes organized into layers with or without attributes attached and directed arcs that connect nodes, as illustrated in Figure 8.3. Each node denotes a part. Different kinds of nodes can be used to differentiate different kinds of parts, if necessary. In this study, circular nodes refer to parts requiring riveting operations only. Triangular nodes indicate that bolting operations are needed (riveting operations might also be needed). The directed arc between two

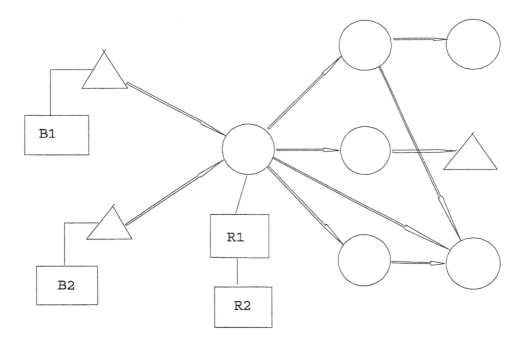

FIGURE 8.3 LDAGA assembly representation scheme.

TABLE 8.2 Fastener Record for the 1-2-5
Joining Group of the Spar Subassembly

Attribute Name	Value
Fastener record	R8
Joining method	Riveting
Joining direction	Z
Fastener quantity	5
Fastener code	R-3.5 × 7
Joining group	1-2-5
Tail part	5

nodes indicates the precedence relationship. The destination node must be assembled after the originating node. The information about the fastener, joining direction, and the parts connected together by the fastener is organized in a fastener record. The collection of parts connected together by the fastener forms a joining group. If the fastener is a rivet, the joining group is more specifically called a riveting group. Likewise, the joining group is more specifically called a bolting group if the fastener is a bolt. The record is always attached to the first component to be assembled that is also called the head part of the joining group. The head part is determined by the assembly direction.

Each fastener record includes information about the fastener record code, joining method, joining direction, fastener quantity, fastener code and specification, joining group, and the last part to be assembled in the joining group (called the tail part). The direction of positioning parts onto a jig/fixture is defined as the assembly direction. On the other hand, the direction of installing fasteners is defined as the joining direction. The joining direction is usually specified in the drawing and could be in the same or opposite direction of the assembly direction. A sample of the fastener record is shown in Table 8.2. The fastener records are used to generate assembly sequence as well as to prepare assembly instruction. A node will be attached with more than one fastener record if there is more than one joining group with the node as the head part. A complete set of attributes for an (sub)assembly should include all the relevant records and associated information.

The LADGA is established from assembly drawing in two stages:

- Construct acyclic directed graphs and their corresponding fastener records for all the joining groups in the drawing. All of the graphs are then combined into one integrated graph for the whole assembly. Each acyclic directed graph with attributes is called an ADGA.
- Convert the ADGA for the assembly into LADGA using an algorithm named Transform.

Construction of ADGA and Its Data Structure

Each joining group is defined as a unique collection of parts connected together with a specific fastener. Each aircraft frame assembly usually has two or more joining groups. It is important that all the joining groups are included in establishing the ADGA for the assembly so that no part is left out. The construction of ADGA takes three steps as follows:

1. Determine the assembly direction:
   The assembly direction is determined based on the considerations of the geometric shape of the assembly and the datum surface. The assembly direction should be kept the same for all the joining groups throughout the construction process. For the spar subassembly shown in Figure 8.2, the assembly direction is determined to be in the Z direction, which means that parts are positioned one by one from the left side of the A-A view to the right.
2. Construct ADGA for each joining group:
   To construct the ADGA corresponding to a joining group in the assembly, the following procedure is used:
   - Find the head part of the group that is the outermost part facing left in the section view if the assembly direction is Z.
   - Identify parts in contact with the foremost part and draw directed arcs between them.
   - Stop if no more parts are left; otherwise, continue to find parts in contact with identified parts in the previous step.
   - Prepare the fastener record for the group and attach it to the head part.

   Following the procedure, 16 joining groups are identified for the spar subassembly shown in Figure 8.2. The results are shown in Figure 8.4. Note that each joining group is denoted as R_i (or B_i) indicating that it is a riveting group (or bolting group).
3. Construct the ADGA for the assembly:
   To construct the ADGA for the assembly, all the ADGAs for joining groups need to be combined. The combining procedure takes the following three steps:
   - Randomly select one ADGA for a joining group as the base for the integration process.
   - Randomly pick one of the remaining ADGAs and combine it with the base (or the intermediate result if not the first combination) by eliminating redundant nodes.
   - Repeat above step until there is no more ADGA to be combined. The result is an integrated ADGA for the assembly.

Following the procedure, the integrated ADGA for the assembly is obtained, as shown in Figure 8.5 (ignoring the fastener record).

A data structure that can be used to represent the integrated ADGA is adjacent lists. Adjacent lists are linked lists, one for each node. Each linked list contains the names of the nodes connected to the node for which the list is created. The heads of these lists are held in an array of pointers. That is, there is a one-dimensional array $V[1,2, ..., n]$ of pointers for a given ADGA, $G_A = (V_A, E_A)$ where V_A is a finite, nonempty list of vertices and E_A is a set of pairs of vertices. Each $v_i \in V_A$ points to a list L of vertices such that vertex v_j is in L if and only if there is an edge $(v_i, v_j) \in E_A$ directed from v_i to v_j in the graph. For the ADGA for the spar subassembly, the adjacent lists are prepared as shown in Figure 8.6. A vertex $v_i \in V_A$ is called the root of G_A if there is no edge (v_j, v_i) in E_A. For the spar subassembly, parts 11 and 12 are the roots of G_A.

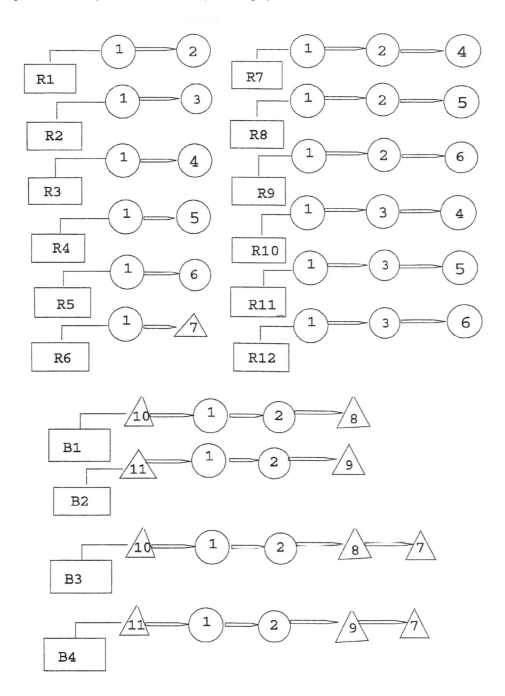

FIGURE 8.4 DAGAs for the joining groups of the spar subassembly.

Construction of LADGA

The assembly precedence relations in an ADGA cannot be easily visualized. This is especially true for aircraft frame assemblies which are usually composed of a large number of components. As has been pointed out before, all the sheet metal parts in an aircraft frame assembly are generally stacked into layers. An algorithm was thus developed to transform an ADGA into a LADGA so that parts are organized into the same layer if they have no precedence relations. Mathematically, a LADGA G_A^L corresponding to a

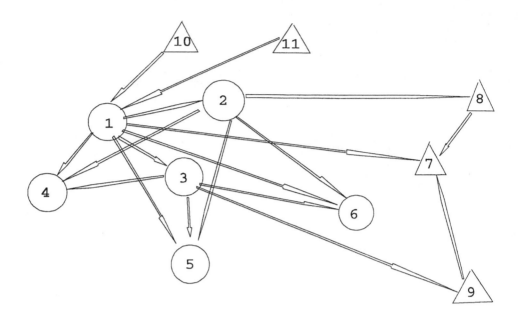

FIGURE 8.5   Integrated DAGA for the spar subassembly.

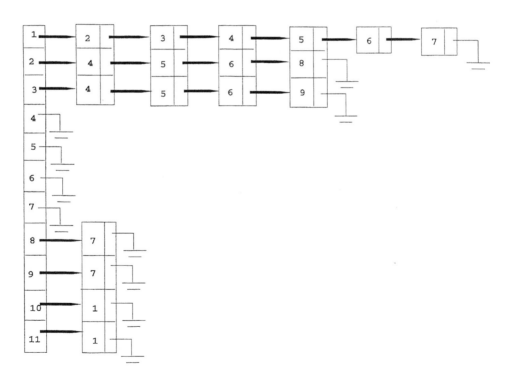

FIGURE 8.6   Linked list for the spar subassembly.

given G_A is a partition of vertices of G_A into m disjoint subsets $P_1, P_2, \ldots, P_m$ so that if v_i and v_j belong to the same P_S, $1 \leq s \leq m$, then there is no edge in G_A from v_i to v_j or from v_j to v_i. We associate a field layer-number with each vertex v_i of G_A. Initially, v_i/layer-number $= 0$ for all v_i in V_A. The algorithm Transform for converting a G_A into a G_A^l follows.

```
Algorithm Transform
begin
        find the set R of roots of G_A;
        for every vertex v_i in R do
        begin:
                v_i / layer-number = 1;
                call Label(v_j, 1);
        end
end
```

The procedure Label used in Transform is as follows:

```
Procedure Label(v_j, current-layer)
begin
        current-layer = current-layer + 1;
        for all v_k in L[j] do
        begin:
                v_k/layer-number = max(v_k/layer-number, current layer);
                call Label(v_k, current-layer);
        end
end
```

After executing Transform, v_i/layer-number of each vertex in G_A is assigned a unique integer. That is, a G_L^A corresponding to G_A is obtained. Each P_s contains all vertices whose layer-number is equal to s, $1 \leq s \leq m$, where m is the maximum number assigned to any vertex. It is not difficult to show that m is the maximum number of layers among all the possible LADGA corresponding to G_A. For the spar subassembly, the LADGA corresponding to the ADGA of Figure 8.5 is obtained as shown in Figure 8.7. The time and storage complexities of Transform are $O(r|F|)$ and $O(|E|)$, respectively, where $r = |R|$, which is the total number of roots, and $|E|$ is the total number of edges in G_A.

Sequence Generation

Two algorithms were developed for the generation of assembly plans. One, named Assembly1, does not consider technological constraints, while the other, named Assembly2, does.

Algorithm Assembly1 is described first. Let $L[i]$ denote layer i in the LADGA, $1 \leq i \leq m$, where the total number of layers m is minimum as obtained by the algorithm Transform. Each fastener record is denoted as a R_{k1}, $k_1 = 1, 2, \ldots, n_r$, for a riveting group or a B_{k2}, $k_2 = 1, 2, \ldots, n_b$, for a bolting group, where n_r and n_b are the total number of riveting groups and bolting groups, respectively. Initially, the set of riveting groups Q_R and bolting groups Q_B are empty. R_{k1}/tail or B_{k2}/tail refers to the tail part for the respective fastener record. The algorithm is presented below.

```
Algorithm Assembly1
begin
        Q_R = ø;
        Q_B = ø;
        for i = 1 to m do
                for all v_j in L[i] do
                begin
                        position part v_j;
                        for all R_{k1} in Q_R s/t/R_{k1}/tail = v_j do
```

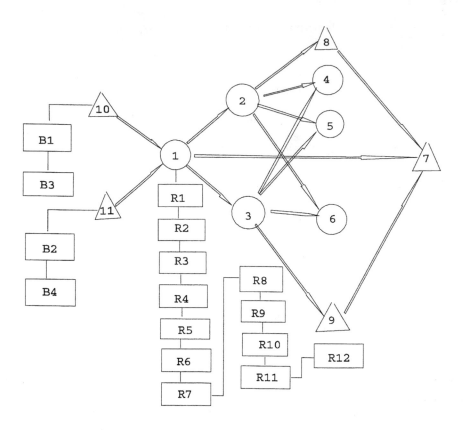

FIGURE 8.7 LDAGA representation for the spar subassembly.

perform riveting operation according to the information of
R_{k1} and remove R_{k1} from Q_R;
for all R_{k1} associated with v_j **do**
$Q_R = Q_R \cup \{R_{k1}\}$;
for all B_{k2} associated with vj **do**
$Q_B = Q_B \cup \{R_{k2}\}$;
end
for $i = 1$ to m **do**
for all B_{k2} in Q_B $s/t/R_{k2}/\text{tail} = v_j$ and v_j is in $L[i]$ **do**
perform bolting operation according to the information of
B_{k2} and remove B_{k2} from QB;
end

Taking Figure 8.7 as input, algorithm Assembly1 is executed to generate an assembly plan as given in Table 8.3. This plan is workable for completing the spar subassembly. However, it is not an optimal plan because technological constraints are not considered.

Technically speaking, the same kind of parts should be positioned together before performing the joining operations in order to reduce tool changing time. Moreover, the positioning and riveting of to-be-riveted parts should be done separately from the positioning and bolting of to-be-bolted parts because of different skills required. It is thus important to distinguish between to-be-riveted parts and to-be-bolted parts and to consider the interactions between their positioning and joining operations. The technological constraints that need to be considered in generating the sequence of positioning and joining

TABLE 8.3   Assembly Plan Generated by Algorithm
Assembly1 for the Spar Subassembly

Operation No.	Description
Step 1	Position part 10
Step 2	Position part 11
Step 3	Position part 1
Step 4	Position part 2
Step 5	Perform riveting operation according to the information of R_1
Step 6	Position part 3
Step 7	Perform riveting operations according to the information of R_2
Step 8	Position part 8
Step 9	Position part 4
Step 10	Perform riveting operations according to the information of R_3, R_7, and R_{10}
Step 11	Position part 5
Step 12	Perform riveting operations according to the information of R_4, R_8, and R_{11}
Step 13	Position part 6
Step 14	Perform riveting operations according to the information of R_5, R_9, and R_{12}
Step 15	Position part 9
Step 16	Position part 7
Step 17	Perform riveting operation according to the information of R_6
Step 18	Perform bolting operation according to the information of B_1
Step 19	Perform bolting operation according to the information of B_2
Step 20	Perform bolting operations according to the information of B_3 and B_4

operations for aircraft frame assemblies include the following:

- Positioning operations should be performed before joining operations.
- Bolting operations should be performed after riveting operations because of tighter tolerance requirements in bolting operations.
- The positioning and joining operations for the riveting groups should be performed and completed before starting the positioning and joining operations for the bolting groups.

To this end, the second algorithm Assembly2 was developed to take the above-mentioned technological constraints into account to generate an assembly plan which, whenever possible, positions to-be-riveted parts first, performs riveting operations second, positions to-be-bolted parts third, and does bolting operations last. Note that jig/fixture selection and design, not discussed here, should be done simultaneously with assembly planning.

In the algorithm Assembly2, a v_j^r refers to a to-be-riveted part which is defined as a part that requires only a riveting operation. On the other hand, a v_j^b refers to a to-be-bolted part which indicates that the part requires bolting operations. A to-be-bolted part might also require a riveting operation. One example is Part 8 of the spar subassembly. Q_P is defined as the set of positioned parts. The algorithm is given below.

Algorithm Assembly2
begin
$\quad Q_P = \varnothing;$
$\quad Q_R = \varnothing;$
$\quad Q_B = \varnothing;$

> **for** $i = 1$ to m **do**
>> **for** all v_j in $L[i]$ **do**
>> **if** v_j is attached with B_{k2} or $v_j = B_{k2}/$tail **then**
>>> label v_j as v_j^b ;
>>
>> **else**
>>> label v_j as v_j^r ;
>>
>> **for** all v_j^r in $L[i]$ **do**
>>> position v_j^r ;
>>> $Q_P = Q_P \cup \{v_j^r\}$;
>>> **for** all R_{k1} associated with v_j^r **do**
>>>> $Q_R = Q_R \cup \{R_{k1}\}$;
>>
>> **for** all R_{k1} in Q_R $s/t/R_{k1}/$tail in Q_P **do**
>>> perform riveting operation according to the information of R_{k1} and delete R_{k1} from Q_R;
>
> **for** $i = 1$ to m **do**
>> **for** all v_j^b in $L[i]$ **do**
>>> position v_j^b ;
>>> $Q_P = Q_P \cup \{v_j^b\}$;
>>> **for** all R_{k1} associated with v_j^b **do**
>>>> $Q_R = Q_R \cup \{R_{k1}\}$;
>>>
>>> **for** all B_{k2} associated with v_j^b **do**
>>>> $Q_B = Q_B \cup \{B_{k2}\}$;
>>
>> **for** all R_{k1} in Q_R $s/t/R_{k1}/$tail in Q_P **do**
>>> perform riveting operation according to the information of R_{k1} and delete R_{k1} from Q_R;
>>
>> **for** all B_{k2} in Q_B $s/t/B_{k2}/$tail in Q_P **do**
>>> perform bolting operation according to the information of B_{k2} and delete B_{k2} from Q_B;

end

Taking Figure 8.7 as input, algorithm Assembly2 is executed to generate an assembly plan as given in Table 8.4. This plan is an optimal one that has taken all the technological constraints into consideration.

TABLE 8.4   Assembly Plan Generated by Algorithm Assembly2 for the Spar Subassembly

Operation No.	Description
Step 1	Position part 1
Step 2	Position part 2
Step 3	Position part 3
Step 4	Position part 4
Step 5	Position part 5
Step 6	Position part 6
Step 7	Perform riveting operations according to the information of R_1, R_2, R_3, R_4, R_5, R_7, R_8, R_9, R_{10}, R_{11}, and R_{12}
Step 8	Position part 10
Step 9	Position part 11
Step 10	Position part 8
Step 11	Position part 9
Step 12	Position part 7
Step 13	Perform bolting operations according to the information of B_1, B_2, B_3, and B_4

Plan Representation

Only one assembly plan is generated by either algorithm. The assembly operations, including positioning parts and performing joining operations, are listed sequentially. The associated details about the operations are also given along with the sequence. In generating the plan, it is assumed that parts in the same layer are arranged in a fixed order. Actually, parts in the same layer can be arranged in any sequence. Therefore, there are many possible sequences for the example discussed. They can all be generated with slight modifications of the algorithms.

8.5 Conclusions

In this chapter, we have reviewed many computer assembly planners that have been developed in the past for mechanical products. The techniques used for their development are classified into four categories: algorithmic, integrated interactive/algorithmic, knowledge-based, and case-based. The representation methods for assembly models and assemblies that have been used are briefly described. The applicability of these techniques in developing planners for aircraft frame assemblies is also discussed.

The second half of the chapter presented a computer-aided assembly planner developed specifically for aircraft frame assemblies. This planner is different from previous planners in assembly model representation, planning algorithms, and plan representation. This system is incomplete in the sense that it focuses only on a small part of the aircraft assemblies. A CAD interface is also needed to convert an assembly drawing into the proposed representation method, i.e., ADGA.

References

1. Baldwin, D. F., Abell, T. E., Lui, M. C. M., De Fazio, T. L., and Whitney, D. E., An integrated computer aid for generating and evaluating assembly sequences for mechanical products, *IEEE Trans. Robotics Automation,* 7(1), 78, 1991.
2. Ben-Arieh, D. and Kramer, B., Computer-aided process planning for assembly. generation of assembly operations sequence, *Int. J. Prod. Res.,* 32(3), 643, 1994.
3. Bourjault, A., Contribution a une approche méthodologique de l'assemblage automatisé: elaboration automatique des séquences opératoires, Doctoral thesis, University of Granche-Comté, 1984.
4. Chakrabarty, S. and Wolter, J., A structure-oriented approach to assembly sequence planning, *IEEE Trans. Robotics Automation,* 13(1), 14, 1997.
5. Chang, T. C. and Terwilliger, J., A rule-based system for printed wiring assembly process planning, *Int. J. Prod. Res.,* 25, 1465, 1987.
6. Chang, T. C., Wysk, R. A., and Wang, H. P., *Computer-Aided Manufacturing,* Prentice-Hall, Englewood Cliffs, NJ, 1991.
7. Chen, C. L. P., Automatic assembly sequences generation by pattern matching, *IEEE Trans. Syst. Man Cybernet.,* 21(2), 376, 1991.
8. Chen, C. L. P. and Pao, Y. H., An integration of neural network and rule-based system for design and planning of mechanical assemblies, *IEEE Trans. Syst. Man Cybernet.,* 23(5), 1359, 1993.
9. Cho, K. K., Ryu, K. R., Oh, J. S., Choi, H. R., and Yun, S. T., Computer-aided process planning system for block assembly shop using CBR, in *Technical Paper of NAMRI/SME XXIV,* Society of Manufacturing Engineers, Dearborn, MI, 1996, 157.
10. De Fazio, T. L. and Whitney, D. E., Simplified generation of all mechanical assembly sequences, *IEEE J. Robotics Automation,* 3(6), 640, 1987.
11. Delchambre, A., A pragmatic approach to computer-aided assembly planning, *Proc. of IEEE Int. Conf. Robotics Automation,* 1600, 1990.
12. Dini, G. and Santochi, M., Automated sequencing and subassembly detection in assembly planning, *Ann. CIRP,* 41(1), 1, 1992.

13. Freedman, R. S. and Frail, R. P., OPGEN: the evolution of an expert system for process planning, *AI Mag.*, 7, 58, 1986.

14. Henrioud, J. M. and Bourjault, A., Computer-aided assembly process planning, *J. Eng. Manuf.*, 206, 61, 1992.

15. Homem de Mello, L. S. and Sanderson, A. C., AND/OR graph representation of assembly plans, *IEEE Trans. Robotics Automation*, 6(2), 188, 1990.

16. Homem de Mello, L. S. and Sanderson, A. C., A correct and complete algorithm for the generation of mechanical assembly sequences, *IEEE Trans. Robotics Automation*, 7(2), 228, 1991.

17. Huang, Y. F. and Lee, C. S. G., A framework of knowledge-based assembly planning, *Proc. IEEE Int. Conf. Robotics Automation*, 599, 1991.

18. Irizarry-Gaskins, V. M. and Chang, T. C., A knowledge-based approach for automatic process plan generation in an electronic assembly environment, *Int. J. Prod. Res.*, 28(9), 1673, 1990.

19. Khosla, P. K. and Mattikali, R., Determining the assembly sequence from a 3D model, *J. Mechanical Working Technol.*, 20, 153, 1989.

20. Ko, H. and Lee, K., Automatic assembling procedure generation from mating conditions, *CAD*, 19(1), 3, 1987.

21. Laperrière, L. and ElMaraghy, H. A., GAPP: a generative assembly process planner, *J. Manufacturing Syst.*, 15(4), 282, 1996.

22. Latombe, J. C., Wilson, R. H., and Cazals, F., Assembly sequencing with toleranced parts, *Comput. Aided Design*, 29(2), 159, 1997.

23. Lee, K. and Gossard, D. C., A hierarchical data structure for representing assemblies: part 1, *Comput. Aided Design*, 17(1), 15, 1985.

24. Lee, S., Subassembly identification and evaluation for assembly planning, *IEEE Trans. Syst. Man Cybernet.*, 24(3), 493, 1994.

25. Lee, S. and Shin, Y. G., Assembly planning based on geometric reasoning, *Comput. Graphics*, 14(2), 237, 1990.

26. Liao, T. W., Wu, X. W., Zheng, S. Q., and Li, S. Q., A computer-aided aircraft frame assembly planner, *Comput. Ind.*, 27, 259, 1995.

27. Lin, A. C. and Chang, T. C., An integrated approach to automated assembly planning for three-dimensional mechanical products, *Int. J. Prod. Res.*, 31(5), 1201, 1993.

28. Liu, D., Utilization of artificial intelligence in manufacturing, *Autofact*, 6, 260, 1984.

29. Maria, A. and Srihari, K., A review of knowledge-based systems in printed circuit board assembly, *Int. J. Adv. Manuf. Technol.*, 7, 368, 1992.

30. Pu, P. and Reschberger, M., Case-based assembly planning, *Proceedings of the Case-Based Reasoning Workshop*, May 1991, 245.

31. Rajan, V. N. and Nof, S. Y., Minimal precedence constraints for integrated assembly and execution planning, *IEEE Trans. Robotics Automation*, 12(2), 175, 1996.

32. Roy, U., Banerjee, P., and Liu, C. R., Design of an automated assembly environment, *Comput. Aided Design*, 21(9), 561, 1989.

33. Roy, U. and Liu, C. R., Establishment of functional relationships between product components in assembly database, *Comput. Aided Design*, 20(10), 570, 1988.

34. Sanderson, A. C. and Homem de Mello, L. S., Task planning and control synthesis for flexible assembly systems, in *Machine Intelligence and Knowledge Engineering for Robotic Applications*, A. K. C. Wong, A. K. C. and Pugh, A., Eds. Springer-Verlag, Berlin, 1987, 331.

35. Sekiguichi, H., Kojima, T., Inue, K., and Honda, T., Study on automatic determination of assembly sequence, *Ann. CIRP*, 32(1), 371, 1983.

36. Supinski, M. R., Egbelu, P. J., and Lehtihet, E. A., Automatic plan and robot code generation for PCB assembly, *Manuf. Rev.*, 4(3), 214, 1991.

37. Swanimathan, A. and Barber, K. S., An experience-based assembly sequence planner for mechanical assemblies, *IEEE Trans. Robotics Automation*, 12(2), 252, 1996.

38. Thomas, J. P., Nissanke, N., and Baker, K. D., Boundary models for assembly knowledge representation, *IEEE Trans. Robotics Automation,* 12(2), 302, 1996.
39. Wesley, M. A., Lozano-Perez, T., Lieberman, L. I., Lavin, M. A., and Grossman, D. D., A geometric modeling system for automated mechanical assembly, *IBM J. Res. Dev.,* 24(1), 64, 1980.
40. Wilson, R. H. and Latombe, J. C., Geometric reasoning about mechanical assembly, *Artif. Intelligence,* 71, 371, 1994.
41. Yokota, K. and Brough, D. R., Assembly disassembly sequence planning, *Assembly Automation,* 12(3), 31, 1992.
42. Zarley, D. K., 1991, A case-based process planner fo small assemblies, *Proceedings of the Case-Based Reasoning Workshop,* May 1991, 363.

9

Petri Net Modeling in Flexible Manufacturing Systems with Shared Resources*

Ke Yi Xing
Xidian University

Bao Sheng Hu
Xi'an Jiaotong University

This chapter develops a Petri net model of production routings and resource allocation in a flexible manufacturing system (FMS) with shared resources, which we call the R^2PN model, and focuses on the deadlock avoidance problems in the system. We introduce the concept of D-structures and explore some of their basic properties, as well as characterize the liveness of the system in terms of D-structures. We address the deadlock avoidance problems by introducing the restrictive controller. In order to synthesize such a controller, we first reduce the R^2PN model. The reduced R^2PN is also an R^2PN model for which we can present an optimal deadlock avoidance controller. Our deadlock avoidance controller for the R^2PN consists of two parts: a Petri net controller and a restrictive policy. The Petri net controller can be reduced from the optimal deadlock avoidance controller for the reduced R^2PN model and can prevent some D-structures from leading to a circular wait relationship which directly causes deadlock in R^2PN. The restrictive policy restricts the utilization of some key resources. In many cases, the Petri net controller can guarantee not only that the controlled R^2PN is live, but also that it is an optimal deadlock avoidance controller. And hence, the controlled R^2PN is an optimal live Petri net model for the system. Some examples are presented to illustrate the results.

---

*This work was supported in part by the National Natural Science Foundation of P. R. China under grant 69574023 and in part by a grant from the Center for Computer Integrated Manufacturing System at Xian Jiaotong University.

9.1   Introduction

A flexible manufacturing system (FMS) is composed of multiple subsystems such as machines, transport systems, and buffers, which must be controlled by a computer to achieve the production goals. Such a system can produce multiple types of products. Various types of products enter the system for processing at discrete points of time. Each type of product has its own prescribed sequence of operations that determines the order in which resources must be used. To effectively utilize the flexibility of an FMS and obtain greater productivity, the most important problem in the FMS is how to assign the resources to different processes. Products are processed concurrently in an FMS and they compete for a finite set of common resources. If the resource requirement is not restricted, these relations of competition can lead to a deadlock. In a deadlock situation, some products are in a circular wait relationship, waiting for resources from each other, and can never advance to completion, while some resources are occupied and can never be used again. Therefore, a deadlock situation is highly undesirable, [1,8] and the live modeling of a system becomes very important.

Petri nets are an effective tool for formulating and solving deadlock problems in FMSs. Banaszak and Krogh [1] considered a class of Petri net models that represent concurrent job flow and dynamic resource allocation in an FMS. They proposed a real-time deadlock avoidance feedback policy and proved that the policy is sufficient for avoiding deadlocks. For a larger class of Petri net models, Hsieh and Chang [4] proposed a new method for synthesizing deadlock avoidance controllers that keeps the FMS live and allows a high resource utilization under any given dispatching policy. Ezpeleta et al. [3] proposed a Petri net based deadlock prevention policy for a class of manufacturing systems. The policy restricts the resource allocation by adding some new places to the Petri net model so that no reachable marking leads to an empty siphon, which guarantees that the controlled system is live. Xing et al. [9] presented a deadlock avoidance Petri net controller for a class of FMSs. In a case where the system contains no key resources, the proposed controller is optimal, that is, minimally restrictive and allows the maximal use of resources; hence, the controlled model is an optimal live Petri net model of the system.

In this chapter we consider a class of FMSs in which a type of products may have different production routings though the system. The different routings may have the same subrouting modeled by the same elements of the Petri net. Our Petri net models that we call R^2PN therefore belong to a larger class than the ones discussed by Banaszak and Krogh., [1] Hsieh and Chang, [4] and Xing et al., [9] but the same class as that presented by Ezpeleta et al. [3].

This chapter begins by demonstrating Petri nets as a model for the dynamic behavior of the production operations and the resource allocation of an FMS. By analyzing Petri net model properties, we formulate a circular waiting concept in the context of Petri net theory: D-structures, and study the conditions under which the Petri net models are live in terms of such structures. Using those results we can define a deadlock avoidance controller for the system. Our controller consists of two parts: a Petri net controller and a restrictive policy. In many cases, the Petri net controller can already avoid all deadlocks, and is an optimal deadlock avoidance controller for the system. Hence, this chapter presents a synthesis method for live modeling of FMSs.

The next section formulates the Petri net models of FMSs. The liveness conditions for Petri net models based on the concept of D-structures are presented in Section 9.3. Section 9.4 gives the definition and the correctness proof of deadlock avoidance controllers for R^2PN s. Section 9.5 introduces two examples of FMSs and illustrates the application of the proposed method and theory. Section 9.6 concludes the chapter.

9.2   Petri Net Models for a Class of FMSs

In this section we use Petri nets to model the dynamic behavior of the production operations and resource allocation of automated manufacturing systems.

Petri Nets

We assume that the reader is familiar with the basic elements of Petri net models. For consistency, basic definitions and notations for Petri nets are summarized as follows.

A marked Petri net is a four-tuple: $G = (P, T, F, m_0)$, where P is a finite set of places, T is a finite set of transitions, and $F \subseteq (P \times T) \cup (T \times P)$ is a set of directed arcs connecting places and transitions of G. $m : P \to N$ is a marking whose ith component represents the number of tokens in the ith place, where N is a set of nonnegative integers. m_0 is an initial marking.

Given a Petri net $G = (P, T, F, m_0)$ and a node $x \in P \cup T$, let $\cdot x$ denote the preset of x, i.e., $\cdot x = \{y \in P \cup T \mid (y, x) \in F\}$. Similarly, $x\cdot$ denotes the postset of x, i.e., $x\cdot = \{y \in P \cup T \mid (x, y) \in F\}$. This notation is extended to a set of nodes as follows: given a set $X \subseteq P \cup T$, $\cdot X = \bigcup_{x \in X} \cdot x$ and $X\cdot = \bigcup_{x \in X} x\cdot$. Given a marking m and a set of places $X \subseteq P$, let $m(X)$ denote the sum of token numbers in all places of X, i.e., $m(X) = \sum_{p \in X} m(p)$.

A transition $t \in T$ is enabled in a marking m if and only if $\forall p \in \cdot t$, $m(p) \geq 1$. A transition t enabled in the marking m can fire-yield a new marking m' defined by

$$m'(p) = m(p) + |\{t\} \cap \cdot p| - |\{t\} \cap p\cdot|.$$

When a marking m' can be reached from the marking m by executing a firing sequence of transitions $\sigma = t_1 t_2 \cdots t_k$, we write $m[\sigma > m'$. We write $m[\sigma >$ to denote that σ can be executed from the marking m. The reachability set $R(G, m_0)$ is the set of all markings reachable from the initial marking m_0 with different sequences of transition firings.

A transition t of a marked Petri net G is live if and only if $\forall m \in R(G, m_0), \exists m' \in R(G, m)$, such that $m'[t >$. A marked Petri net is live if and only if every transition is live; otherwise it contains some deadlocks. A transition t is dead in a marking m if and only if there exists no marking $m' \in R(G, m)$ such that $m'[t >$.

A set of places $S \subseteq P$ is a siphon if and only if $\cdot S \subseteq S\cdot$. A siphon is minimal if it is not a superset of any other siphon.

A sequence of places and transitions $\sigma = x_1 x_2 \cdots x_n$ is called a path of G if and only if no place or transition except x_1 and x_n appears more than once, and $x_{i+1} \in x_i\cdot$ for every i. A path $x_1 \cdots x_n$ is a cycle if $x_1 = x_n$. Let P_1 be the subset of places. A path is called a P_1 path if its place nodes are in P_1.

Given two Petri nets $G_1 = (P_1, T_1, F_1, m_{10})$ and $G_2 = (P_2, T_2, F_2, m_{20})$, the concurrent composition of G_1 and G_2, denoted as $G = G_1 * G_2$, is a marked Petri net defined by

$$G_1 * G_2 = (P_1 \cup P_2, T_1 \cup T_2, F_1 \cup T_2, m_0),$$

where $m_0(p) = m_{10}(p)$, $\forall p \in P_1$, and $m_0(p) = m_{20}(p), \forall p \in P_2$.

Using Petri Nets to Model FMSs

We consider an FMS consisting of k types of resources R, which can produce n different types of product Q. Each type of product has its own production routings. A production routing is a sequence of operations. Assume that each operation requires a single resource and a resource can only be assigned to one operation at a time. Modeling the production routings with a Petri net, we use a place in the Petri net to represent either an operation or a resource status. To establish a cyclic model, we first assign a place denoted as p^0, which represents the operation of replacing completed products with raw products, which requires no system resources. Tokens in place p^0 represent products to be produced or completed. We then model a production routing by a directed cycle: $p^0 t_1 p_1 t_2 p_2 \cdots t_n p^0$, where $n - 1$ is the number of operations in the production routing. Every place p_i represents an operation which requires a resource. For any two directed cycles $p^0 x_1 x_2 \cdots x_n p^0$ and $p^0 y_1 y_2 \cdots y_m p^0$ have the same nodes, and if $x_{i+k} = y_{j+k}, k = 1, 2, \cdots l$, $x_i \neq y_j$ and $x_{i+l+1} \neq y_{j+l+1}$, then $x_{i+1} = y_{j+1}$ and $x_{i+l} = y_{j+l}$ are places. Then the production routing Petri net for an FMS is a state machine,

$$G_p = (P, T, F, m_{p0}),$$

where

1. $P\backslash\{p^0\} \neq \emptyset$ is the set of operation places.
2. For every $t \in T$, $|{}^{\cdot}t| = |t^{\cdot}| = 1$.
3. Every transition and place are contained in some directed cycles, and every directed cycle contains place p^0.
4. In the initial marking m_{p0}, only place p^0 is marked; that is, $\forall p \in P\backslash\{p^0\}$, $m_{p0}(p) = 0$ and $m_{p0}(p^0) \geq 1$. Here we assume that the initial marking of p^0 is an integer C which is larger than the total number of resources for analyzing all deadlocks in the system.

Given a transition $t \in T$, let ${}^{(p)}t$ denote the place in P that is an input place for t. Similarly, $t^{(p)}$ denotes the output place in P for t.

For each resource type $r \in R$, we assign a resource place, also denoted as r. Let R also denote the set of resource places. If an operation modeled by $p \in P\backslash\{p^0\}$ requires a type-r resource, denoted by $R(p) = r$, then the requirement of resource r by place p is modeled by introducing arcs from the resource place r to every input transition of p in G_p, and the release of resource r is modeled by introducing arcs from every output transition of p to r. Since no resource is required for the operation p^0, there are no resource places connected to a transition in ${}^{\cdot}(p^0)$ or $(p^0)^{\cdot}$. The resource Petri net model for the FMS is a Petri net

$$G_r = (R, T, F, m_{r0}),$$

where $F_r = \{(t, r)|t \in T, r \in R, R({}^{(p)}t) = r\} \cup \{(r, t) \mid t \in T, r \in R, R(t^{(p)}) = r\}$, and $m_{r0}(r) = C_r \geq 1$, $\forall r \in R$ is the capacity of type-r resources.

Then the complete Petri net model for the FMS is the composition of the routing Petri net and the resource Petri net

$$G = G_p * G_r = (P \cup R, T, F, m_0),$$

where $F = F_p \cup F_r$, $m_0(p) = m_{p0}(p), \forall p \in P$, and $m_0(p) = m_{r0}(p), \forall p \in R$

This model is called a routing/resource Petri net R^2PN. Similar Petri net models for FMSs have been developed by other researchers.[1,3,9] We illustrate the modeling methodology using the following simple manufacturing system.

Example 1

Consider a manufacturing system in which two types of products, q_1 and q_2, are produced by six types of machines. The machines are serviced by a transport system TS, which can pick up and deliver products between any pair of machines. TS can hold one product at a time. The machine types are denoted by M_i, $i = 1, 2, \cdots, 6$, and each type has 5 machine units. Hence the resource-type set of the system is given by $R = \{M_1, \cdots, M_6\}$. Suppose that a product of type q_1 is produced by the machine sequences: $M_1M_2M_1M_2M_6$ or $M_1M_3M_4M_5M_2M_6$. A product of type q_2 is produced by the machine sequence $M_5M_4M_3M_6$. The R^2PN model of such a system is shown in Figure 9.1.

Given a transition $t \in T$, let ${}^{(r)}t$ denote the input place in R for t. Similarly, $t^{(r)}$ denotes the output place in R for t. These notations are extended to a set of transitions $X \subseteq T$ in natural way, for example, ${}^{(r)}X = \cup_{x \in X}X^{(r)}t$, etc. Given a marking m, we will say a transition t is process enabled if $m({}^{(p)}t) \geq 1$, which means a product is currently in the production operation preceding the transition t. A transition t is resource enabled if ${}^{(r)}t = \emptyset$ or $m({}^{(r)}t) \geq 1$, which means a resource is currently available for the production operation succeeding the transition t.

9.3   Liveness Conditions for R^2PN Models

In this section we study some structural properties of R^2PN models and introduce the concept of D-structures. We then characterize deadlocks in terms of D-structures and prove a necessary and sufficient condition for an R^2PN model to be live. These characterizations are used in the next section to synthesize the deadlock avoidance controller for an R^2PN model.

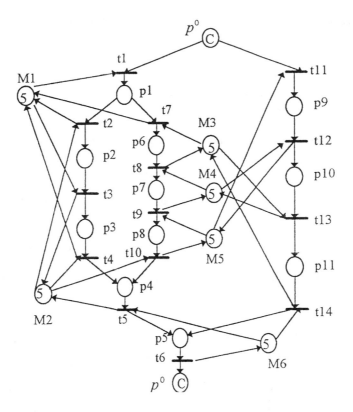

FIGURE 9.1 The R^2PN model for a manufacture system.

Given a marked R^2PN G and a reachable marking $m \in R(G, m_0)$, let $DT(m)$ denote the set of transitions in $T\backslash\{(p^0)^{\cdot} \cup^{\cdot} (p^0)\}$ which is process enabled, but dead in the marking m. Therefore, $t \in DT(m)$ means that t is not resource enabled in the marking m; i.e., $m(^{(r)}t) = 0$. If $DT(m) = \emptyset$, $\forall m \in R(G, m_0)$, then G is live. Hence, assume that the R^2PN considered in the sequel contains deadlocks.

Theorem 1

Let $G = (P \cup R, T, F, m_0)$ be a marked R^2PN and $m \in R(G, m_0)$. If $DT(m) \neq \emptyset$, then there exists a nonempty transition set $D \subseteq DT(m)$ such that $^{(r)}D = D^{(r)}$, $(^{(p)}D)^{\cdot} \subseteq D$, and $m(^{(p)}D) = m_0(^{(r)}D)$.

Proof

First we prove that $^{(r)}DT(m) \subseteq DT(m)^{(r)}$. Let $r \in {}^{(r)}DT(m)$. There exists a transition $t \in DT(m)$ such that $^{(r)}t = r$ and $m(r) = 0$. Therefore, there exists a transition t_1 such that $m(^{(p)}t_1) \geq 1$ $t_1^{(r)} = r$. Let $T_0 = \{t \in T \mid m(^{(p)}t) \geq 1, t^{(r)} = r\}$. Then $T_0 \neq \emptyset$ and $m(^{(p)}T_0) = m_0(r)$. Thus $T_0 \subseteq DT(m)$, $r \in T_0^{(r)} \subseteq DT(m)^{(r)}$ and $^{(r)}D T(m) \subseteq DT(m)^{(r)}$.

Let $D_1 = DT(m)$. If $D_1^{(r)} = {}^{(r)}D_1$, then let $D = D_1$. Now suppose that $D_1^{(r)} \neq {}^{(r)}D_1$. For any resource $r \in D_1^{(r)}\backslash^{(r)}D_1$, let $D_2 = D_1\backslash\{t \in d_1 \mid t^{(r)} = r\}$, then $D_2 \neq \emptyset$. Repeat the above procedure if $^{(r)}D_2 \neq D_2^{(r)}$. Since G is finite, a finite sequence of nonempty transition sets exists: $D_k \subset D_{k-1} \subset \cdots \subset D_1 = DT(m)$ such that, $^{(r)}D_i \neq D_i^{(r)}$, $D_i \neq \emptyset$, $i = 1, \cdots, k - 1$, and $D_k \neq \emptyset$ and $^{(r)}D_k = D_k^{(r)}$. Let $D = D_k$, and $r \in {}^{(r)}D = D^{(r)}$, $t \in DT(m)$, $t^{(r)} = r$, then $t \in D$ and $m(^{(p)}D) = m_0(^{(r)}D)$.

Finally we prove that $(^{(p)}D)^{\cdot} \subseteq D$. Let $t \in D$. If $|(^{(p)}t)^{\cdot}| = 1$, then $(^{(p)}t)^{\cdot} = t \in D$. Now let $|(^{(p)}t)^{\cdot}| \geq 2$, $t^{(r)} = r$, $t_1 \in (^{(p)}t)^{\cdot}$, $t_1 \neq t$. Then $t_1 \in DT(m)$, otherwise t_1 can be fired in a reachable marking $m' \in R(G, m)$, $m'[t_1 > m''$, $m''(r) = 1$. By $^{(r)}D = D^{(r)}$, there exists a transition $t' \in D$ such that $^{(r)}t' = \tau$, and t' can be fired in the marking m''. This is impossible since $t' \in D \subseteq DT(m)$. Hence $t_1 \in DT(m)$, and therefore $m(^{(r)}t_1) = 0$. Let $^{(r)}t_1 = r_1$. Then $m(r_1) = 0$ and there exists $t_2 \in DT(m)$, $t_2^{(r)} = r_1$ and $^{(r)}t_2 = r_2$

such that $m(^{(p)}DT(m) \cap {}^{\cdot}\tau_2) = m_0(r_2)$. Similarly there exists $t_3 \in DT(m)$, $t_3^{(r)} = r_2$, $^{(r)}t_3 = r_3$ and $m(^{(p)}(DT(m) \cap {}^{\cdot}r_3) = m_0(r_3)$. In this manner, we construct a transition sequence in $DT(m)$: $t_1, t_2, t_3, \cdots$, such that $t_i^{(r)} = r_{i-1}$, $^{(r)}t_i = r_i$, $m(^{(p)}(DT(m) \cap {}^{\cdot}r_i) = m_0(r_i)$. Since R is finite, there exist resource places r_k and r_l with $k > l$ such that $r_k = r_l$. Then $r_i \in {}^{(r)}D T(m) \cap DT(m)^{(r)}$ and $t_i \in D, i = 1, \cdots, k$. Thus $(^{(p)}D)^{\cdot} \subseteq D$.

The conditions $^{(r)}D = D^{(r)}$ and $(^{(p)}D)^{\cdot} \subseteq D$ imply that the set of resources used by the operations in $^{(p)}D$ is the same set of resources required for firing transitions in D. $m(^{(p)}D) = m_0(^{(r)}D)$ means that at marking m, all resources in $^{(r)}D$ are held by the operations in $^{(p)}D$. Hence, in marking m, the operations in $^{(p)}D$ are in a circular wait chain in which each operation is waiting for a resource held by the next operation in the chain. This circular wait relation leads to deadlock in the system.

Example 2

In the marked R^2PN shown in Figure 9.1, $D = \{t_8, t_{13}\}$ satisfies the conditions $^{(r)}D = D^{(r)}$, $(^{(p)}D)^{\cdot} \subseteq D$. For any reachable marking m at which $m(p_6) = m(p_{10}) = 5$, $m(^{(p)}D) = m_0(^{(r)}D) = 10$, the operation p_6 holds all type M_3 resources and waits for a type M_4 resource, while the operation p_{10} holds all type M_4 resources and waits for a type M_3 resource. Hence p_6 and p_{10} form a circular wait chain and cannot be completed. Transitions t_8 and t_{13} are in deadlock.

Observations and study of the above theorem and example have motivated us to define the following D-structures.

Definition 1

A transition set $D \subseteq T \setminus \{ {}^{\cdot}(p^0) \cup (p^0)^{\cdot}\}$ is called a D-structure if $^{(r)}D = D^{(r)}$ and $(^{(p)}D)^{\cdot} \subseteq D$. Let $\Psi(G)$ denote the set of all D-structures in the R^2PN G, that is,

$$\Psi(G) = \{D \subseteq T \setminus \{ {}^{\cdot}p^0 \cup (p^0)^{\cdot}\} | {}^{(r)}D = D^{(r)} \qquad \text{and} \qquad (^{(p)}D)^{\cdot} \subseteq D\}.$$

In an R^2PN model, a D-structure can lead to a circular wait under a marking m satisfying $m(^{(p)}D) = m_0(^{(p)}D)$, and hence cause deadlock in the system. Now we can characterize the liveness of R^2PN models in terms of D-structures.

Theorem 2

Let G be a marked R^2PN model. Then G is live if and only if $\forall m \in R(G, m_0)$ and $\forall D \in \Psi(G), m(^{(p)}D) \leq m_0(^{(r)}D) - 1$.

Proof

If there exist a marking $m \in R(G, m_0)$ and a D-structure $D \in \Psi(G)$ such that $m(^{(p)}D) = m_0(^{(r)}D)$, then for any resource place $r \in {}^{(r)}D$, $m(r) = 0$. Therefore $\forall m' \in R(G, m)$, $m'(^{(p)}D) = m_0(^{(r)}D)$ and every transition $t \in D$ cannot be enabled in the marking m'. All transitions in D are dead in marking m. The rest is the same as for Theorem 1.

The following theorem proves that if a marked R^2PN contains some D-structures, then it must contain deadlocks.

Theorem 3

Given a marked R^2PN $G = (P \cup R, T, F, m_0)$, for any D-structure $D \in \Psi(G)$, there exists a reachable marking $m \in R(G, m_0)$ such that $m(^{(p)}D) = m_0(^{(r)}D)$. Therefore, the set of transitions in D are dead in marking m.

Proof

Let $D_0 = D \in \Psi(G)$. The marking m satisfying conditions in the theorem can be constructed from D_0 and m_0 in the following way:

- For $i = 0, 1, 2, \cdots$, repeat the following steps until $D_k = \emptyset$.
- Let $t \in D_i$ be a transition such that $t^{(p)} \notin {}^{(p)}D_i$ and there exists no P path from $^{(p)}t$ to $D_i \setminus \{t\}$. Let $D_{i+1} = D_i \setminus \{t\}$ and $t^{(r)} = r$. Consider the following two cases:

Case 1: $r \notin D_{i+1}^{(r)}$

Consider a P path from p^0 to $^{(p)}t$ in which the sequence of transitions is $\sigma = t_0 \, t_1 \, t_2 \, \cdots \, t_k$, $t_i^{(p)} = {}^{(p)}t_{i+1}$, $i = 0, 1, \cdots, k - 1$, $t_k^{(p)} = {}^{(p)}t$. Then $m_i \, [\sigma^{Cr} >$ is well defined, and let $m_i[\sigma^{Cr} > m_{i+1}$. Then $m_{i+1}({}^{(p)}t) = m_0(r) = C_r$.

Case 2: $r \in D_{i+1}^{(r)}$

Let $m_{i+1} = m_i$. Then, using D and m_d, with $d = |D|$, satisfy the condition $m_d({}^{(p)}D) = m_0({}^{(r)}D)$, and all transitions in D are dead in marking m_d.

As an immediate result of Theorem 3 we have the following corollary.

Corollary 1

A marked R^2PN G is live if and only if G contains no D-structures, i.e., $\Psi(G) = \emptyset$.

The value of $m_0({}^{(r)}D) - 1$ is the greatest number of tokens held in the places of $^{(p)}D$ while no deadlock results from D-structure D. This value is the token capacity in $^{(p)}D$ and represents the capacity to hold the products in a group of processes. This token capacity plays a key role in the deadlock avoidance controller synthesis in the next section.

9.4 Deadlock Avoidance Controllers for R^2PN Models

Let us consider an R^2PN where some deadlock can arise. Our objective is to design a controller that guarantees that a deadlock situation will not occur in the system. Using Theorem 1, deadlock occurs only when there exist a D-structure D and a reachable marking m such that $m({}^{(p)}D) = m_0({}^{(r)}D)$. Therefore, the condition

$$m({}^{(p)}D) \leq m_0({}^{(r)}D) - 1, \forall m \in R(G, m_0), \forall D \in \Psi(G)$$

is necessary to avoid deadlocks, but not sufficient. In this chapter we use a Petri net controller to achieve the necessary condition.

Example 3

Consider the marked R^2PN G shown in Figure 9.2, in which there are three D-structures: $D_1 = \{t_2, t_7\}$, $D_2 = \{t_3, t_6\}$, and $D_3 = \{t_2, t_3, t_6, t_7\}$. We can easily present a Petri net controller C shown in Figure 9.3,

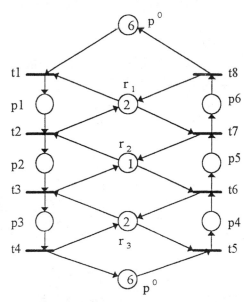

FIGURE 9.2 An R^2PN model with a cyclic chain and a key kind of resource.

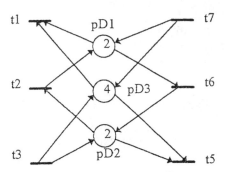

FIGURE 9.3   A Petri net controller for R^2PN in Figure 9.2.

so that for each reachable marking m of the controlled R^2PN $C * G$,

$$m(^{(p)}D_i) \leq m_0(^{(r)}D_i) - 1, \quad i = 1, 2, 3.$$

In the controlled R^2PN $C * G$, the following marking m_c is reachable from the initial marking m_{c0}:

$$m_c(p) = \begin{cases} 2, & \text{if } p \in \{p^0, p_1, p_4\}, \\ 1, & \text{if } p = r_2, \\ 0, & \text{otherwise.} \end{cases}$$

In the marking m_c, transitions t_2 and t_6 are process and resource enabled, but they cannot fire since they are not control enabled, i.e., $m_c(pD_1) = m_c(pD_2) = 0$. A new deadlock occurs. Hence, we need to avoid not only deadlocks in the R^2PN, but also deadlocks in the controlled R^2PN.

Notice that $D_3 = D_1 \cup D_2$ and pD_3 does not exert an influence on the R^2PN. $^{(r)}D_1 \cap {}^{(r)}D_2 = \{r_2\}$, $m_0(r_2) = 1$.

Given a D-structure D of G, let us denote

$$I(D) = {}^{\cdot}(^{(p)}D)\backslash D$$

$$L(D) = {}^{\cdot}(D^{(p)}\backslash^{(p)}D) \cap D.$$

The firings of transitions in $I(D)$ add tokens to $^{(p)}D$ and the firings of transitions in $L(D)$ withdraw tokens from $^{(p)}D$.

Definition 2

A subset of $\Psi(G)$, $V = \{D_1, D_2, \cdots, D_n\}$, is called a cyclic chain if for every $D_k \in V$, there exist $D_i, D_j \in V$ and $t_i, t_j \in T$ such that $t_i \in D_i \cap I(D_k)$, $t_j \in I(D_j) \cap D_k$.

Let $V = \{D_1, D_2, \cdots, D_n\}$ is a cyclic chain. A resource in $^{(r)}I(D_1) \cap {}^{(r)}I(D_1) \cap \cdots \cap {}^{(r)}I(D_n)$ is called the key resource if it has capacity $C_r = 1$. Let R_K denote the set of all key resource kinds. A P path $\sigma = t_1 p_1 \cdots p_{n-1} t_n$, $n \geq 1$ is a key path of G if $R(p_i) \in R_K$, $i = 1, 2, \cdots, n$. A key path $\sigma = t_1 p_1 \cdots p_{n-1} t_n$ is maximal if $t_1^{(r)} = \emptyset$ or $t_1^{(r)} \notin R_k$ and $^{(r)}t_n = \emptyset$ or $^{(r)}t_n \notin R_k$.

For example, in the marked R^2PN shown in Figure 9.2, D-structures $D_1 = \{t_2, t_7\}$ and $D_2 = \{t_3, t_6\}$ form a cyclic chain, and r_2 is a key resource; $t_2 p_2 t_3$ and $t_5 p_6 t_7$ are the maximal key paths.

Although the Petri net controller designed in the above manner cannot guarantee that the controlled R^2PN model is live, we can prove in the following that if no key resource exists in the R^2PN model, then the Petri net controller can avoid all deadlocks and hence the controlled model is a live Petri net. Motivated by this fact, we will introduce a method for reducing R^2PN such that the reduced R^2PN is also R^2PN, but contains no key resources. Another aim of reducing the R^2PN is to lower the complexity for synthesizing a deadlock avoidance controller for the R^2PN. In the following, we first define the reduced R^2PN models and then present an optimal deadlock avoidance controller for the reduced R^2PN model. Finally, we define a controller for the R^2PN, which guarantees that the controlled model is live.

Reducing R^2PN Models

The greater the number of D-structures in an R^2PN, the larger the opportunity for deadlock and the more complex the design of the deadlock avoidance controller. To lower the complexity of controller synthesis, we first present the following method for reducing R^2PN models.

Definition 3

Let $G = (P \cup R, T, F, m_0)$ be an R^2PN and r be a resource place. Let $H(r) = \{p \in P \mid R(p) = r\}$. The r-reduced R^2PN G is a Petri net $G(r)$ constructed from G by the following steps.

1. Remove the place r from G and all arcs which are incidental to or from r.
2. Repeat the following steps for each place $p \in H(r)$. Select and remove $p \in H(r)$ and all transitions in $\cdot p \cup p \cdot$ from G. For every pair of transitions (t_1, t_2) of $\cdot p \times p \cdot$ in G:

   - if $t_1^{(r)} \neq {}^{(r)}t_2$ in G, then add a new transition, denoted by $t_1 + t_2$, and some arcs which are incidental from or to $t_1 + t_2$ so that $\cdot(t_1 + t_2) = (\cdot t_1 \cup \cdot t_2) \backslash \{p, r\}$ and $(t_1 + t_2)\cdot = t_1 \cdot \cup t_2 \cdot \backslash \{p, r\}$.
   - if $t_1^{(r)} = {}^{(r)}t_2 = \{r_1\}$ in G, then add a new transition, denoted by $t_1 + t_2$, and some arcs which are incidental from or to $t_1 + t_2$ so that $\cdot(t_1 + t_2) = (\cdot t_1 \cup \cdot t_2) \backslash \{p, r, r_1\}$ and $(t_1 + t_2)\cdot = (t_1 \cdot \cup t_2 \cdot) \backslash \{p, r, r_1\}$.

We will call $t_1 + t_2$ τ-transition, and let T_τ denote the set of all τ-transitions in $G(r)$. Notice that in the second situation above ${}^{(r)}(t_1 + t_2) = (t_1 + t_2)^{(r)} = \emptyset$ and $R({}^{(p)}(t_1 + t_2)) = R((t_1 + t_2)^{(p)}) = r_1$. and so we can consider ${}^{(p)}(t_1 + t_2)$ and $(t_1 + t_2)^{(p)}$ as the same operation places requiring the same resource r_1. In this manner, the r-reduced R^2PN model may also be considered an R^2PN model, and any R^2PN model can be reduced by any resource place.

Let G be an R^2PN and r_1 and r_2 be two resource places. For the r_1-reduced R^2PN $G(r_1)$, we can do an r_2-reduced procedure for $G(r_1)$ and obtain an $\{r_1, r_2\}$-reduced R^2PN $G(r_1, r_2)$ in the same way. $G(r_1, r_2)$ is an R^2PN model. In general, for any set of resource places R', we can construct R'-reduced R^2PN, denoted by $G(R')$.

Example 4

Consider the R^2PN G shown in Figure 9.1. The M_2-reduced R^2PN $G(M_2)$ is given in Figure 9.4. The firing of τ-transition $t_2 + t_3$ requires no resources; $R({}^{(p)}(t_2 + t_3)) = R((t_2 + t_3)^{(p)}) = M_1$. The operations ${}^{(p)}(t_2 + t_3)$ and $(t_2 + t_3)^{(p)}$ can be considered the same operation as in $G(M_2)$, and $G(M_2)$ is an R^2PN. G contains at least ten D-structures and $G(M_2)$ has only three D-structures.

Let R_K be the set of key resource places and $G(R_K) = (P' \cup R', T', F', m'_0)$ be the R_K reduced R^2PN. Then $G(R_K)$ is an R^2PN model in which there are no key resources. The set of transitions in $G(R_K)$ can be divided into two parts, T_0 and T_1, where T_0 is a set of τ-transitions and every τ-transition $t_1 + t_2 + \cdots + t_k$ in T_0 corresponds to a maximal key path $t_1 p_1 t_2, p_2 \cdots p_{k-1} t_k$, $K \geq 2$, in G. $T_1 = T' \backslash T_0$. Let T_2 denote the set of transitions of G which are in some key path. Then $T = T_1 \cup T_2$.

The complexity for reducing an R^2PN by the set of key resources R_K is linear with $|\{p \in P \mid R(p) \in R_K\}|$. Since $\{p \in P \mid R(p) \in R_K\}$ is finite, the procedure for reducing an R^2PN model is efficient, and any R^2PN admits this reduction.

Optimal Deadlock Avoidance Petri Net Controllers for a Class of R^2PNs

Let G be a marked R^2PN and R_K be the set of key resource places. Then $G(R_K)$ is an R^2PN which contains no key resources. For such a special class of R^2PNs, we can first present the following deadlock avoidance Petri net controller:

Definition 4

Let $G = (P \cup R, T, F, m_0)$ be a marked R^2PN, $R_K = \emptyset$. A controller for G is a marked Petri net defined by

$$C' = (P_c, T, F_c, m_0^c),$$

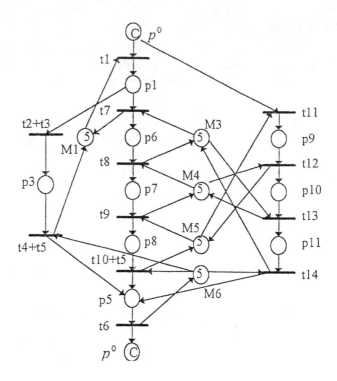

FIGURE 9.4   The M_2-reduced R^2PN of Petri net model in Figure 9.1.

where $P_c = \{P_D \mid D \in \Psi(G)\}$ is a set of control places so that there exists a bijective mapping from $\Psi(G)$ to it,

$$F_c = \{(p_D, t) \mid t \in I(D), D \in \Psi(G)\} \cup \{(t, p_D) \mid t \in L(D), D \in \Psi(G)\},$$

$$m_0^c(p_D) = m_0(^{(r)}D) - 1, \qquad \forall p_D \in P_c.$$

The marked R^2PN G with the controller C' can be modeled by the composition of G and C' given by

$$C' * G = (P \cup R \cup P_c, T, F \cup F_c, m_{c0}),$$

where $m_{c0}(p) = m_0(p), \forall p \in P \cup R$, and $m_{c0}(p) = m_0^c(p), \forall p \in P_c$.

The controller C' guarantees simply that

$$m(^{(p)}D) \le m_0(^{(r)}D) - 1, \forall D \in \Psi(G), \forall m \in R(C' * G, m_{c0}).$$

This is necessary for avoiding deadlocks in G. Hence if $C' * G$ is live, then C' is the optimal deadlock avoidance controller, and $C' * G$ is the optimal live Petri net model of the system.

Example 5

Consider the R^2PN G shown in Figure 9.5. Let m_0 be an initial marking of G and $m_0(r_i) \ge 1, i = 1, 2, 3, 4$. Let C' be a Petri net controller for G as mentioned above. $D_1 = \{t_2, t_3\}$ and $D_2 = \{t_2, t_3, t_4\}$ are D-structures of G. $^{(r)}D_1 = {}^{(r)}D_2 = \{r_1, r_2\}$ and $^{(p)}D_1 = \{p_1, p_2\} \subseteq {}^{(p)}D_2 = \{p_1, p_2, p_3\}$. Therefore, for any reachable marking $m \in R(C' * G, m_{c0})$, if

$$m(^{(p)}D_2) \le m_{c0}(^{(r)}D_2) - 1,$$

then

$$m(^{(p)}D_1) \le m_{c0}(^{(r)}D_1) - 1.$$

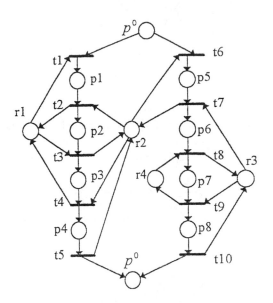

FIGURE 9.5 An R^2PN model for Example 5.

$D_3 = \{t_8, t_9\}$ and $D_4 = \{t_2, t_3, t_4, t_7, t_8, t_9\}$ are two D-structures. D_u, $D_3 \subseteq D_4$, $(D_4 \backslash D_3)^{(r)} = \{r_1, r_2\}$, $^{(r)}D_3 = \{r_3, r_4\}$, hence $(D_4 \backslash D_3)^{(r)} \cap {}^{(r)}D_3 = \emptyset$, and for any marking m, if

$$m(^{(p)}D_3) \le m_{c0}(^{(r)}D_3) - 1,$$

then

$$m(^{(p)}D_4) \le m_{c0}(^{(r)}D_4) - 1.$$

Therefore, in the definition of Petri net controller C', we need not consider D-structures D_1 and D_4. This is an example of the case that we will consider below.

Let $D \in \Psi(G)$. If $\exists D \in \Psi(G)$ such that $D \subset D_0$ and $^{(r)}D = {}^{(r)}D_0$ then for any reachable marking $m \in R(C' * G, m_{c0})$,

$$m(^{(p)}D_0) \le m_{c0}(^{(r)}D_0) - 1$$

implies that

$$m(^{(p)}D) \le m_{c0}(^{(r)}D) - 1$$

Hence, in the definition of Petri net controller C', we may remove the control place p_D and arcs relevant to or from p_D.

Let $\Psi_M(G)$ denote the set of all D-structures D for which there exists no D-structure D_0 such that $D \subset D_0$ and $^{(r)}D = {}^{(r)}D_0$; that is, D is the maximum D-structure with the set of resources $^{(r)}D$.

Let $D \in \Psi_M(G)$. If $\exists D_0 \in \Psi(G)$ such that $D_0 \subset D$ and $(D \backslash D_0)^{(r)} \cap {}^{(r)}D_0 = \emptyset$, then for any reachable marking $m \in R(C' * G, m_{c0})$,

$$m(^{(p)}D_0) \le m_{c0}(^{(r)}D_0) - 1.$$

implies that

$$m(^{(p)}D) \le m_{c0}(^{(r)}D) - 1.$$

Hence, in the definition of Petri net controller C', we may remove the control place p_D and arcs relevant to or from p_D.

Let $\Psi_N(G)$ denote the set of all D-structures D that contain a D-structure D_0 such that $(D \backslash D_0)^{(r)} \cap {}^{(r)}D_0 = \emptyset$. Let us denote $\Psi_B(G) = \Psi_M(G) \backslash \Psi_N(G)$. A D-structure in $\Psi_B(G)$ is called a basic D-structure. For example, in the R^2PN in Figure 9.5, only $D_2 = \{t_2, t_3, t_4\}$ and $D_3 = \{t_8, t_9\}$ are basic D-structures.

Considering the above results, we can replace C' with the following Petri net controller:

$$C = (P_B, T, F_B, m_0^B),$$

where $P_B = \{p_D \mid D \in \Psi_B(G)\}$ is a set of control places so that there exists a bijective mapping from $\Psi_B(G)$ to it:

$$F_B = \{(p_D, t) \mid t \in I(D), D \in \Psi_B(G)\} \cup \{(t, p_D) \mid t \in L(D), D \in \Psi_B(G)\},$$

$$m_0^B(p_D) = m_0(^{(r)}D) - 1, \forall D \in \Psi_B(G).$$

The controlled R^2PN is a composition Petri net of G and C,

$$C * G = (P \cup R \cup P_B, T, F \cup F_B, m_{B0})$$

where $m_{B0}(p) = m_0(p), \forall p \in P \cup R$, and $m_{B0}(p) = m_0^B(p), \forall p \in P_B$.

In the controlled R^2PN $C * G$, let $^{(c)}t$ and $t^{(c)}$ denote $^.t \cap P_B$ and $t^. \cap P_B$, respectively.

Given a reachable marking $m \in R(C * G, m_{B0})$, let $DT_c(m)$ denote the set of transitions which is dead, but process enabled in the marking m. If $DT_c(m) = \emptyset, \forall m \in R(C * G, m_{B0})$, then $C * G$ is live. Now we need to prove that the proposed controller leads to a live Petri net model.

Lemma 1

Given an R^2PN G, $R_K = \emptyset$, and its controller C is as above. If $\exists m \in R(C * G, m_{B0})$ such that $DT_c(m)$ is not empty, then $\exists t \in DT_c(m)$ and $D \in \Psi_B(G)$ such that $m(^{(r)}t) \ge 1$, $p_D \in {}^{(c)}t$, and $m(p_D) = 0$.

Proof

Let $t \in DT_c(m)$. Then $m(^{(p)}t) \ge 1$ and one of the following two cases must hold:

1. $m(^{(r)}t) = 0$.
2. $m(^{(r)}t) \ge 1$ and $\exists D \in \Psi_B(G)$ such that $p_D \in {}^{(c)}t$, $m(^{(p)}D) = 0$. In this case, $m(p_D) = m_0(^{(r)}D) - 1$.

Suppose the lemma is not true. Then for every transition $t \in DT_c(m)$, $m(^{(r)}t) = 0$. Let $t_1 \in DT_c(m)$, $^{(r)}t_1 = r_1$ and let $A(r_1) = \{t_1 \in DT_c(m) \mid t^{(r)} = r_1\}$. Then $A(r_1) \ne \emptyset$ and $m(^{(p)}A(r_1)) = m_0(r_1)$. For a transition $t_2 \in A(r_1)$, let $^{(r)}t_2 = r_2$ and $A(r_2) = \{t \in DT_c(m) \mid t^{(r)} = r_2\}$. Then $A(r_2) \ne \emptyset$ and $m(^{(p)}A(r_2)) = m_0(r_2)$. In this manner, we construct a transition sequence $t_1, t_2, \ldots$, such that $^{(r)}t_i = r_i = t_{i+1}^{(r)}, t_{i+1} \in A(r_i) = \{t \in DT_c(m) \mid t^{(r)} = r_i\} \ne \emptyset, m(^{(p)}A(r_i)) = m_0(r_i), i = 1, 2, \ldots$. Then there exist t_j and t_k so that $^{(r)}t_j = {}^{(r)}t_k$ and $j < k$. Let $Y = \{t_j, \ldots, t_k\}$. Similarly, for $t_i \in Y$ we can construct another sequence $t_{i1}, t_{i2}, \ldots, t_{il}$ so that $^{(r)}t_{ij} = r_{ij} = t_{ij+1}^{(r)}, t_{i(j+1)} \in A(r_{ij}) = \{t \in DT_c(m) \mid t^r = r_{ij}\} \ne \emptyset, {}^{(r)}t_{ik} = {}^{(r)}t_{il}, h < l \ m^p A(r_{ij})) = m_0(r_{ij}),$. Then let $Y = Y \cup \{t_{i1}, t_{i2}, \ldots, t_{il}\}$. Repeat the above procedure until all sequences starting from Y are in Y. Then Y is a D-structure and $m(^{(p)}Y) = m_0(^{(r)}Y)$, producing a contradiction and completing the proof.

Theorem 4

Let G be a marked R^2PN and $R_K = \emptyset$. Let C be a Petri net controller defined as above. Then the controlled R^2PN $C * G$ is live. Hence, C is the optimal deadlock avoidance controller and $C * G$ is an optimal live Petri net model for the system.

Proof

Suppose the theorem is not true. Then there exists a marking $m \in R\ (C * G, m_{B0})$ such that $DT_c(m)$ is not empty. Using the above Lemma, $\exists t_1 \in DT_c(m)$, $D_1 \in \Psi_B(G)$ and $p_{D1} \in^{(c)} t_1$ such that $m(^{(r)}t_1) \geq 1$, $m(p_{D1}) = 0$. Then $m(^{(p)}D_1) = m_0(^{(r)}D_1) - 1$, and $t_1 \in I(D_1)$. We denote $r_1 =^{(r)}t_1$ and $A(r_1) = \{t \in D_1 \mid^{(r)}t = r_1\}$, then $m(r_1) = 1$ and $A(r_1) \neq \emptyset$ since $^{(p)}D_1 = D_1^{(r)}$ and $t_1 \notin D_1$. Here we first prove the claim: there exists a transition $t_2 \in A(r_1)$ such that $m(^{(p)}t_2) \geq 1$.

Suppose that $\forall t \in A(r_1)$, $m(^{(p)}t) = 0$. Let $E_1 = D_1 \backslash A_1$, where $A_1 \equiv \{t \in D_1 \mid t \in (p^{\cdot} \cup {}^{\cdot}p), R(p) = r_1\}$, then $E_1 \neq \emptyset$. Since $m(^{(p)}t) = 0$ for any transition $t \in A_1 \cap r_1^{\cdot}$, $R(^{(p)}t) \in E_1^{(r)}$ and hence $^{(r)}E_1 \subseteq E_1^{(r)}$. If $^{(r)}E_1 \neq E_1^{(r)}$, then $\exists x_1 \in E_1$ so that $x_1^{(r)} \in E_1^{(r)} \backslash^{(r)}E_1$. Thus $E_2 \equiv E_1 \backslash \{x_1\} \neq \emptyset$ and $^{(r)}E_2 \subseteq E_2^{(r)}$. In this manner, we can construct a sequence $E_1, E_2, \ldots$, so that $E_1 \supset E_2 \supset E_3 \supset \ldots$, $E_i \neq \emptyset$ and $E_i^{(r)} \backslash^{(r)}E_i \neq \emptyset$. Since E_1 is finite, the sequence is also finite and there exists $E_k \neq \emptyset$ so that $E_k^{(r)} =^{(r)}E_k$. Thus, E_k is a D-structure and $m(^{(p)}E_k) = m_0(^{(r)}E_k)$, which produces a contradiction.

For $m(^{(r)}t_2) = m(r_1) = 1$, there exists a D-structure $D_2 \in \Psi_B(G)$ such that $t_2 \in I(D_2)$ and $m(^{(p)}D_2) = m_0(^{(p)}D_2) - 1$. Similar to obtaining t_2 and D_2 from t_1 and D_1, we can get a transition t_3 and a basic D-structure D_3 from t_2 and D_2 such that $m(^{(p)}t_3) \geq 1$, $^{(r)}t_3 = r_1$, $t_3 \in I(D_3) \cap D_2$ and $m(^{(p)}D_3) = m_0(^{(r)}D_3) - 1$. In this way, we construct a basic D-structure sequence $D_1, D_2, \ldots$, and a transition sequence $t_1, t_2, \ldots$, such that $r_1 =^{(r)}t_i \in^{(r)}D_i$, $i = 1, 2, \ldots$. And there exists a cyclic chain in $D_1, D_2, \ldots$, say $\{D_1, D_2, \ldots, D_k\}$, and $r_1 \in^{(r)} I(D_1) \cap^{(r)}I(D_2) \cap \ldots \cap^{(r)} I(D_k)$. Since $m_0(r_1) \geq 2$, there exists a transition $v_1 \in D_1$ such that $v_1^r = r_1$, $m(^{(p)}v_1) \geq 1$. Then $v_1 \in D_i$, $i = 1, 2, \ldots, k$. Let $^{(r)}v_1 = r_2$, then $r_2 \in D_1^{(r)} \cap \ldots \cap D_k^{(r)}$ and $m(r_2) = 0$. For any transition $v_2 \in D_1$, if $v_2^{(r)} = r_2$, and $m(^{(p)}v_2) \geq 1$, then $v_2 \in D_i$, $i = 1, 2, \ldots, k$. Since for any resource $r \in D_1^{(r)}$, there exists a resource sequence $r_1, r_2, \ldots, r_k = r$ in $^{(r)}D_1$ and a transition sequence $t_1, t_2, \ldots, t_k$ in D_1 such that $r_i = t_i^{(r)}$, $^{(r)}t_i = r_{i+1}$, $m(^{(p)}t_i) \geq 1$. Hence, $D_1^{(r)} \subseteq D_2^{(r)} \cap \ldots \cap D_k^{(r)} \cdot DT_c(m) \cap D_1 \subseteq D_2 \cap \ldots D_k$. In a like manner, it can be proved that $D_i^{(r)} \subseteq D_1^{(r)} \cap \ldots \cap D_{i-1}^{(r)} \cap D_{i+1}^{(r)} \cap \ldots \cap D_k^{(r)}$ and $D_i \cap DT_c(m) \subseteq D_1 \cap \ldots \cap D_{i-1} \cap D_{i+1}D_k$, that is, $D_i \cap DT_c(m) = \{t \in D_i \mid m^{(p)}(t) \geq 1\}$ is a same set for every i. But $t_2 \in D_1$, $t_2 \in I(D_2)$, $m(^{(p)}t_2) \geq 1$ and $t_2 \notin D_2$, producing a contradiction and completing the proof.

Example 6

Consider the application of the synthesis method of deadlock avoidance Petri net controller synthesis to the R^2PN G shown in Figure 9.1.

$$\Psi_B(G) = \{D_1, D_2, D_3, D_4, D_5\},$$

where

$$D_1 = \{t_3, t_4\}, \qquad D_2 = \{t_8, t_{13}\},$$
$$D_3 = \{t_9, t_{12}\}, \qquad D_4 = \{t_8, t_9, t_{12}, t_{13}\},$$
$$D_5 = \{t_2, t_3, t_4, t_7, t_8, t_9, t_{10}, t_{12}, t_{13}\}.$$

We compute for each $D \in \Psi_B(G)$ the sets $I(D)$, $L(G)$ and the number $m_0(^{(r)}D)$ to synthesize the Petri net controller for G as follows:

$$I(D_1) = \{t_2\} \qquad L(D_1) = \{t_4\} \qquad m_0(^{(r)}D_1) = 10$$
$$I(D_2) = \{t_7, t_{12}\} \qquad L(D_2) = \{t_8, t_{13}\} \qquad m_0(^{(r)}D_2) = 10$$
$$I(D_3) = \{t_8, t_{11}\} \qquad L(D_3) = \{t_9, t_{12}\} \qquad m_0(^{(r)}D_3) = 10$$
$$I(D_4) = \{t_7, t_{11}\} \qquad L(D_4) = \{t_9, t_{13}\} \qquad m_0(^{(r)}D_4) = 15$$
$$I(D_5) = \{t_1, t_{11}\} \qquad L(D_5) = \{t_4, t_{10}, t_{13}\} \qquad m_0(^{(r)}D_5) = 25.$$

Then the optimal deadlock avoidance Petri net controller C can be synthesized as shown in Figure 9.6.

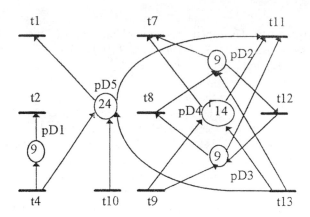

FIGURE 9.6   The optimal deadlock avoidance Petri net controller for the R^2PN in Figure 9.1.

A Petri net controller for the R^2PN G is given by Ezpeleta et al. [3]. In the case where the R^2PN model contains no key resources, a comparison of our controller with those of Ezpeleta et al. shows that the performance achieved by our controller is better than the one achieved by the controller used by Ezpeleta et al.

Deadlock Avoidance Controllers for the R^2PNs

Let G be an R^2PN model and $R_K \neq \emptyset$. Then we can reduce G by R_K and obtain a reduced R^2PN model $G(R_K)$ which contains no key resources. Hence an optimal deadlock avoidance Petri net controller C for $G(R_K)$ can be defined. In this subsection we present a deadlock avoidance controller for the R^2PN model G by using the liveness of $C * G(R_K)$.

Definition 5

Let $G = (P \cup R, T, F, m_0)$ be a marked R^2PN model and $R_K \neq \emptyset$. C is the optimal deadlock avoidance Petri net controller for the reduced R^2PN $G(R_K)$. A restriction controller ρ for G is defined as follows.
 ρ makes G operate in synchrony with $C * G(R_K)$, that is,

1. A transition $t \in T_1$ fires in G and $C * G(R_K)$ at the same time.
2. A transition $t \in T_2$ fires once in G only if some maximal key path containing t fires once; any maximal key path $\sigma = t_1 p_1 t_2 \ldots p_{k-1} t_k$ can be fired if and only if the transition $t_1 + t_2 + \cdots + t_k$ corresponding to σ can be fired in $C * G(R_K)$.

The R^2PN G under the control of ρ is denoted as ρ/G. Let $\sigma = t_1 p_1 \ldots t_n$ be a maximal key path in G. If the τ-transition $t_1 + t_2 + \cdots + t_k$ corresponding to σ can fire once in $C * G(R_K)$, then $t_1, t_2, \ldots, t_k$ can be fired once in order, and if t_1 is fired, then $t_2, \ldots, t_k$ must be fired once, respectively in ρ/G. Let $\sigma = x_1 x_2 \ldots x_n$ be a sequence of transitions in $C * G(R_K)$ and let $f(\sigma)$ be a sequence of transitions in G constructed from σ by replacing $x = t_1 + t_2 + \cdots + t_k$ with a subsequence $t_1 t_2 \ldots t_k$. Then σ can be fired from m_{B0} in $C * G(R_K)$ if and only if $f(\sigma)$ can be fired from m_0 in G under the controller ρ.
 The following theorem establishes the liveness of the controlled system ρ/G.

Theorem 5

Let $G = (P \cup R, T, F, m_0)$ be a marked R^2PN model and $R_K \neq \emptyset$. ρ is the restriction controller for G as in Definition 5. Then the controlled system ρ/G is live.

Proof

Let σ be a sequence of transitions in $C * G(R_K)$ which can be fired from the initial marking m_{B0} of $C * G(R_K)$ such that $m_{B0}[\sigma > m_c$ and let $m_0[f(\sigma) > m$ in ρ/G. Let t be a transition of G. Then $t \in T_1$ or t is in some key path $t_1 p_1 t_2 \ldots t_n$ which corresponds to τ-transition $x = t_1 + t_2 + \cdots + t_n$ and $t = t_i$ for some i.

By the liveness of $C * G(R_K)$, there exists a marking $m'_c \in R(C * G(R_K), m_c)$ such that t or x can be fired in the marking m'_c. Let $\delta = x_1 x_2 \ldots x_k$ be a sequence of transitions in $C * G(R_K)$ such that $m_c[\delta > m'_c$. For simplicity, we ensure that $\delta = 1$ empty string. Then t or x can be fired in the marking m_c. If $t \in T_1$, then t can be fired in the marking m. If t is in the key path $t_1 p_1 \ldots p_{n-1} t_n$ and $t = t_i$, then $m(^{(p)}t_1) = m_c(^{(p)}x)$ ≥ 1. $m(R(p_1)) = m(R(p_2)) = \cdots = m(R(p_{n-1})) = 1$. If $R(^{(P)}t_1 + \cdots + t_n)) \neq R(t_1 + \cdots t_n)^{(P)}$, then $m(R(t_n^{(p)})) = m_c(^{(r)}x) = m(R(x^{(p)})) \geq 1$ and $\beta = t_1 t_2 \ldots t_n$ can be fired from the marking m; if $R(^{(p)}(t_1 + \cdots + t_n)) = R(t_1 + \cdots + t_n)^{(p)}$, then $R(^{(p)}t_1) = R(t_n^{(P)})$ and $\beta = t_1 t_2 \ldots t_n$ can be fired from the marking m. Thus $t = t_i$ is live and hence the R^2 PN G with the controller ρ is live.

The control function of ρ to G is dependent on the behavior of $C * G(R_K)$. Now we can present another form of ρ which is independent to $C * G(R_K)$. This controller consists of two parts: a Petri net controller and a restrictive policy.

Definition 6

Let $G = (P \cup R, T, F, m_0)$ be a marked R^2 PN model. $R_K \neq \emptyset$. $C = (P_B, T, F_B, m_0^B)$ is the optimal deadlock avoidance Petri net controller for the reduced R^2 PN $G(R_K)$. A Petri net controller for G is defined by

$$C_P = (P_B, T, F'_B, m_0^B),$$

where $F'_B = \{(p_D, t_n) \mid (p_D, x) \in F_B, x = t_1 + t_2 + \cdots + t_n\}; \cup \{(t_1, p_D) \mid (x, p_D) \in F_B, x = t_1 + \cdots + t_n\}$.

The restrictive policy ρ_0 is defined for the composition Petri net of G and C_p, $C_p * G$, as follows. For each marking $m \in R(C_p * G, m_{Bo})$,

1. $t \in T_1$ can be fired if t can be fired in $C_p * G$.
2. Any key path $t_1 p_1 t_2 p_2 \cdots p_{n-1} t_n$ can be fired if $m(^{(p)}t_1) \geq 1$, $m(R(p_i)) = 1$, $i = 1, \cdots, n - 1$, $m(R(t_n^{(P)})) \geq 1$ if $R(^{(p)}t_1) \neq R(t_n^{(p)})$, and $m(p_D) \geq 1, \forall p_D \in {}^{(c)}t_n$ in $C_p * G$. If t_1 is fired, then $t_2, \ldots t_n$ must be fired once in order.

The Petri net controller C_p has the same function of condition 1 in ρ, and the function of ρ_0 is equivalent to condition 2 of ρ in Definition 5. Hence, C_p together with ρ_0 is a deadlock avoidance controller for the R^2PN G.

Corollary 2

Let G be a marked R^2PN model. C_p and ρ_0 are defined as above. Then $C_p * G$ under the control of ρ_0 is live.

The key to synthesis of the deadlock avoidance controllers mentioned above is to compute the set of basic D-structures, $\Psi_B(G)$. Xing and Li [11] established a one to one corresponding relationship between the set of basic D-structures and the set of minimal siphons which may be empty as follows.

Lemma 2

Let $G = (P \cup R, T, F, m_0)$ be an R^2 PN model. Γ is the set of minimal siphons of G which can be empty in some reachable marking. For a minimal siphon $S \in \Gamma$, define

$$h(S) = {}^{\cdot}(S \cap R) \backslash (S \cap P)^{\cdot}$$

Then h is a one-to-one mapping from Γ to $\Psi_B(G)$ and

$$\Psi_B(G) = \{h(S) \mid S \in \Gamma\}.$$

For instance, in the R^2 PN shown in Figure 9.5, the set of basic D-structures $\Psi_B(G) = \{D_1, D_2\}$ where $D_1 = \{t_2, t_3, t_4\}$ and $D_2 = \{t_8, t_9\}$ and the set of minimal siphons $\Gamma = \{S_1, S_2\}$, where $S_1 = \{r_1, r_2, p_4, p_5\}$ and $S_2 = \{r_3, r_4, p_8\}$. Then $D_i = h(S_i)$, $i = 1, 2$, can be verified.

Since the efficient algorithms to compute the set of minimal siphons can be found in the literature, we can compute the set of basic D-structures $\Psi_B(G)$ by the bijective mapping h and computing the set of minimal siphons.

9.5 Examples

We illustrate the $R^2 PN$ methodology and the application of the controllers of the previous section via two examples.

Example 7

Consider a work cell which consists of three robots r_1, r_2, and r_3 and four kinds of machines M_1, M_2, M_3, and M_4. Each robot can hold one product at a time. Each machine type has two units of machines, and each unit of machine can process one product at a time. The system can produce three types of products q_1, q_2, and q_3. For a type q_1 product there are two production routings through the system resources: $r_1 M_1 r_2 M_2 r_3$ and $r_1 M_3 r_2 M_4 r_3$; the production routing for q_2 is $r_2 M_2 r_2$ and the production routing for q_3 is $r_3 M_4 r_2 M_3 r_1$.

The production routings for q_1 are modeled by the directed cycles:

$$p^0 t_1 p_1(r_1) t_2 p_2(M_1) t_3 p_3(r_2) t_4 p_4(M_2) t_5 p_5(r_3) t_6 p^0$$

and

$$p^0 t_1 p_1(r_1) t_7 p_6(M_3) t_8 p_7(r_2) t_9 p_8(M_4) t_{10} p_5(r_3) p^0,$$

respectively. The routing for q_2 is modeled by

$$p^0 t_{11} p_9(r_2) t_{12} p_{10}(M_2) t_{13} p_{11}(r_2) t_{14} p^0 .$$

The routing for q_3 is modeled by

$$p^0 t_{15} p_{12}(r_3) t_{16} p_{13}(M_4) t_{17} p_{14}(r_2) t_{18} p_{15}(M_3) t_{19} p_{16}(r_1) t_{20} p^0.$$

A token in the place $p_i(r_j)$ or $p_i(M_j)$ represents a product which is held by r_j or processed on M_j.

Then the $R^2 PN$ model resulting from the system is shown in Figure 9.7, where $p_i(M_j)$ or $p_i(r_j)$ are written as p_i, for brevity. The places M_i and r_j model the available stats of machine M_i and robot r_j, respectively. The initial marking is shown in Figure 9.7, where $m_0(p^0) = C \geq 11$.

In the $R^2 PN$ model G there are 18 minimal siphons which can be empty [3] and hence 18 basic D-structures.

$$D' = \{t_8, t_{18}\}, \qquad D'' = \{t_9, t_{17}\}$$

are two basic D-structures. $V = \{D', D''\}$ is a cyclic chain.

$$^{(r)}I(D') = \{^{(r)}t_7, {}^{(r)}t_{17}\} = \{M_3, r_2\}$$
$$^{(r)}I(D'') = \{^{(r)}t_8, {}^{(r)}t_{16}\} = \{M_4, r_2\},$$

where r_2 is a key resource and is used by five operations modeled by p_3, p_7, p_9, p_{11}, p_{14}. The r_2-reduced $R^2 PN$ $G(r_2)$ is shown in Figure 9.8. $G(r_2)$ contains only four basic D-structures and

$$\Psi_B(G(r_2)) = \{D_1, D_2, D_3, D_4\},$$

where

$$D_1 = \{t_{10}, t_{16}\}$$
$$D_2 = \{t_8 + t_9, t_{17} + t_{18}\}$$
$$D_3 = \{t_8 + t_9, t_{10}, t_{16}, t_{17} + t_{18}\}$$
$$D_4 = \{t_2, t_3 + t_4, t_5, t_7, t_8 + t_9, t_{10}, t_{16}, t_{17} + t_{18}, t_{19}\}.$$

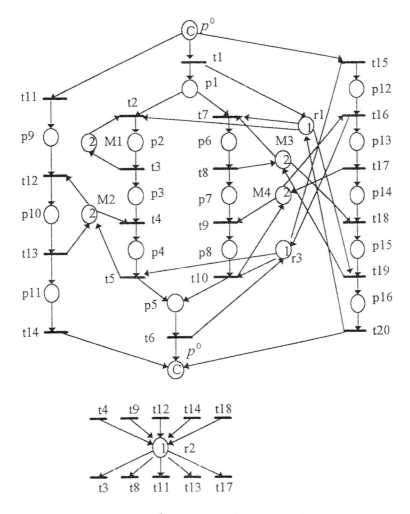

FIGURE 9.7 The R^2PN modeling of a manufacturing system.

For each D-structure $D \in \Psi_B(G(r_2))$, $I(D)$, $L(D)$, and $m_0(^{(r)}D)$ are computed as follows.

$$I(D_1) = \{t_8 + t_9, t_{15}\} \qquad L(D_1) = \{t_{10}, t_{16}\}$$
$$I(D_2) = \{t_7, t_{16}\} \qquad L(D_2) = \{t_8 + t_9, t_{17} + t_{18}\}$$
$$I(D_3) = \{t_7, t_{15}\} \qquad L(D_3) = \{t_{10}, t_{17} + t_{18}\}$$
$$I(D_4) = \{t_7, t_{15}\} \qquad L(D_4) = \{t_5, t_{10}, t_{19}\}$$
$$m_0(^{(r)}D_1) = m_0(\{r_3, M_4\}) = 3 \qquad m_0(^{(r)}D_2) = m_0(\{M_3, M_4\}) = 4$$
$$m_0(^{(r)}D_3) = m_0(\{r_3, M_3, M_4\}) = 5 \qquad m_0(^{(r)}D_4) = m_0(\{r_1, r_3, M_1, M_2, M_3, M_4\}) = 10.$$

$G(r_2)$ contains no key resources. Hence, an optimal deadlock avoidance Petri net controller can be synthesized as in Figure 9.9a. Figure 9.9b shows the Petri net part C_p of the controller for G.

In the R^2PN G, there are five maximal key paths. If this $R^2 PN$ G is controlled by the Petri net controller C_p and the restrictive policy ρ_0, then each maximal key path can be fired only when all of its transitions are resource enabled, its first transition is process enabled, and its last transition is control enabled. For example, the maximal key path $\sigma = t_8 p_7 t_9$ can be fired in a marking m of $C_p * G$ only if $m(^{(r)}t_8) = m(r_2) = 1$, $m(^{(r)}t_9) = m(M_4) \geq 1$, $m(^{(p)}t_8) = m(p_6) \geq 1$, and $m(^{(c)}t_9) = m(p_{D_2}) \geq 1$. And if t_8 is fired, then t_9 must be fired before M_4 or p_{D_2} becomes empty. A Petri net controller, given Ezpeleta et al. [3] contains 18 control places.

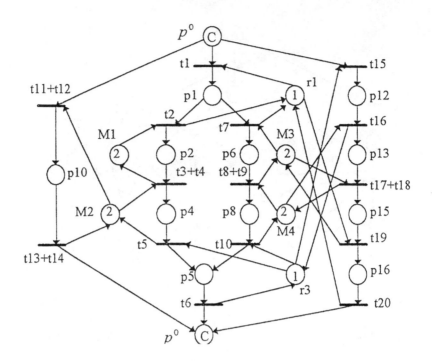

FIGURE 9.8   The R^2-reduced R^2PN of the Petri net model in Figure 9.7.

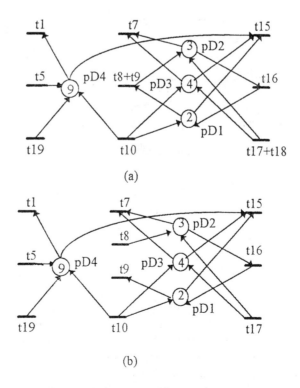

FIGURE 9.9   (a) The optimal Petri net controller for the R^2PN in Figure 9.8. (b) The Petri net conroller part for the R^2PN in Figure 9.7.

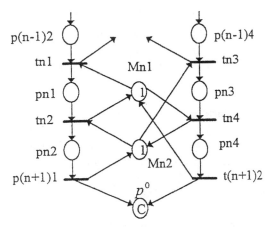

FIGURE 9.10   The R^2PN model for the manufacturing system in Example 8.

Example 8

Consider a flexible manufacturing system which consists of n workcells WC_i, $i = 1, 2, 3, \cdots, n$. The workcell WC_i has two machines M_{i1} and M_{i2}. Suppose two types of products q_1 and q_2 are processed through the workcells $WC_1, WC_2, \cdots, WC_n$, in order. Then the set of resources in the system can be given by $R = \{M_{i1}, M_{i2}, i = 1, 2, \cdots, n\}$ and $C_{M_{ij}} = 1$. Deadlocks occur only in some workcells. For example, in the workcell WC_n, a type q_1 product is processed by the machine sequence M_{n1}, M_{n2} and a type q_2 product by M_{n2}, M_{n1}. A deadlock occurs if M_{n1} is processing a type q_1 product and M_{n2} is processing a type q_2 product. Figure 9.10 illustrates the $R^2 PN$ model of the system.

In the system, there is no key resource, and a D-structure is only some set $\{t_{i2}, t_{i4}\}$ if it exists. For this $R^2 PN$, we can introduce an optimal deadlock avoidance Petri net controller C. For a D-structure $D_i = \{t_{i2}, t_{i4}\}$, the controller C restricted only the number of tokens in places p_{i1} and p_{i3} not greater than $m_0(\{M_{i1}, M_{i2}\}) - 1 = 1$. That is, the controller only restricts the number of products processed in the workcell WC_i where deadlock can occur. Hence, the function of our controller is local for the system.

Suppose that in the workcell WC_n deadlock can occur, that is, $D_n = \{t_{n2}, t_{n4}\}$ is a D-structure and Ezpeleta et al.'s Petri net controller [3] is used for this $R^2 PN$, then the number of products processed in the whole system is at most two even if all $D_i = \{t_{i2}, t_{i4}\}$, $i \neq n$, are not D-structures. In any case, our controller allows at least n products to be processed and the maximal use of resources in the system. This has a clear implication for improving the resource utilization and system productivity.

9.6   Conclusion

This chapter has (1) formulated a circular wait concept in the context of Petri net models: D-structure; (2) characterized the liveness conditions of Petri net models in terms of such structures; and (3) proposed a method for synthesizing deadlock avoidance controller for the $R^2 PN$ models.

A D-structure is defined to capture the characteristics of the circular wait chain in the system. Such a structure can lead to deadlock when a large number of products are dispatched in it. To avoid such phenomena, the token capacity in a D-structure is proposed, which plays a key role in the synthesis of the deadlock avoidance controller. We combined the Petri net controller with the restrictive policy to generate valid and resource utilization maximizing control for the FMS. The computation of the controller is carried out off-line, and the respond time of the controlled system is short.

This chapter presents results for a linear manufacturing system. The limitation of the $R^2 PN$ model is that, for each transition, there can only be one input place that is an operation place. That is, an $R^2 PN$ cannot model an assembly operation, which occurs when several different parts are assembled into one product. For such a system, the related necessary and sufficient conditions for a live Petri net model and the deadlock avoidance controller will be more complex.[12] Future research will focus on the extension of $R^2 PN$ to solve the deadlock problems in manufacturing/assembly systems.

References

1. Banaszak, Z. and Krogh, B., Deadlock avoidance in flexible manufacturing systems with concurrently competing process flows. *IEEE Trans. Robotics and Automation,* 6(6), 724, 1990.
2. Ezpeleta, J., Couvreur, J. M., and Silva, M., A new technique for finding a generating family of siphons, traps, and st-components. *Lecture Notes on Computer Science,* (674). Rozenberg, G., Ed. Springer-Verlag, New York, 126, 1993.
3. Ezpeleta, J., Colom J., and Martinez, J., A Petri net based deadlock prevention policy for flexible manufacturing systems, *IEEE Trans. Robotics and Automation,* 11(2), 173, 1995.
4. Hsieh, F. S., and Chang, S. C., Dispatching-driven deadlock avoidance controller synthesis for flexible manufacturing systems, *IEEE Trans. Robotics and Automation,* 10(2), 196, 1994.
5. Murata, T., Petri nets: properties, analysis and applications, in *Proc. IEEE,* 77(4), 541, 1989.
6. Peterson, J. L., *Petri Net Theory and the Modeling of Systems,* Prentice-Hall, Englewood Cliffs. NJ, 1981.
7. Lautenbach, K., Linear algebraic calculation of deadlocks and traps, in *Concurrency and Nets,* Voss, Genrich, and Rozonberg, Eds., Springer-Verlag, New York, 315, 1987.
8. Viswanadham, N., Narahari, Y., and Johuson, T., Deadlock prevention and deadlock avoidance in flexible manufacturing systems using Petri net models, *IEEE Trans. Robotics and Automation,* 6(6), 713, 1990.
9. Xing, K. Y., Hu, B. S., and Chen, H. X., Deadlock avoidance policy for Petri net modeling of flexible manufacturing systems with shared resources, *IEEE Trans. Automation Contr.,* 41(1), 1996.
10. Xing, K. Y., Xing, K. L., and Hu, B. S., Deadlock avoidance controller for a class of manufacturing systems, *IEEE International Conference on Robotics and Automation,* 1996.
11. Xing, K. Y., and Li, J. M., Correspondence relation between two kinds of structure elements in a class of Petri net models, *J. of Xidian University,* 24(1), 11, 1997 (in Chinese).
12. Xing, K. Y., and Hu, B. S., The Petri net modeling and liveness analysis of manufacturing/assembly systems, in *Proceedings of the Second Chinese World Congress on Intelligent Control and Intelligent Automation* Xi'an, China, 1997.
13. Zhou, M. C., and DiCesare, F., Parallel and sequential mutual exclusions for Petri nets modeling for manufacturing systems with shared resources, *IEEE Trans. Robotics and Automation,* 7(4), 515, 1991.

Index